PHILOSOPHY IN MIND

PHILOSOPHICAL STUDIES SERIES

*Founded by Wilfrid S. Sellars and Keith Lehrer*




VOLUME 60

# PHILOSOPHY IN MIND

## THE PLACE OF PHILOSOPHY IN THE STUDY OF MIND

*Edited by*

MICHAELIS MICHAEL

and

JOHN O'LEARY-HAWTHORNE

*The University of New South Wales, Sydney, Australia*

Springer-Science+Business Media, B.V.

A C.I.P. Catalogue record for this book is available from the Library of Congress.

ISBN 978-94-010-4438-7 ISBN 978-94-011-1008-2 (eBook)
DOI 10.1007/978-94-011-1008-2

*Printed on acid-free paper*

CONTENTS

# PREFACE

This collection grew out of a mini-conference on the place of philosophy in the study of mind, held at The University of New South Wales during August 1992. We would like to thank both the other contributors and members of the large audience for making the conference so successful, thereby confirming the continuing strength of interest in philosophy of mind in Australia. Ten of the papers included here — by André Gallois, Gilbert Harman (two papers), Frank Jackson, Jeanette Kennett and Michael Smith, Michaelis Michael, John O'Leary-Hawthorne, Lee Overton, Huw Price and Graham Priest — are descendants of the papers delivered at that conference and the discussion thereof. Also included here are solicited contributions from Karl-Otto Apel, Annette Baier, José Benardete, Robert Brandom, Gideon Rosen and John Searle, whose involvement with this volume we very much appreciate. With the exception of the contributions by Apel, Priest and Searle, the papers that appear here are previously unpublished. Apel's paper appears for the first time in English. We are most grateful to Lisabeth During for translating the original German text.

We would like to thank the Faculty of Arts and Social Sciences at the University of New South Wales for financial assistance, our School of Philosophy for their support, and both Rowland Hilder and Paul Hannigan for their technical assistance.

THE EDITORS

MICHAELIS MICHAEL AND JOHN O'LEARY-HAWTHORNE

# INTRODUCTION: PHILOSOPHY IN MIND

Even in this post-Idealist era, the study of mind remains centrally important to philosophy. A number of prevalent views support this contention: that problems about intentionality and mental content are among the most basic philosophical problems; that modern philosophy began and is even constituted by Descartes' discovery of subjectivity; that mental phenomena constitute the most serious problem for the naturalistic ambition.

Yet what is the place of philosophy in the study of mind? The question is a difficult one, not least because there is no agreed conception as to what counts as a distinctively philosophical method or a distinctively philosophical subject matter. Nor is there any consensus about exactly how a study of mind might shed light on other issues on the philosophical agenda, including central issues in metaphysics, ethics, philosophy of science, philosophy of language and so on.

While our main question is inevitably rather open-ended, it nevertheless strikes many of us with some force. As things now stand, one finds a marked insecurity about the continuing importance, even credibility, of philosophical inquiry into the nature of mind. In the Anglo-American tradition from which these papers are largely drawn, the perceived threat has rarely been the poet, novelist or literary critic, but invariably the scientist. This disquiet has taken the form of a critique of various philosophical distinctions — such as those between analytic and synthetic truths, essential and accidental properties, and observation and theory — accompanied by a concern that philosophers may have obscured the empirical, scientific nature of the central questions about the mind. Some philosophers have sought security by aligning themselves with science in the form of a *cognitive* science; others continue to reaffirm more traditional pictures of the philosopher's activity in this area. It is hoped that this collection will provide food for thought for all those who are seriously concerned about how the philosopher's inquiries in this area should be integrated with other sorts of investigation.

Let us remind ourselves of some of the many overlapping pictures that a so-called philosopher of mind might have of her own activities: One who studies the essence of mind — a student of mind *qua* mind; one who analyses our ordinary mentalistic concepts or mentalistic language; one who articulates common sense; one who exercises wisdom concerning the human condition; one who, in defending a theory of mind, runs the gauntlet of objections, taking

*Michaelis Michael and John O'Leary-Hawthorne (eds.), Philosophy in Mind, pp.1-7.*

on a range of challenges that are considerably wider than those that arise in ordinary affairs or even scientific investigation; one who engages in *a priori* speculation about the mind; one who is putting a protoscience of mind into place; one who discovers links between theories of mind and other, fairly diverse subject matters; one who attends carefully to the phenomenology of our experience of ourselves and each other; one who finds arguments in defense of this or that view about the mind that begin with generally accepted premises and which yield surprising and/or interesting conclusions about the mind; one who tidies up some of the conceptual mess left by scientists as they proceed on with their work at the coalface; one who works at the most theoretical end of cognitive science alongside the scientists.

The aim of this collection is not to focus on any one picture concerning the nature and importance of philosophy of mind, let alone provide a cumulative defense of some such view. Our aim is rather to offer a wide variety of perspectives that might shed light on the contribution and debt that philosophy has had and will have with respect to the study of mind. In what follows, we shall offer a partial introduction to the papers by identifying some questions and topics that provide common threads.

### *Philosophy of Mind: The Track Record*

One obvious place to begin when thinking about philosophy and the study of mind is to ask 'What discoveries concerning the mind have been peculiar to philosophy?' At least some of us will find plenty of reason to be optimistic here. We might begin by citing various problems that have been uncovered by philosophers: qualia, free will, other minds, intentionality and so on. We might then make a list of fundamental insights about the mind that have been achieved from the armchair. Candidates addressed (though not necessarily defended) in this collection include: that the propositional has primacy; that the mental doesn't supervene on the neural; that the basic problems about the mind are empirical; that there is no privileged access; that cognition is fundamentally social; that mental states are identical to brain states; that intentionality is fundamentally normative; that computers can't think; that the world is, in a deep way, mind-dependent; that there is a fundamental link between ethics and reason. If one is interested in the place of philosophy in the study of mind, one must surely reflect on the level and nature of philosophy's achievements here. One feels that if many of these achievements are genuine, then philosophers must have been doing something right.

There are, of course, ways of approaching the history of philosophical study of mind that do not take it as a history of discovery. One such approach claims to detect a history of pseudo-problems and pseudo-discoveries produced by speculative excess, giving itself the critical task of uncovering the basis of such illusions so that they do not recur. In this respect, at least, Kant and Wittgenstein can both be seen as allies. Following Wittgenstein's example, the modern way of trying to make good on this critical role for philosophy is to shift from the material to the formal mode: one tries to see philosophical problems about the mind as pseudo-problems by reflecting on the linguistic tasks performed by mental language. Huw Price's paper is very much in this critical tradition. There, he argues that, as philosophers of mind, our primary goal should be that of laying bare the point of our mental concepts rather than that of "conceptual analysis".

Another approach to philosophical thought about the mind is to focus on its self-constitutive role. This role for philosophy is discussed in Lee Overton's paper. On Overton's view, it is important to see how philosophical theorizing about the nature of deliberation shapes the sort of deliberators, and so agents, we become.

*The A Priori*

The nature of mind will clearly have a bearing upon our capacity, as philosophers, to study it. Whether and to what extent we can have *a priori* insight is particularly pertinent here. Logical Positivism is no longer with us; yet Anglo-American philosophy has largely acquiesced in the positivist's insistence that there is no synthetic *a priori* knowledge. While the term 'metaphysics' has reassumed some measure of respectability, the idea that we have some special intuitive ability to acquire substantive metaphysical truths that will forever elude empirical inquiry has remained one to be embraced with some embarrassment, if at all. Meanwhile, the project of conceptual analysis, heralded by many as the activity to keep philosophers in business, has itself become the target of much critical scrutiny. The crucial issue here is whether we can viably distinguish the analytic from the synthetic in a way that is epistemologically significant and *a priori* determinable. The major figure in this movement against *a priorism* has, of course, been Quine. His work has aimed towards sweeping away the *a priori* altogether, along with the attendant distinctions of necessary and contingent, essence and accident. The only broadly philosophical projects that can be reconciled with the Quinean vision are those of formalizing scientific theories and articulating scientific method; and this is because the task of regimenting the language of science and reflecting on what we do when we engage in scientific inquiry is continuous with austere science itself.

A priorism, while perhaps shaken, has nevertheless retained a loyal following. The philosophical literature abounds with armchair arguments for important conclusions about the mind. André Gallois' paper 'Deflationary Self-Knowledge' discusses a cluster of *a priori* arguments which alternatively attempt to subvert or endorse Descartes' important view that we have privileged access to our own minds. These arguments run quite orthogonal to the psychological literature on this matter. For more on privileged access, see Annette Baier's 'How to get to Know One's Own Mind: Some Simple Ways', though there we find a style of philosophising that is far less a prioristic.

How, in general, a synthetic *a priori* investigation into the mind might proceed is described in José Benardete's 'AI and the Synthetic *A Priori*'. Benardete resurrects an avowedly Kantian conception of the synthetic *a priori,* claiming not merely that it can be grafted onto various research programmes in cognition, but rather that even such sober projects as 'hard' AI may require it.

Meanwhile, Frank Jackson's 'Armchair Metaphysics' reaffirms a commitment to the project of analyzing our psychological concepts as the proper task for philosophers of mind. In that paper we find the striking view that physicalism requires the analytic reduction of all concepts, including the normative and the sensate, to those of natural science. Gilbert Harman's response to the Jackson piece offers us a statement of Quinean suspicions about the analytic project. There we find the equally striking conclusion that since we cannot distinguish the analytic from the synthetic, physicalism is an incoherent doctrine.

Whether logic, that *a priori* discipline par excellence, might tell us anything important about the mind, is discussed in Harman's 'Epistemology and the Diet Revolution' and Graham Priest's 'Gödel's Theorem and the Mind . . . Again'. While Harman's piece considers how reasoning, in general, relates to logic, Priest consider perhaps the most notorious attempt to derive interesting results about the mind from logico-mathematical premises: Lucas' use of Gödel's Theorem to show that the mind is not a machine. On Priest's view, Lucas' argument will ultimately be of use not to transcendental human psychology, but to natural theology.

*Science and Common Sense*

What is philosophy's relationship to science? One view of the matter here is the Quinean one that we have already alluded to. On this view, legitimate philosophy is continuous with theoretical science. Such a view is exemplified in Gilbert Harman's 'Epistemology Today' where epistemology problems are argued to be continuous with a scientific study of reasoning processes and where philosophy is contrasted with scientific investigations only on the score of its having 'no clear experimental paradigm'. For those who are more reluctant to align themselves so closely with science, another ally needs to be be sought. The layman might expect that ally to be wisdom or the light of (natural) *a priori* reason: however, the most popular ally in Anglo-American philosophy these days is common sense. We thus, for example, find John Searle claiming of his firmest philosophical convictions about the mind — including his recipe for a solution to the mind-body problem — that they are nothing but a form of common sense ('The Problem of Consciousness').

The Quinean view on common sense is clear enough: common sense tells us that science is a better bet than common sense. Yet other philosophers have reckoned a sort of epistemological privilege to be enjoyed by common sense that can anchor philosophical reflection. From this idea emerges a vision of the philosophical enterprise as articulating common sense and on that basis, casting out theories where they conflict with common sense, endorsing theories where they are endorsed by common sense. Jeanette Kennett and Michael Smith's 'Philosophy and Common Sense: The Case of Weakness of Will' offer us this sort of picture.

*Objectivity and the Mind*

There was a time when the idea that world was mind-dependent was subscribed to at face value by a large segment of the philosophical community. Today, few believe that the world is, literally, all mental. Yet some still find talk of a mind-dependent reality as suggestive of some important insight. There is, on their account, something deeply right about the idea that the mind permeates the world (or at least some segments of it), so long as that is not construed as denying that there are rocks or else as insisting that rocks are collections of ideas. Could this be one area where, by reflecting on mind and its interface with the world, we will shed light on the deep metaphysical issues? If so, it will surely fall within the distinctive province of philosophy to carry out this project. There have been various ways of cashing out the idea that the mind somehow projects itself onto the world. One way has been to focus on the notion of perspective, provoking certain realistically minded philosophers to argue for a non-perspectival conception of knowledge. The notion of perspective and its bearing on issues to do with objectivity is discussed in contributions by Michaelis Michael, Gideon Rosen and Huw Price. Another has been to see certain sorts of discourse as playing the role of expressing an attitude other than belief. This idea is discussed in John O'Leary-Hawthorne's 'Belief and Truth-Aptness'. Another has been to subscribe to a response-dependent view of some concept according which what a concept is true of is linked to our dispositions in some salient set of circumstances. This is also discussed in Rosen's and Price's papers.

*Philosophy of Mind and Ethics*

While purported links between the study of mind and ethics will not figure very highly on the agenda of those philosophers oriented towards cognitive science, we should not overlook the segment of the philosophical community that seeks to shed light on the nature of ethics by reflecting on those aspects of emotion, motivation and cognition that bear on it. The project of moral psychology, set in motion by Aristotle, continues to figure as one way that philosophers may play a role in the study of mind that is importantly distinct from straightforwardly empirical disciplines. Jeanette Kennett and Michael Smith's 'Philosophy and Common sense: The Case of Weakness of Will' and Lee Overton's 'Unprincipled Decisions' are contributions to this project.

Contemporary hermeneutical philosophy in Germany has an especially strong commitment to the practical involvement of philosophy in ethics and politics. One might thus expect to find moral psychology alive and well in that tradition. Yet, importantly, the hermeneutic tradition tends to seek insight into the nature of ethics not by reflecting on mind, but rather upon language. On that view, language matters a lot more to philosophy than does mind. (And, of course, this echoes one important strand of Anglo-American philosophy). Karl-Otto Apel's 'Can the Philosophy of Language Provide the Key to the Foundations of Ethics?' provides us with a useful account of why one prominent philosopher in the hermeneutic tradition has found reason to ground ethics in language rather than mind.

*Mind and Its Place in Society*

The task of elucidating the relationship between the individual mind and societal structures is, of course, hardly the distinctive task of philosophy. One of the acknowledged tasks of the social sciences is precisely to catalogue the causal influence of society upon the individual mind's development. Yet there may be relations, some of a constitutive nature, between mind and society that are frequently overlooked or insufficiently appreciated by the social sciences that the philosopher can usefully explain. Or so one important strand of philosophical thought maintains at any rate. Philosophers this century (most famously, Wittgenstein and Davidson) have often taken to arguing that thought is impossible unless certain specified relations hold between the individual and society. Work along these lines has typically relied on a purported philosophical discovery — that thought is fundamentally normative — and has then gone on to somehow ground the normative in the social. Robert Brandom's 'Reasoning and Representing' can be usefully located in this tradition. Brandom asks whether we should understand socially articulated inference in terms of representation or representation in terms of socially articulated inference, and opts for the latter. Related themes are explored in Apel's paper.

Philosophers inclined towards some interesting or surprising thesis about the individual and society will inevitably find themselves reacting to the Cartesian picture according to which we begin with direct access to the contents of our our minds and work out towards the external world from those foundations. This Cartesian inheritance is very much the target of Annette Baier's 'How To Get To Know One's Own Mind: Some Simple Ways'. Her reaction attempts to stand Descartes on his head by showing how, in some important respects, the real epistemic problem is not so much that of other minds but our own.

*The University of New South Wales*

JOSÉ BENARDETE

# AI AND THE SYNTHETIC A PRIORI

## I

'The questions surrounding the nature of mind and consciousness', writes Paul Churchland, 'have slowly come to be represented as empirical questions, high-level perhaps, but solvable eventually by the methods of theoretical science' (1992, p. xi). Is that really true? And if so, how did 'this dramatic new course' come about? Churchland explains it as owing to certain 'seminal thinkers' where it turns out that 'Feigl, Smart, Sellars and Feyerabend spring immediately to mind', philosophers one and all. So if two centuries from now mind proves to be at last integrated into the scope of the natural sciences — Churchland speaks only of 'theoretical science' — philosophers will have to be credited in no small measure with that striking achievement?

Whether philosophy played an important role in the work of Galileo and Newton at the serious outset of modern science, I take to remain a bone of scholarly contention even today; but in any event we have been taught to believe that, once having been securely launched on its way by the end of the 17th century, modern science would henceforth be required never again to call philosophy to its aid. Until today, that is, when the natural sciences are readying themselves to pursue the issue of mind as a research imperative? Maybe so! But then there really must be something pretty special about the nature of mind after all (one expects tough-minded philosophers like Churchland to deny that) if the sciences can secure access to it only through the mediation of philosophy; and one would then want to get very clear about what that difference might consist in. Long chastened by the so-called critical philosophy of Kant not to mention, more recently but in the same vein, the challenge of the Logical Positivists, the professional philosopher today is bound to hesitate before lending credence to Churchland's flattering program, lest in allowing his traditional, indeed almost pre-Socratic pretensions to be reactivated he should only expose himself to further humiliation in the future.

One of the more pressing questions surrounding the nature of mind and consciousness might be crisply summed up in the words, 'Can a robot really think?' And there is at least one philosopher today whose express 'intent' is not merely 'to begin the process of providing a recipe for building a person' in robot form. Much more than that, the larger intent of John Pollock's project is 'to implement that recipe' in artificial intelligence, as he is persuaded that 'many of the problems involved are essentially philosophical, while re-

*Michaelis Michael and John O'Leary-Hawthorne (eds.), Philosophy in Mind, pp.9-22.*

searchers in AI have not usually been trained in philosophy' (1989, p. ix). Two centuries from now, did I say? Rather, from Pollock's words added to those of Paul Churchland, we might suppose ourselves even today to be well into the early dawn of a Golden Age of cognitive science where the alienation of philosophy from mathematical physics above all, enshrined in the Cartesian distinction between *res cogitans* and *res extensa*, can be seen to be in the process of being overcome.

Of two minds in this matter, I have hinted at some cynical misgivings only as a prelude to climbing aboard the cognitive science bandwagon with a proposal of my own that is at least as visionary as any other I have found in this area. To wit: the proposition that robots can think has a synthetic *a priori* status, where half of my task will consist in motivating interest in this moribund theme of the synthetic *a priori* (impatient readers are invited to skip ahead to Section 2) that has scarcely been mentioned in the literature during the past —a conservative estimate —25 years. How to justify linking that theme to the seemingly remote issue of robots will supply my further burden. Reversing Churchland then, the answer to at least one question surrounding the nature of mind and consciousness is thereby represented in no merely empirical (whether high - or low-level) terms. No great surprise, surely, when one considers that if the natural sciences remain committed to the a posteriori, philosophy by contrast probably continues to be affiliated largely if not entirely with the a priori.

How Churchland's heuristic retains its value can be brought out in connection with a genuinely surprising discovery that it enabled me to make, involving the subtext of a discussion by Quine where he is engaged in fending off Noam Chomsky. Significantly entitled 'Quine's Empirical Assumptions', Chomsky's challenge consists above all in inviting Quine, who is on record as rejecting all *a priori* knowledge, to identify the empirical and *a fortiori* falsifiable premises that underlie his thesis as to Indeterminacy of Radical Translation. Passing by in a flash, the moment is easy to miss when Quine in his response to Chomsky allows that his position presupposes the absence of mental telepathy (1969, p. 306). Think about it! Quine must then be conceding that there is a determinate mental content that we associate with the word 'rabbit'. In the absence of telepathy, however, that content fails to count as the meaning of the word, being rather filtered out in the behavioral give-and-take of our public language, in a fashion reminiscent of Wittgenstein's beetle in the box. Notice above all that the truth or falsity of Quine's Indeterminacy Thesis is herewith represented by him as depending on the truth or falsity of a fairly low-level empirical hypothesis, namely the hypothesis that there is no mental telepathy, which might be readily defeated tomorrow by humdrum experiments.

Another example in the same vein is beautifully supplied by Neil Tennant when he proposes 'extra means of bridging the gap between possibly inconclusive behavior "on the surface" and the rich content that we uncritically accord our reports of pain.' Tennant can thus ask, 'Could it be...that when one is "really" in pain one emits pheromones of a particular kind and utters one's words with limbic overtones from an otherwise unused range?' And maybe then ' an evolutionary story could build them in as part of what is tacitly known by someone who understands that another is in pain' (1987, p. 17).

The issue of meaning (public and private) being acknowledged to be one of the questions surrounding the nature of mind and consciousness, we have seen twice over, in Tennant as well as Quine, how it might well be represented in contingent, empirical terms. Even so, I am surely not alone in holding that there is another, rather more aprioristic and certainly more familiar version of Quine's Indeterminacy Thesis that is perfectly compatible with the widespread even ubiquitous presence of telepathy. And in much the same way I take there to be an aprioristic version of strong AI, to which I am especially attracted, as well as the probably more familiar, aposterioristic variety.

'The only *theoretical* reason to take contemporary Artificial Intelligence more seriously than clockwork fiction', according to John Haugeland, 'is the powerful suggestion that our own minds work on computational principles' (1985, p.5, his emphasis). Is that really true? The only theoretical reason? But then Churchland's heuristic proves to be confirmed afresh in the case of the 'robot' question where in Haugeland's formulation it is represented as turning on the empirical hypothesis that our minds work on computational principles à la Alan Turing. Advertising aposterioristic AI, Haugeland can even continue, 'In other words, we're really interested in AI as part of the theory that *people* are computers — and we're all interested in people' (his emphasis).

Who 'we'? Probably not the Marvin Minsky of 1955 who kept me almost mesmerized (talk about visions!) late into the night one evening in Cambridge, Massachusetts as he projected the destiny of AI well into the century ahead. If some in the field were interested in modelling their robots on the actual workings of the human brain which they were accordingly keen to investigate, Minsky made it loftily clear that he was not in their pedestrian ranks. How the wetware of the brain might produce cognition, could be readily seen to be far removed from the fastidious interests of this professional mathematician. There was thus a high as well as low road to AI. The thought that the human brain might be in fact close to the optimum solution to the problem of organizing a limited supply of material into a thing capable of thinking at our level, came rather as a surprise to Minsky a decade later when I heard him bemusedly retreat from his earlier attitude in a public lecture.

Notice, however, that the original point of principle remained intact. Although a study of the brain might well provide a short-cut toward our building an intelligent robot, it was by no means needed to provide independent validation of our success. No giant step has had to be taken now by me in order to construe the low road to artificial intelligence as an explicit case of aposterioristic AI while the high road arguably offers the prospect of being an implicit case of a more or less aprioristic AI. I say 'more or less' here only to acknowledge that even the high road may be infected with aposterioristic considerations somewhere down the line. Although Minsky at one point in our conversation in 1955 did say, ' Few people are as materialistic as I am' , my first thought the next day in marvelling over his project can almost be quoted as follows, 'Can this really be materialism when the material constitution of the only thing in nature known by us to exercise intelligence fails to interest this pure mathematician? Why, Minsky is none other than Socrates the Younger!' And here with this shock of recognition I was harking back not to the famous snub-nosed figure but to the ancient mathematician reproached by Aristotle in Chapter 11 of book Zeta of his *Metaphysics* (at 1036b 20-30) for regarding man and circle as being on a par, as if all men's being made of flesh was to be likened to a possible world where all circles were made of bronze, the merest accident in the one case as in the other.

Especially apposite today, this conceit of Minsky's avatar in antiquity is very much on our minds when, against the background of artificial hearts, lungs, eyes, etc. fashioned out of inorganic materials, we propose to enter the brain itself, replacing one by one each neuron as it decays with an artificial substitute made (say) of silicon. A metamorphosis (recalling caterpillar and butterfly) of man into robot? And is the resulting robot, presumed to be at any rate behaviorally equivalent (in the relevant respects) to a man, properly to be described as also a man, as I take Socrates the Younger to have maintained? So that only some men fall in the province of biology? The most compelling case of aposterioristic AI that one could imagine, our robot will be no more than a zombie lacking consciousness, according to John Searle who can readily appear to be the only genuine materialist working today in the neighborhood of cognitive science. What we are made of really does matter. Matter matters even when it comes to mind.

More precisely, Searle allows as a point of logic that, most improbably, our silicon robot might well enjoy consciousness though a relevantly equivalent copper model should fail to do so (1992, p. 66).Which is not to concede that the improbable consciousness of the former robot would be merely epiphenomenal when it comes to its conduct. If the movements (internal as well as external) of the two robots may be allowed to be the same *mutatis mutandis*, the consciousness produced by the silicon brain will cause some of the former's movements to constitute actions, being invested with intentionality (e.g., raising one's arm by way of signalling someone to halt). If there is now the awkward question as to how to tell apart conscious robots from the zombie variety, Daniel Dennett can be indulged when he has this bit of serious fun at Searle's expense: 'who knows, maybe zombies have green brains' (1991, p. 405). The party is over, however, when he goes on to say, 'Maybe women aren't really conscious! Maybe Jews! What pernicious nonsense' (1991, p. 406). For shame! That even such a good fellow as our very own Dan Dennett should poison the wells with this foray in 'political correctness', indicates that a nerve has probably been struck hereabouts.

In a lighter vein where Dennett dismisses 'the possibility of zombies' as 'preposterous', Pollock writes, ' It seems proposterous to suppose that what a creature is made of can have anything to do with what mental states it is in, except indirectly by influencing what structures can be built out of that stuff', observing that 'it makes no difference whether a can opener is built out of steel or aluminum'. Just the sort of thing that Socrates the Younger was doubtless prepared to say! And so when Pollock rests his case on 'a distinction between matter and form' (1989, p. 66), we know that we are back with Aristotle in Zeta, thereby explaining how it is that our two mathematicians, ancient and modern, might undertake to abstract from man's flesh and blood his functional form (presumably instantiated in our silicon and copper robots) rather in the fashion that they were accustomed in their science to abstract the form of circularity from bronze and gold circles.

How Minsky's program might connect with the natural sciences, I found particularly baffling, for again it smacked more of pure mathematics than empirical research in that — witness Winograd's SHRDLU — computer simulations obviate any need to implement one's program on a nuts and bolts level. That Minsky was a materialist if the relevant contrast was with Cartesian mentalism, I could hardly ignore if only because I noticed in his apartment a copy of *The Concept of Mind* by Gilbert Ryle whose critique of 'the ghost in the machine' could only please him, even as Ryle's airy dismissal of man's being *just* a machine at the end of the work I knew (so why mention it to him?) that Minsky would be entitled to dismiss in turn.

Far from dismissing Searle's wetware approach to mind as proposterous, I esteem *The Rediscovery of the Mind* as being at least as worthy a candidate for our assent as *Consciousness Explained* with its up-to-date cognitive science spin given to its disqualification of qualia via a protracted session in neo-Wittgensteinian therapy that, even so, might well be extended with profit much further. Taking these two works pretty much to define the opposite poles in philosophy of mind today, I have invoked Searle in particular largely as a foil enabling me to press the claims of aprioristic AI as against those of the aposterioristic variety. In parody form the nub of my argument goes then as follows. If Searle at any rate will find the former version of AI to be hardly more implausible than the latter, one might as well be hanged for a sheep as for a lamb by opting for the radical version. So I am thus to be seen as catering first and foremost to the cognitive science crowd for whom Searle is a favorite whipping-boy? Well, it must be admitted that the expression 'aprioristic AI' is bound to excite distrust, precisly by conflating almost in an oxymoronic mode the tender-mindedness of the *a priori* with the tough-mindedess of AI, and I must thus be allowed some leeway in easing my reader along.

Answering to Ned Block's (1980) distinction between analytic and synthetic functionalism the difference between the two versions of AI is brought out with especial vividness when we focus on our ex-flesh-and-blood-person-now-transformed-into-a-silicon-robot whose adventures we are free to follow either in real life or merely in various scenarios of computer simulation where it interacts intelligently (*pace* Searle) with its environment. Does it really matter now if we have only been hallucinating all along, deceived perhaps by Descartes' Evil Genius into believing that there is a physical world? Just as it would not matter when we abstract the form of circularity from our bronze and gold circles, if they were figments of our imagination. Circularity would still be grasped by us. By contrast of course the computer program defining the functional organization of our robot, in conjunction with that of its world, would be so formidably complex that we could not possibly internalize it. So let us simply take that infirmity of ours as a bit of human 'friction' from which we are entitled to abstract in the idealizing spirit of high theory. Then we have two paradigms that compete for our allegiance in the present instance, namely the aprioristic circularity case and the aposterioristic electrolysis case where someone merely hallucinates water being decomposed into hydrogen and oxygen, thereby being right only by accident and hence failing to achieve a cognitive grasp of anything whatever. In the other case we standardly suppose that even an immaterial spirit might grasp what it is for something to be circular simply on the basis of hallucinatory experience, and I shall simply help myself to that supposition.

Designing then a robot and its world in the aprioristic imagination (= computer simulation), we shall put it through its paces in its environmental niche (also imagined). From the outset the movements of our robot in its niche must satisfy the HD (or hypothetico-deductive) requirement whereby hypothesizing these movements as constituting the actions of an intelligent agent allows an observer to deduce further movements (some actual, others diversely counterfactual) of the robot which prove to be in fact confirmed as the scenario plays itself out counterfactually as well as actually. So we have here the sequence 'movements → actions → movements' or MAM, enshrined in Dennett's discussion of the Intentional Stance.

A second, but much more intimate, requirement takes us to the central insight motivating the entire project. If the first requirement is expressly conceptual, the second can be described without exaggeration as involving a pure *a priori* spatio-temporal intuition reminiscent of Kant himself, for it trades on the most important use to which Wittgenstein's remarks on seeing-as (prompted by Jastrow's duck-rabbit) could be put. This has to do with seeing the movements of a body *as* (constituting) the actions of a person, and though many such cases have an *a posteriori* character (e.g., ink-inscribing pen-in-hand movements viewed as someone's entering into a commercial contract) there are other cases (e.g., the action of a horse in leaping over a fence or in avoiding the branch of a tree) that I take to involve an *a priori* synthesis of physical and psychological concepts. Call this then the FY connection where leaping counts as a paradigm of mentality insofar as it involves belief and desire. It is precisely this FY *a priori* synthesis that will be found to issue in a synthetic *a priori* science of mind that is as alien to the naturalistic juggernaut of today (Searle and Dennett are bedfellows here) as it is to traditional Cartesian mentalism.

Highly behavioristic even as it incongruously resists being absorbed into the aposterioristic framework of the natural sciences, aprioristic AI is prepared to welcome the primary imperative of the new post-Skinnerian cognitive sciences when they mandate that the movements of our robot be content-driven by representations ensconced inside it that may even on occasion misrepresent its environment. No need then to insist on a black box behaviorism that respects only sensory input and motor output! And we may thus model these internal representations on Denis Stampe's tree rings that represent (and misrepresent) the age of the tree (1979). Thereby invoking a naturalistic account of representation? Not at all. Suppose once more that we are being deceived by Descartes' Evil Genius. So Stampe's account of representation is accordingly discredited? By no means. We can still say that in any relevant world his aprioristic account still applies, and my only assumption here is that a properly naturalistic account must always consist principally of contingent propositions.

Welcoming as well a compositional semantics for our robot's representations in line with Jerry Fodor's 'only game in town' (the more recent Connectionist approach may be harder to handle), we may be advised to adopt the crudest version, by installing in our robot two boxes, one labelled 'beliefs', the other 'desires', each filled with strips of paper inscribed with English sentences, the internal processing of which (as activated by sensory input and causing in its turn motor output) will constitute the thought processes of our robot as it engages in Mentalese. And there are other, even more naturalistically oriented suggestions that aprioristic AI is fully entitled to appropriate. Take, for example, the deepest moment in early Dennett's hermeneutics of content (it flits by much too quickly) when discussing 'intelligent storage' he writes, 'For information to be *for* a system, the system must have some *use* for the information, and hence the system must have *needs*' (1969, p. 46, his emphasis).

Suggestive surely, if only because we characteristically think of animals but not robots as having needs, and we may thus, prior to building our robot proper, and recalling Aristotle's stratified soul (rational, sensitive and *nutritive*) undertake to build an artificial *plant*, needing water and sunlight (say) in order to grow and flourish let alone merely survive, only later equipping it with the resources for sensory input and motor output, for we recall moreover that even in the bad old days of SR (i.e., stimulus-response) psychology hard-line researchers like C. L. Hull did not hesitate to speak of 'need reduction' in connection with their rats quenching their thirst as a 'reward' for running a maze. Pursuing this theme still deeper in a quasi-metaphysical if not altogether physicalistic vein, aprioristic AI might arguably even profit from Peter van Inwagen's (1990, pp. 86-87) account of an organisms's life as being first and foremost a self-maintaining event like a flame or an ocean wave, by providing for such homeostatic processes at the outset.

## II

However enjoyable these more speculative spin-offs of the theory, the thematic primacy of the synthetic *a priori* cannot be evaded, for it defines the status of its axioms, which will characteristically take the form 'If $\Phi$ then $\Psi$', that is to say, if such and such a complicated physical state of affairs obtains (and here a robot's design is specified in exhaustive detail, along with such a description of its environment as well as the laws of nature presiding over it), then such and such a psychological state supervenes upon it, e.g., the robot's leaping into the air in order to warn of fire in the neighborhood. Played out in the mode of computer simulation, the whole scenario of seeing-as by involving an *a priori* synthesis of physical and psychological concepts issues in propositions that can only be described as being themselves at once synthetic and *a priori* in character.

How now to provide an *independent* justification of what would otherwise strike most philosophers today as being hardly more than an antiquarian oddity, namely trotting out the synthetic *a priori* at this late date. I say independent because when it comes to a taxonomy of positions in current philosophy of mind my own proposal may be felt to be hardly inferior to others more familiar, like Dennett's and Searle's, with which it might be profitably contrasted; and I can even imagine someone's suddenly acquiring — to her surprise — some sympathy for the synthetic *a priori* if only in the present context. Hardly encouraging is the last time the synthetic *a priori* made a pronounced appearance on the philosophic scene, which I take to have been Roderick Chisholm's invoking it in order to characterize such basic principles of his epistemology as that something's looking red to a person justifies her in believing it to be red, in the absence of evidence to the contrary (1957, pp. 105 and 112). Even here, however, Chisholm mentions the synthetic *a priori* only in passing, probably still a little embarrassed by the ill repute with which the Logical Positivists invested it. Yet his point has not lost its force even for those who in a more up-to-date vein have opted for a reliabilist account of justification. For it is not as if they expect to show that a belief's being a reliable (even fool-proof) indicator of truth logically entails that holding it is justified. Nor do they suppose that reliabilism is a well-confirmed empirical hypothesis, presumably based on the 'fact' that in case after case people holding reliable beliefs have been independently certified to be justified in so doing. Slotting it now into the niche of the synthetic *a priori* may now begin to have some plausibility to it.

My prize exhibit featuring Saul Kripke, which I propose to examine in some detail, takes us rather closer to the contemporary scene than Chisholm's 1957 first edition of *Perceiving*, though the synthetic *a priori* is never so much as once mentioned in *Naming and Necessity* where its powerful presence is felt rather in its subtext. Precisely by advertising his startling rediscovery of necessary *a posteriori* propositions that one only dimly recalled from Aristotle, Kripke performed a sleight-of-hand whereby the synthetic *a priori* could be smuggled in for the ride. Yet I am sure that very few recognized it even when it virtually cried out for recognition in his branding as 'self-evidently absurd.... the view that the *very pain I now have* could have existed without being a mental state at all' (1980, p. 147). Alerted by Kripke's own emphasis to an expressly *de re* thesis, we can generalize it in (1), thereby signalling its difference from its *de dicto* counterpart in (2) which (at least for purposes of argument) I could be brought around to accepting as analytic.

(1) (x) (x is a pain ⊃Nec(x is a mental state)).

(2) Nec(x) (x is a pain ⊃ x is a mental state).

Having elsewhere entertained possible worlds where Kripke's pain token is neither a pain nor a mental state (1989, pp. 180-1), I can readily vouch for Kripke's tacit assignment to (1) of a synthetic as well as *a priori* status.

If the example is probably less than fully satisfactory for my present purposes, which is only to show that the reader, having already acquiesced albeit unwittingly in the synthetic a priori, need not recoil from my explicit appeal to it, that is doubtless to be explained by the fact that Kripke's discussion of pain arises in connection with what was widely felt to be the least convincing part of his book, where he challenges the mind-body Identity Thesis. Much more compelling was Kripke's doctrine as to the Necessity of Origins, so that when David Wiggins styled himself as 'agnostic' in its regard, he was taken to be quite bold and independent in resisting the force of received opinion. And it is precisely in connection with this doctrine above all, which features an interplay between the necessary *a posteriori* and the synthetic a priori, that aprioristic AI and its contrast with the aposterioristic version can begin to take on an almost reassuringly familiar look. The difficulty to be overcome of course is that it was not enough for the necessary a posteriori, with its great novelty, to engross the lion's share of attention. If that was properly to be expected with regard to Kripke, the absence of any awareness of the synthetic *a priori* can only be marvelled at.

The two themes are precisely related in the 'Necessity of Origins' doctrine as particular to universal, with the former instantiating the latter. The particular here is supplied by (3) and the universal by (4), and where (quoting Kripke) we ' let 'B' be a name (rigid designator) of a table, let 'A' name the piece of wood from which it actually came', while we 'let 'C' name another piece of wood' (1980, p. 114).

(3) (x) (y) x = B & y = C ⊃ Nec ¬(x was made out of y).

(4) (x) (y) x is a table & y is a block of wood &
¬(x was made out of y) ⊃ Nec ¬(x was made out of y).

That (3) can only be known to be true on *a posteriori* grounds is evident enough, for one must learn through experience that B was in fact made out of A and not from C. Because (3) ranges over all possible worlds, insisting that in none of them is B made out of C, a further condition for knowing it to be true lies in knowing that the general principle underlying it is true, namely (4), and that can only be accomplished a priori. At which point, encouraged by Kripke's remark that (4) is 'perhaps susceptible of something like proof' based on little more than 'the principle of the necessity of identity', one could readily suppose that (4) is at any rate quasi-analytic. Not so, though only a short detour through pure haecceitism can make the relevant point expeditiously.

Pure haecceitism is the doctrine that (just about) the only essential property of Socrates is his Socrateity, i.e., his being identical with Socrates, entailing that in some possible world Socrates originates precisely as a table, and not even a thinking table, while B correspondingly originates in some world as a human embryo, acknowledging which one hardly need take a big step in order to locate another where B originates once more as a plain table but now made out of C and not A. An absurd line of suggestions? (Albeit not without attraction to the vulgar anti-essentialist who is ever keen to expose any putatively essential property of a thing as being really only accidental to it.) I should like to think so, but as yet I have failed to find it infected with logical incoherence strictly so-called, leading me to conclude that in accepting (4) philosophers have tacitly subscribed to the synthetic *a priori* partly on the strength of which they can then thematically espouse (3) as being a necessary *a posteriori* truth.

My apologies for digressing so far afield from my principal thesis! The excursion has performed more than one service, however, if it has shown how the old and the new, in the form, respectively, of the synthetic *a priori* and the necessary a posteriori, have in Kripke issued in a fruitful union, which we may then emulate afresh in the form now of the synthetic *a priori* and Artificial Intelligence, the latter juxtaposition being surely no more *outré* than the former. In a more positive vein one notices how something like visualization figures in both unions, as when in Kripke's case one imagines a world where two tables are made, one out of A, the other out of C, and where most people can be brought to 'see' that in those circumstances B is surely found to be embodied again in A. A much less metaphorical instance of seeing will be found in our own case where bodies in motion (kinematics) are *seen* as persons in action (cinematics) in the synthesis, reminding us that is is in connection with *Anschauung* above all that the synthetic *a priori* emerges in Kant. Someone's recognizing that a straight line between two points — where 'straight' falls under the category of Quality on a par with 'curved' or 'looped' or 'zig-zagged' — is identical with one between them than which none such is shorter — 'shorter' falls under the category of Quantity — Kant in effect (at B16 in the second edition of the first *Critique*) takes to involve an *a priori* synthesis of two types of concepts requiring therewith the exercise of a faculty of pure *Anschauung*. Something of the sort appears now to go on as well in the projected synthetic *a priori* science of pure robotics ('pure robotics', for short), reminding us moreover of the great importance Kant attached to the construction of figures in geometry. No mere visual aid for him! (See 'Construction in the index of Friedman, 1992.)

Starting out then with an infinite supply of Democritean atoms obeying the laws of Newtonian mechanics, doubtless the simplest case though one early refinement might allow for a touch of causal indeterminacy through some such device as the Lucretian *clinamen* or swerve, pure robotics as the geometry of mind proceeds to 'construct' all possible configurations — diachronic as well as synchronic — of those atoms. Since the portfolio of kinematic world-configurations can be shown to have the daunting cardinality of the real numbers, it will be no routine task for the transcendental imagination to scan them all cinematically in a search for worlds where mind emerges in the form of intelligent robots. No matter. Pure robotics is content to insist on the existence of a master function, Ψ, that picks out the robots in such worlds answering to Dennett's 'real patterns' (1991A), and it is in the light of Ψ that various scaled-down versions of pure (and hybrid or aposterioristic) robotics open up opportunities for the kind of effective research in which Minsky and Pollock are engaged.

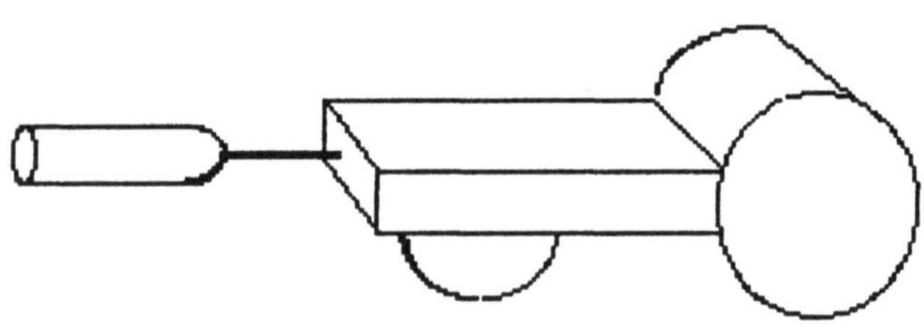

Figure 1: A Simple Braitenbergian Vehicle

If as a long-time student of infinity (see especially Benardete 1964), I have a special relish for the more uninhibited forms of pure robotics, on a sober level we may descend into ' the systems built inthe *qedanken* lab' (1989, p. 89) of Dan Lloyd, notably Squint in fig. 1, designed by Valentino Braitenberg (1984), whose ' simple photocell at the left....is connected to a motor that drives a single fat wheel at the right', thereby 'driving the vehicle in whatever direction it happens to be facing'. Squint 'whirrs in light, rests in darkness' and serves Lloyd as the starting point for an exercise in 'synthetic psychology' (1989, p. 52), inspired by Dennett when he writes that ' if the point is to uncover very general, very abstract principles [of] cognitive organization...why not make up a whole cognitive creature, a Martian three-wheeled iguana, say, and an environmental niche for it to cope with?' (1978, p. 103-4). Well, isn't that precisely the 'point' of my paper albeit dressed up in the jargon of the synthetic *a priori*? Just about, though my emphasis has been more in the line of coming upon Martian three-wheeled iguanas, in the course of leafing through the Democritean portfolio, than in uncovering any very abstract principles of cognitive organization.

Recall that I posited *axioms* with the form 'If Φ then Ψ', thereby indicating scepticism as to their derivability as theorems. Yet it is eminently arguable that any such holistic, cinematic grasp of these low-level 'axioms', which remove the sting from Davidson's anomalous monism, do presuppose an implicit, antecedent grasp of highly general and abstract principles of cognitive organization from which, taken in conjunction with the raw physical scenarios, the 'axioms' can be deduced. How these (presumably aprioristic) principles are to be brought to bear on the raw data, how — in a word — they are to schematize it, may well require something like Kant's doctrine of schematism, and with Kant (at A141/B180-81) one may even fear that ' this schematism of our understanding, in its application to appearances and their mere form, is an art concealed in the depths of the human soul, whose real modes of activity nature is hardly likely ever to allow us to discover, and to have open to our gaze.' Oh, ye of little faith. Cognitive science today positively rejoices in the prospect of exposing to view the neural mechanisms that underly Kant's schematism, and if the aposterioristic features of cognitive science inevitably loom largest (as in the acquired schematism of Dennett's piano tuners[25]), once we allow for different sub-disciplines of cognitive science, depending on how neurological or psychological each may be, there is no great cause now to be surprised by one of them turning out to have a synthetic *a priori* status.

A case in point is already suggested by Lloyd after he has intriguingly characterized his project in terms of 'asking questions very like those of science'. So there is a difference that we need to get hold of, almost certainly connected with the fact that Lloyd's results are expected to be only 'partly contingent'. Much more gritty than the aprioristic psychology of Squint prove to be the T5(2) cells in the optic tectum of the toad whose representational content on being fired Lloyd identifies twice over, once non-extensionally and again extensionally, seeing that 'toads like their dinners to be "wormlike" — fairly small and longer parallel to the direction of movement.' [26] Yet even this thick aposterioristic scenario will be readily seen to have 'artificial' counterparts that the function will pull out of the Democritean portfolio. If the synthetic *a priori* will characterize the latter scenarios, the necessary *a posteriori* may well be found to have a role to play in the former.

*Syracuse University*

## REFERENCES


Benardete, J.A. (1964), *Infinity: An Essay in Metaphysics* (Oxford).
Benardete, J.A. (1989), *Metaphysics: The Logical Approach* (Oxford).
Block, N. (1980), 'Troubles with Functionalism', in *Readings in the Philosophy of Psychology*, Vol. 1, N. Block, ed., (Cambridge, Mass.).
Braitenberg, V. (1984), *Vehicles* (Cambridge, Mass.).
Chisholm, R. (1957), *Perceiving* (Ithaca).
Churchland, P. (1992), *A Neurocomputational Perspective: The Nature of Mind and the Structure of Science* (Cambridge, Mass.).
Dennett, D. (1969), *Content and Consciousness* (London).
Dennett, D,.(1978), 'Why Not the Whole Iguana?' in *Behavioral and Brain Sciences*, 1.
Dennett, D. (1991), *Consciousness Explained* (Boston).
Dennett, D. (1991A), 'Real Patterns', *Journal of Philosophy*, 89, pp. 27-51.
Friedman, M. (1992), *Kant and the Exact Sciences* (Cambridge, Mass.).
Haugeland, J. (1985), *Artificial Intelligence: The Very Idea* (Cambridge, Mass.).
Kant, I. *Critique of Pure Reason.*
Kripke, S. (1980), *Naming and Necessity* (Cambridge, Mass.).
Lloyd, D. (1989), *Simple Minds* (Cambridge, Mass.).
Pollock, J. (1989), *How to Build a Person: A Prolegomenon* (Cambridge, Mass.).
Quine, W.V. (1969), 'Response to Chomsky' in D. Davidson and J. Hintikka, eds., *Words and Objections: Essays on the Work of W.V. Quine* (Dordrecht).
Searle, J. (1992), *The Rediscovery of the Mind* (Cambridge, Mass.).
Stampe, D. (1979), 'Towards a Causal Theory of Linguistic Representation', in P.A. French et al., eds., *Midwest Studies in Philosophy*, (Minneapolis).
Tennant, N. (1987), *Anti-Realism and Logic* (Oxford).
Van Inwagen, P. (1990), *Material Beings* (Ithaca).

FRANK JACKSON

# ARMCHAIR METAPHYSICS

What role if any is there for conceptual analysis in metaphysics? On the face of it, very little. Metaphysics is to do with what is in the world and what it is like, not with concepts and semantics.[1] We would expect science in the wide sense to be highly relevant, but not the armchair deliberations of the philosopher concerned with the analysis of concepts. However, traditionally metaphysicians have paid at least as much attention to questions of conceptual analysis, and to related questions of logical interconnections (to what entails or fails to entail what) as they have to what science tells us about the world. David Armstrong, for example, while rightly and famously insisting that what is said in the philosophy of mind must be answerable to what science tells us about the role of the brain in the causation of behaviour, spends most of *A Materialist Theory of the Mind* (1968) doing conceptual analysis.[2] It is understandable that recently many philosophers writing under the banner of 'naturalism' have declared the traditional preoccupations of metaphysicians with such armchair matters as conceptual analysis and entailments to be a mistake.

Naturalism comes in an extreme and in a moderate form. The extreme form rejects conceptual analysis and its ilk altogether. The extremists sometimes describe the history of conceptual analysis as the history of failure. I think that they are forgetting about biased samples. True, the well known and much discussed examples of putative analyses in the philosophy journals are highly controversial but that is why they are much discussed in the philosophy journals. The more moderate form of naturalism accepts that there is a legitimate activity properly called 'conceptual analysis': for example, conceptual analysis is what mathematicians did when they elucidated the notions of infinity, convergence on a limit, irrational numbers, and so on, and that was a good thing to do. The moderate naturalist view is that conceptual analysis and its ilk are all very well in their place, but their place is not in metaphysics (*qua* speculative cosmology), or at least not at the heart of metaphysics. Questions of analysis and of what entails what belong to semantics and logic, not to ontology and metaphysics. Here is a characteristically straightforward statement by Kim Sterelny of moderate naturalism as it applies to the metaphysics of mind

My approach [to the mind] is not just physicalist, it is naturalist. Naturalists are physicalists . . . [b]ut naturalists have methodological views about philosophy as well; we think philosophy is

*Michaelis Michael and John O'Leary-Hawthorne (eds.), Philosophy in Mind, pp.23-42.*

continuous with the natural sciences. On this view, philosophical theories are conjectures whose fate is ultimately determined by empirical investigation. . . . An alternative conception is to see philosophy as an investigation into conceptual truths proceeding by thought experiments that probe the way we understand our own concepts. . . . There very likely are 'conceptual truths'; truths that depend only on the way we understand concepts and thus depend not at all on how the world is. But I doubt that there are any very interesting conceptual truths about the mind, or about thinking (1990, p. ix).[3]

This paper is a reply to moderate naturalism. I will argue that issues to do with conceptual analysis broadly conceived are inevitably central to any serious metaphysics.

I will start by explaining what I mean by a 'serious metaphysics' and why any such metaphysics brings with it a problem I will call the *placement problem*. I will then argue that there is only one way to solve that problem, namely, by embracing a doctrine I will call *the entry by entailment* doctrine. But this doctrine, I will argue, inevitably makes matters of conceptual analysis central in metaphysics.

## *Serious metaphysics and the placement problem*

Some physical objects are true. For example, if I were to utter a token of the type 'Snow is white', the object I would thereby bring into existence would be true. The object I would thereby bring into existence would also have a certain mass, be caused in a certain way, be of a type the other tokens of which have characteristic causes and effects in my mouth and from my pen and in the mouths and from the pens of my language community, have a certain structure the parts of which have typical causes and effects, and so on and so forth. How is the first property, the semantic property of being true, related to the host of non-semantic properties of the sentence? How can we find a place or location for the semantic in the physical story about the sentence?

Some who have puzzled about this question have been tempted by a sceptical position on truth, and correspondingly on meaning and reference. They feel that the list of non-semantic properties — suitably expanded of course — is an in principle complete one. Sentences are, when all is said and done, a species of physical object, and we know that science can in principle tell us the whole story about physical objects. And though we are not yet, and may never be, in a position actually to give that whole story, we know enough as of now to be able to say (a) that it will look something like the story I gave a glimpse of, and (b) that in any case it will not contain terms for truth, reference and meaning. But if the whole story does not contain truth, reference and meaning, then so much the worse for truth, reference and meaning.[4]

A quite different response is to distinguish what appears explicitly in a story from what appears implicitly in a story. I tell you that Jones weighs 70 kilos and Smith weighs 80 kilos. That is something I tell you explicitly. Do I also tell you that Jones weighs less than Smith? Not in so many words, but of course it is implicit in what I said in the following sense: what I said entails that Jones weighs less than Smith. Likewise, runs the alternative response, truth, reference and meaning are implicit in the story completed science will tell about our sentence and the world in which it is produced: that story will entail that the sentence is true, that it has a certain meaning, and that its parts refer to certain things, including snow. This response locates the semantic properties of sentences within the picture completed science tells about sentences and the world by arguing that they are entailed by that story. The semantic gets a place in the physical story by being entailed by the physical story.

I have just described a familiar example of the placement problem, and two responses to it. But we can generalise. Metaphysics, we said, is about what there is and what it is like. But of course it is concerned not with any old shopping list of what there is and what it is like. Metaphysicians seek a comprehensive account of some subject matter – the mind, the semantic, or, most ambitiously, everything – in terms of a limited number of more or less basic notions. In doing this they are following the good example of physicists. The methodology is not that of letting a thousand flowers bloom but rather that of making do with as meagre a diet as possible. Some who discuss the debate in the philosophy of mind between dualism and monism complain that each position is equally absurd. We should be *pluralists*. Of course we should be pluralists in some sense or other. However, if the thought is that any attempt to explain it all, or to explain it all as far as the mind is concerned, in terms of some limited set of fundamental ingredients is mistaken in principle, then it seems to me that we are being, in effect, invited to abandon serious metaphysics in favour of drawing up big lists. But if metaphysics seeks comprehension in terms of limited ingredients, it is continually going to be faced with the problem of location. Because the ingredients *are* limited, some putative features of the world are not going to appear explicitly in the story. The question then will be whether they nevertheless figure implicitly in the story. Serious metaphysics is simultaneously discriminatory and putatively complete, and the combination of these two facts means that there is bound to be a whole range of putative features of our world up for either elimination or location.

What is it for some putative feature to have a place in the story some metaphysic tells in its favoured terms? One answer, already mentioned, is for the feature to be entailed by the story told in the favoured terms. I take it that few will quarrel with this as a sufficient condition. One good way to get a place in the story is by entailment. I will argue that not only is it one way, it is the only way. The one and only entry ticket into the story told in the preferred terms is by being entailed by that story.

In order to focus and make familiar the discussion, I am going to develop the argument for the entry by entailment view in terms of the particular example of physicalism and the psychological. I will be arguing, that is, that the psychological appears in the physicalists' story about our world if and only if that story entails the psychological. Similarly, when I discuss the connection between entry by entailment and conceptual analysis I will focus on the example of physicalism and the psychological. It may help in following the argument that is to come to assume the truth of physicalism (an assumption I take, reluctantly, to be contrary to fact). Although the considerations will be deployed in this familiar and particular context, it will be clear, I trust, that they would apply generally.

*Placing the psychological in the physicalists's picture*

Physicalism is the very opposite of a 'big list' metaphysics. It is highly discriminatory, operating in terms of a small set of favoured properties and relations, typically dubbed 'physical'; and it claims that a complete story, or anyway a complete story of everything contingent, about our world can in principle be told in terms of these properties and relations alone. It is miserly in its basic resources while being as bold as can be in what it claims.

A fair question is how to specify precisely the notion of a physical property. I am not going to answer this fair question. Roughly, I will mean what is typically meant: the kinds of properties that figure in, or are explicitly definable in terms of, those that figure in physics, chemistry, biology, and neuroscience. This rough characterisation leaves it open why those sciences, rather than say psychology or politics, are chosen to settle the favoured class, and it says nothing about how committed this approach is to those sciences being roughly right in the kinds of properties they need for their own internal purposes. Nevertheless, I think that the rough characterisation will do for our purposes here. As far as I can see, nothing in what follows turns on the answers to these two controversial matters. What is important here is that there is a favoured list, not how a property or relation gets to be on that favoured list.[5]

What will be important is the notion of a complete story. Our argument for the entry by entailment thesis as it applies to physicalism and the psychological, will be that it is the physicalists' claim to have a complete story about the nature of our world which commits them to our world having a psychological nature if and only if that nature is entailed by the world's physical nature.

Physicalists variously express their central contention as that the world is entirely physical; as that when you have said all there is to say about physical properties and relations you have said all there is to say about everything, or anyway everything contingent including psychology; that the world is nothing but or nothing over and above the physical world; that a full inventory of the instantiated physical properties and relations would be a full inventory *simpliciter*; and so on and so forth. What does all this come to?

We can make a start by noting that one particularly clear way of showing *incompleteness* is by appeal to independent variation. What shows that three co-ordinates do not provide a complete account of location in space-time is that we can vary position in space-time while keeping any three co-ordinates constant. Hence, an obvious way to approach completeness is in terms of the lack of independent variation. Four co-ordinates completely specify position in space-time because you cannot have two different positions with the same four co-ordinates. But of course lack of independent variation is supervenience: position in space-time supervenes on the four co-ordinates. So the place to look when looking for illumination regarding the sense in which physicalism claims to be complete, and, in particular, to be complete with respect to the psychological, is at various supervenience theses.[6]

Now physicalism is not just a claim about the completeness of the physical story concerning the individuals in our world. It claims completeness concerning the world itself. Accordingly, we need to think of the supervenience base as consisting of worlds, not merely of individuals in them. *Intra*-world supervenience theses will not capture what the physicalists have in mind. We need to look to *inter*-world *global* supervenience theses, an example of which is

(A) Any two possible worlds that are physical duplicates (physical property and relation for physical property and relation identical) are duplicates *simpliciter*.[7]

But (A) does not capture what the physicalists have in mind. Physicalism is a claim about our world to the effect that its physical nature exhausts all its nature, and our world is nowhere mentioned in (A). It is, though, mentioned in

(B) Any world that is a physical duplicate of our (the actual) world is identical *simpliciter* with our world.

But (B) is false, and for a reason that does not appear to count against physicalism. Physicalists are typically happy to grant that there is a possible world physically exactly like ours but which contains as an addition a lot of mental life sustained in non-material stuff. Indeed, what they grant as possible is what some theists believe to be the truth about our world. These theists hold that physicalism is the correct account of earthly existence but it leaves out of account the after-life. When we die our purely material psychology is reinstated in purely non-material stuff (which fortunately is considerably more durable than the material variety).(See John Hick, 1964.)

The trouble with (B) is that it represents physicalists' claims as more wide ranging than they in fact are. What we need is something like (B) but which limits itself to worlds more nearly like ours, or at least more nearly like ours on the physicalists' conception of what our world is like. I suggest

(C) Any world which is a *minimal* physical duplicate of our world is a duplicate *simpliciter* of our world.

What is a minimal physical duplicate? Think of a recipe for making scones. It tells you what to do, but not what *not* to do. It tells you to add butter to the flour but does not tell you not to add dirt to the flour. Why doesn't it? Part of the reason is that no-one would think to add dirt unless explicitly told to. But part of the reason is logical. It is impossible to list all the things not to do. There are indefinitely many of them: don't add bats wings; don't add sea water; don't add . . . . Of necessity the writers of recipes rely on an intuitive understanding of an implicitly included 'stop' clause in their recipes. A minimal physical duplicate of our world is what you would get if you — or God, as it is sometimes put — used the physical nature of our world (including of course its physical laws) as a recipe in this sense for making a world.

We noted earlier the physicalists' talk of everything being nothing over and above, nothing more than, the physical; of the full physical story being complete, being the full story *simpliciter*, and so on. The hope has sometimes been that the key notion of, as it is sometimes called, 'nothing buttery' might be susceptible of a precise, non-circular definition in terms of supervenience. If (C), or something like it, is the best we can do by way of specifying what physicalism is committed to in terms of supervenience, then this hope has been dashed. The notion of minimality that features in (C) is too close to nothing buttery. We have I trust made matters clearer, but we have not reached what might have been hoped for.

We arrived at (C) by eliminating alternatives. But we can give a positive argument for the conclusion that the physicalist is committed to (C). Suppose, first, that (C) is false; then there is a difference in nature between our world and some minimal physical duplicate of it. But then either our world contains some nature that the minimal physical duplicate does not, or the minimal physical duplicate contains some nature that our world does not. The second is impossible because the extra nature would have to be non-physical (as our world and the duplicate are physically identical), and the minimal physical duplicate contains no non-physical nature by definition. But if our world contains some nature that the duplicate does not, this nature must be non-physical (as our world and the duplicate are physically identical). But then physicalism would be false, for our world would contain some non-physical nature. Hence, if (C) is false, physicalism is false. We now show that if physicalism is false, (C) is false. If physicalism is false, our world contains some non-physical nature. But that nature cannot be present in any minimal physical duplicate of our world, as any such duplicate is a minimal duplicate. But then any such world is not a duplicate *simpliciter* of our world, and hence (C) is false.

There is a debatable step in this little proof that physicalism is true if and only if (C) is true. At the last stage I assumed that if our world contains some non-physical nature, there is a minimal physical duplicate which lacks that nature. But suppose that some complex of physical properties in this world is necessarily connected in the strongest sense of 'necessary' to some quite distinct, non-physical property. In that case the assumption will be false. One response to this objection would be to invoke the Humean ban on necessary connections between distinct existences, but a less controversial response is available. I think that we know enough about what the physical properties are like — despite the fact that we ducked the question of how to specify them precisely — to be confident that the connection in question is not possible. The kinds of properties and relations that figure in the physical sciences do not have mysterious *necessary* connections with quite distinct properties and relations. In any case, this is what physicalists should say. To say otherwise is to abandon physicalism (in favour of something far more mysterious than interactionist dualism). So physicalists must, it seems to me, accept (C).

I said before that (C) does not provide the sometimes hoped for non-circular account of 'nothing buttery'. If we eschew the Humean dismissal, we must allow that (C) does not even capture the content of physicalism. (C) is something the physicalist is committed to by virtue of claiming completeness, but (C) does not express what she is committed to *qua* physicalist. Someone who thought that there are complexes of physical properties in this world which necessarily connected with irreducibly psychic properties could accept that every minimal physical duplicate of our world is a duplicate *simpliciter* — either on the ground that there are no *minimal* physical duplicates because it is impossible to have the physical properties without the associated but distinct psychic ones, or on the ground that 'minimal physical duplicate' means as minimal as possible, in which case there are minimal physical duplicates but every one of them has the associated psychic properties and so is a duplicate *simpliciter* — but she would not thereby be a physicalist (obviously!). I suspect that the only way to say what physicalists hold is to use the kinds of terms they have used traditionally. Appropriate supervenience theses illuminate what they are committed to, but do not capture what they believe.[8]

*A side issue: supervenience and singular thought*[9]

To accept (C) is *ipso facto* to accept

(C*) Any world which is a minimal physical duplicate of our world is a psychological duplicate of our world.

Consider a minimal physical duplicate of the actual world. It will contain a duplicate of Bush. It might be urged that our Bush's psychology, while being very similar to his doppelgänger's, will not be quite the same as his doppelgänger's. Their singular thoughts will be different by virtue of being directed to different objects. Only our Bush is thinking about our Clinton. Thus, if physicalism is committed to (C*), physicalism is false.

One response to this putative disproof of physicalism would be to challenge the view about singular thought that lies behind it, but I think we can steer clear of that issue for our purposes here.[10] The disproof as stated is trading on the counterpart way of thinking about objects in possible worlds: the way according to which no object appears in more than one world, and what makes it the case that an object which is *F* might have failed to be *F* is the fact that its counterpart in some possible world is not *F*.[11] However, the duplicate of our Bush is thinking about the very same person as our Bush in the only sense that the counterpart theorist can take seriously. If I had scratched my nose a moment ago, I would still have had the very same nose that I actually have. Noses are not that easy to remove and replace. The counterpart theorist has to say that what makes that true are certain facts about the nose of my counterpart in a world where my counterpart scratched his nose a moment ago. If that is good enough for being the very same nose, then the corresponding facts about Bush's counterpart are good enough for it to be true that he is thinking about the very same person and hence having the same singular thought.

The putative disproof might be developed without trading on the counterpart way of looking at matters. A believer in trans-world identity typically holds that whether an object in our world is literally identical with an object in another is not a qualitative matter. Such a believer might well hold that though Bush, our Bush, and Clinton, our Clinton, eyeball each other in more worlds than this one, it is nevertheless true that in some minimal physical duplicates of our world, our Bush thinks about a qualitative duplicate of our Clinton who nevertheless is not our Clinton. On some views about singular thought, Bush will count as having a different thought in such a world from the thought he has in our world. In this case the steering clear requires a modification of (C). The physicalist will need to require that minimal physical duplicates of our world be ones which, in addition to being physical property and relation identical with our world, are haecceity-associated-with-physical-property-and-relation identical with our world.

*From (C) to entry by entailment*

Given that (C) follows from physicalism, there is a straightforward and familiar argument to show that if physicalism is true, then the psychological story about our world is entailed by the physical story about our world. (C) entails that any psychological fact about our world is entailed by the physical nature of our world.

We can think of a statement as telling a story about the way the world is, and as being true inasmuch as the world is the way the story says it is. Let Ø be the statement (an infinite disjunction of very long conjunctions) which tells the rich, complex and detailed physical story which is true at the actual world and all the minimal physical duplicates of the actual world, and false elsewhere. Let $\prod$ be any true statement entirely about the psychological nature of our world: $\prod$ is true at our world, and every world at which $\prod$ is false differs in some psychological way from our world. If (C) is true, every world at which Ø is true is a duplicate *simpliciter* of our world, and so *a fortiori* a psychological duplicate of our world. But then every world at which Ø is true is a world at which $\prod$ is true — that is, Ø entails $\prod$.

We have thus derived the entry by entailment thesis for the special case of physicalism and the psychological. A putative psychological fact has a place in the physicalists' world view if and only if it is entailed by some true, purely physical statement. Any putative psychological fact which is not so entailed must be regarded by the physicalist as either a refutation of physicalism or as *merely* putative. Although the argument was developed for the special case of physicalism and the psychological, the argument did not depend crucially on matters local to that special case. We could have argued in the same general way in the case of physicalism and the semantic, or in the case of cartesian dualism and the semantic, or in the case of berkelean idealism and physical objects.[12] Our argument essentially turned on just two facts about any serious metaphysics: it is discriminatory, and it claims completeness. It is those two features of serious metaphysics which mean that it is committed to views about what entails what.

How does entry by entailment show the importance of conceptual analysis? If Ø entails $\prod$, what makes Ø true also makes $\prod$ true (at least when Ø and $\prod$ are contingent). But what makes Ø true is the physical way our world is. Ø needs nothing more than that. Hence, the physicalist is committed to each and every psychological statement being made true by a purely physical way our world is. The analytical functionalist has a story about how this could be. It comes in two stages. One stage — the most discussed stage — is about how certain functional cum causal facts make it true that a subject is in one or another psychological state. The other stage is about how certain physical facts make it true that the appropriate functional-cum-causal states obtain. The story is a piece of conceptual analysis. Analytical functionalism is defended by the *a priori* methods characteristic of conceptual analysis. For us the important point is that the physicalist must have *some* story to tell; otherwise how the purely physical makes psychological statements true is rendered an impenetrable mystery. But it is the very business of conceptual analysis to explain how matters framed in terms of one set of terms and concepts can make true matters framed in a different set of terms and concepts. When we seek an analysis of knowledge in terms of truth, belief, justification, causation and so on, we seek an account of how matters described in terms of the latter notions can make true matters described in terms of the former. So far we have been unsuccessful (I take it). When we seek a causal account of reference, we seek an account of the kinds of causal facts which make it true that a term names an object. When and if we succeed, we will have an account of what makes it true that 'Moses' names Moses in terms of, among other things, causal links between uses of the word and Moses himself. Compatibilists about free will seek accounts of what it is to act freely in terms of facts about the agent's abilities and the causal role of the agent's character. If they have succeeded, they have told us how the way things are described in terms of abilities and causal origins make true the ways things are described in terms of free action. And so on and so forth.

I take the understanding of conceptual analysis just outlined to be pretty much the standard one, but there seems to be another understanding around. For example, some who discuss, and indeed advocate, some version or other of the causal theory of reference also say that they oppose conceptual analysis. But the causal theory of reference *is* a piece of conceptual analysis in our sense. It is defended by *a priori* reflection on, and intuitions about, possible cases in the manner distinctive of conceptual analysis. And, indeed, it is hard to see how else one could argue for it. The causal theory of reference is a theory about the conditions under which, say, 'Moses' refers to a certain person. But that is nothing other than a theory about the possible situations in which 'Moses' refers to that person, and the possible situations in which 'Moses' does not refer to that person. Hence, inuitions about various possible situations — the meat and potatoes of conceptual analysis — are bound to hold centre stage simply because it is better to say what is intuitively plausible than what is intuitively implausible. Or consider the book by Sterelny from which we quoted the passages saying that 'an alternative [to the methods of this book] is to see philosophy as an investigation into conceptual truths proceeding by thought experiments' and 'I doubt that there are any very interesting conceptual truths about the mind, or about thinking'. Later in the book we find the following passage:

> . . . representational notions play a central role in psychology. Yet representational kinds are not individualistic. There is a standard parable that makes this point. Imagine that, far away in space and time, there is a world just like ours, Twin Earth. Twin Earth really is a twin to Earth; the parallel is so exact that each of us have a doppelganger on Twin Earth, our twinself. A person's twinself is molecule-for-molecule identical to himself or herself. . . . you and your twin are individualistically identical, for what is going on inside your heads is molecule-for-molecule the same. But the representational properties of your thoughts are not always the same. For consider your thoughts about your mother. These are thoughts about *your* mother. That individual's motherish thoughts are about another individual. An individual amazingly similar to your mother, but none the less a different person . . . . So the representational properties of thoughts about particular people don't supervene on brain states; they are not individualistic. Variations of the Twin Earth theme have been run by Putnam and Burge to show that thoughts about natural kinds and socio-legal kinds are not individualistic either (pp. 82 - 83).

Surely in this passage we are being offered a defence of a conceptual claim about the representational properties of our thoughts, namely, to the effect that they are not individualistic, and moreover the defence is via a thought experiment (a 'parable')? Clearly, Sterelny, and at least some other critics of conceptual analysis, must mean by conceptual analysis something different from what we mean here.[13]

Many will feel that there is a major objection to the argument given from (C) to the conclusion that Ø entails Π. The objection is that the argument neglects a crucial distinction between entailment and *fixing* which arises from the recognition of the necessary *a posteriori* .[14]

*Entailment and the necessary a posteriori*

'Water = $H_2O$' is necessarily true (modulo worlds where there is no water) despite being *a posteriori.*[15] This means that conditionals which say that if $H_2O$ has a certain distribution, so does water — for example, 'if $H_2O$ is *L*-located then water is *L*-located' — are necessarily true though *a posteriori.* Does it follow that '$H_2O$ is *L*-located' entails 'Water is *L*-located'? You might say yes on the ground that every world where $H_2O$ is *L*-located is a world where water is *L*-located. You might say no on the ground that the conditional is not *a priori* knowable. The second response might be bolstered by noting that we would not normally assess the argument

> $H_2O$ is *L*-located.
> Therefore, water is *L*-located.

as valid. The second response might be spelt out by insisting on a distinction between fixing and entailment proper.[16] If every *P*-world is a *Q*-world, then *P* fixes *Q*, but in order for *P* to entail *Q* it must in addition be the case that 'if *P* then *Q*' is *a priori.*

The decision between the two responses turns, it seems to me, on the decision between two different ways of looking at the distinction between necessary *a posteriori* statements like 'Water = $H_2O$' and necessary *a priori* ones like '$H_2O$ = $H_2O$'. You might say that the latter are analytically or conceptually or logically (in some wide sense not tied to provability in a formal system) necessary, whereas the former are metaphysically necessary, meaning by the terminology that we are dealing with two senses of 'necessary' in somewhat the way that we are when we contrast logical necessity with nomic necessity (see, for example Forrest, 1992).[17] Thus, the class of nomic possibilities is a proper subset of the class of logical possibilities – every nomic possibility is a logical possibility, but some logical possibilities (for example, travelling faster than the speed of light) are not nomic possibilities. Similarly, the idea is that the class of metaphysical possibilities is a proper subset of the class of logical possibilities. Every metaphysically possible world is logically possible, but some logically or conceptually possible worlds – for instance, those where water is not $H_2O$ – are metaphysically impossible. On this approach, the reason the necessity of water's being $H_2O$ is not available *a priori* is that though what is logically possible and impossible is available in principle to reason alone given sufficient grasp of the relevant concepts and logical acumen, what is metaphysically possible and impossible is not so available.

I think, as against this view, that it is a mistake to hold that the necessity possessed by 'Water = $H_2O$' is different from that possessed by 'Water = water' (or that possessed by '2 + 2 = 4'). Just as Quine insists that numbers and tables exist in the very same sense, and that the difference between numbers existing and tables existing is a difference between numbers and tables, I think that we should insist that water's being $H_2O$ and water's being water are necessary in the very same sense. The difference lies, not in the kind of necessity possessed, but rather where the labels '*a priori*' and '*a posteriori*' suggest it lies: in our epistemic access to the necessity they share.

My reason for holding that there is one sense of necessity here relates to what it was that convinced us that 'Water = $H_2O$' is necessarily true. What convinced us were the arguments of Saul Kripke and Hilary Putnam to the conclusion that the terms 'Water' and '$H_2O$' are rigid designators, and so that it follows from the fact that 'Water is $H_2O$' is true that it is necessarily true. This was the discovery of a semantic fact about certain terms, not the discovery of a new sort of necessity. Kripke and Putnam taught us something important about the semantics of certain singular terms that means that we must acknowledge a new way for a statement to be necessary, a way which cannot be known *a priori*; but it is a new way to have the old property (Kripke, 1980; Putnam, 1975).

If the relevant statements are necessary in the same sense but differ with regard to the epistemological status of their shared necessity, there is an obvious response to the question, Does '$H_2O$ is *L*-located' entail ' Water is *L*-located'? It is to say yes, but hasten to distinguish *a priori* from *a posteriori* entailment, and hold that the entailment in this case is *a posteriori*. In short, the response is to adopt the following definitions:

*P a priori* entails *Q* iff (a) every *P*-world is a *Q*-world, and (b) the conditional 'If *P* then *Q*' is *a priori*.
*P* entails (or fixes) *Q* iff every *P*-world is a *Q*-world.
*P a posteriori* entails *Q* iff (a) *P* entails *Q*, and (b) *P* does not *a priori* entail *Q*.

Hence it seems to me that our result that physicalism is committed to Ø's entailing ∏ stands. (From here on we will think of ∏ as some arbitrary true psychological statement; it might even be the huge one that says all there is to say about our world's psychology.) But there are two objections that the moderate naturalist might well make all the same.

First, she might object that our argument assumed that there are psychological facts. The entailment result was obtained from the combination of (C) with the assumption that there are true psychological statements. A physicalist of the eliminativist persuasion can reject one of our premises. This is right, but does not affect the overall issue. Eliminativism about the mental is implausible. Eliminativism across the board about matters described in any terms other than the austerely physical is incredible. We could have developed the argument in terms of the relationship between the physical story and any true statement S about our world not framed in the austere terms in order to show that the physicalist is committed to Ø's entailing S. The general point is that a serious metaphysics is committed to there being entailments between the full story told in its favoured terms and each and every truth told in other terms. The only physicalistic eliminativism which would avoid commitment to entailments from Ø to statements in other terms would be one which said that there are no truths tellable in other terms. But such a metaphysics would no longer be a serious one in our sense. For it would not effect a partition of the truths acknowledged by it into those pertaining most directly to the properties and relations viewed as the fundamental ingredients of everything, and the others; it would not be discriminating.[18]

Secondly, she might very reasonably object that the result that Ø entails ∏ is not a result which supports the importance of conceptual analysis in anything like the traditional sense. Conceptual analysis in the traditional sense, as we noted earlier, is constituted by *a priori* reflection on concepts and possible cases with an aim to elucidating connections between different ways of describing matters. Hence, it might be objected, if we allow that some entailments are *a posteriori*, to demonstrate an entailment is conspicuously not to demonstrate the importance of conceptual analysis. In order to reply to this objection I need to say more about the necessary *a posteriori*.

*The a priori part of the story about the necessary a posteriori*

We urged that the explanation of the *a posteriori* nature of the necessary *a posteriori* does not lie in the special necessity possessed; where then does it lie? I think that the answer is best approached if we have before us from the beginning two central facts about the issue. First, the issue is an issue about sentences, or if you like statements or stories in the sense of assertoric sentences in some possible language, and not about propositions, or at least not propositions thought of as sets of possible worlds.[19] By the argument lately rehearsed, the set of worlds where water is water is the very same set as the set where water is $H_2O$, so by Leibnitz's Law there is no question of the proposition that water is water differing from the proposition that water is $H_2O$ in that one is, and one is not, necessary *a posteriori*. Secondly, the puzzle about the necessary *a posteriori* is not how a sentence can be necessary and yet it take empirical work to find this out. Russians utter plenty of sentences which are necessarily true and yet it takes, or would take, many of us a lot of empirical work to discover the fact. The puzzle is how a sentence can be necessarily true and understood by someone, and yet the fact of its necessity be obscure to that person. And the reason this is a puzzle is because of the way we use sentences to tell people how things are.

Consider what happens when I utter the true sentence 'There is a coin in my pocket'. I tell you something about how things are, and to do that is precisely to tell you which of the various possibilities concerning how things are is actual. My success in conveying this simple bit of information depends on two things: your understanding the sentence; and your taking the sentence to be true. We have here a folk theory that ties together understanding, truth, and information about possibilities; and the obvious way to articulate this folk theory is to identify, or at least essentially connect, understanding a sentence with knowing the conditions under which it is true, that is, knowing the possibilities at which it is true and the possibilities at which it is false.[20] Indeed, if we could not connect understanding a sentence, accepting it, and knowing about possibilities in this kind of way, there would be little point in asserting sentences to one another. There are many possibilities concerning when the bus leaves. Your uttering a sentence I understand and accept, namely, 'The bus leaves at six' is exactly what I need to tell me which of the many possibilities is actual, and that fact constitutes the essential point of your utterance. But it seems then that understanding a necessarily true sentence should, at least in principle, be enough to reveal its necessary status. For understanding it would require knowing the conditions under which it is true, and how could you know them and yet fail to notice that they are universal? The puzzle is particularly pressing when, as in the cases we are concerned with, the statements are about relatively accessible, contingent features of our world.

I think – unoriginally[21] – that the way out of our puzzle is to allow that we understand some sentences without knowing the conditions under which they are true, but to do this in a way which retains the central idea of the folk theory that the three notions of: understanding a sentence, of a sentence's being true, and of the information carried by a sentence, are very closely connected. It is just that the connection is not always so simple as that understanding requires knowledge of the truth-conditions of what is expressed.

Here is an illustrative example, familiar from discussions of two-dimensional modal logic, of understanding without knowing truth-conditions.[22] Suppose I hear someone say 'He has a beard'. I will understand what is being said without necessarily knowing the conditions under which what is said is true, because I may not know who is being spoken of. As it is sometimes put, I may not know which proposition is being expressed. If I am the person being spoken of, the proposition being expressed is that Jackson has a beard, and what is said counts as true if and only if Jackson has a beard; if Jones is the person being spoken of, the proposition being expressed is that Jones has a beard, and what is said counts as true if and only if Jones has a beard. Hence, if I don't know whether it is Jackson, Jones or someone else altogether, I don't know which proposition is being expressed, and I don't know the conditions under which what is said is true. But obviously I do understand the sentence. However granting this does not require us to abandon the folk theory that understanding, truth and information are very closely connected. For I am much better placed than the Russian speaker who hears 'He has a beard'. Unlike the Russian speaker, I know how context determines the conditions under which what is said is true. I know how to move from the appropriate contextual information, the information which in this case determines who is being spoken of, to the truth-conditions of what is said. Thus, uttering the sentence has for me but not for the Russian speaker, the potential to convey information about which possibilities are realised, and it has this potential precisely because I understand the sentence.[23]

A similar point can be made about 'water' sentences. The doctrine that meanings ain't in the head means that the truth-conditions of our 'water' sentences depend on matters outside our heads that we may be ignorant of. But someone who does not know that it is $H_2O$ that falls from the sky and all the rest, may, as we have already noted, understand 'water' sentences. Hence, understanding, say, 'There is water hereabouts' does not require knowing the conditions under which it is true, though it does require knowing how the conditions under which it is true depend on context, on how things are outside the head.[24] But this means that although understanding alone does not give truth-conditions, understanding alone does give us the way truth-conditions depend on context, and that fact is enough for us to move *a priori* from, for example, statements about the distribution of $H_2O$ *combined with the right context-giving statements*, to information about the distribution of water. For example, suppose that the right account of the semantics of 'water' is that it is a rigidified definite description meaning roughly 'stuff which actually falls from the sky, fills the oceans, is odourless and colourless, is essential for life, is called "water" by experts,. . . , or which satisfies enough of the foregoing', and consider the inference

(1) $H_2O$ is *L*-distributed.
(2) $H_2O$ is the stuff which falls from the sky, fills the oceans, etc.
(3) Therefore, water is *L*-distributed.

Although, as noted earlier, the passage from (1) to (3) is *a posteriori* — (2) being the *a posteriori*, contingent fact that needs to be known in order to make the step from (1) to (3) — the passage from (1) *and* (2) to (3) will be *a priori*. And it will be so because although our understanding of the word 'water' does not determine its reference-conditions, it does determine how the reference-conditions of the word depend on context, and (2) gives that context. It gives the relevant fact about how things are outside the head.[25] Our understanding of the relevant terms plus logical acumen is enough to enable us to go from (1) combined with (2) to (3). Indeed, this is the *a priori* fact which reveals the *a posteriori* entailment. We did not know that (1) entailed (3) until we learnt (2). But as soon as we learnt (2) we had the wherewithal to move *a priori* to (3). We could put it this way. Conceptual analysis tells us that we can move from the combination of (1) and (2) to (3). For conceptual analysis — what is in principle possible *a priori* from understanding plus logical acumen — tells us how truth-conditions depend on context, and that is all we need to go from the combination of (1) and (2) to (3).

Earlier I drew the connection between the entry by entailment thesis which commits physicalists to holding that $\emptyset$ entails $\Pi$, and the centrality of conceptual analysis to physicalism, by saying (a) that physicalism, on pain of making mysteries, had better tell us how physical nature makes true, in the sense of necessitating, psychological nature, and (b) that conceptual analysis tells us how matters framed in one set of terms can make true matters framed in different terms. One way of putting the objection under discussion is that $H_2O$ being *L*-distributed makes it true that water is *L*-distributed, but conceptual analysis cannot tell us about this. That $H_2O$ facts make true water facts is *a posteriori*. Our reply is that there *is* an *a priori* story to be told about how $H_2O$ facts make true water facts.

More generally, the two-dimensional modal logic way of looking at the necessary *a posteriori* shows that even if the entailment from $\emptyset$ to $\prod$ is *a posteriori*, there is still an *a priori* story tellable about how the story in physical terms about our world makes true the story in psychological terms about our world. Although understanding may not even in principle be enough to yield truth-conditions, it is enough to yield how truth-conditions depend on context. But of course the context is, according to the physicalist, entirely physical. Hence, the physicalist is committed to there being an *a priori* story to tell about how the physical way things are makes true the psychological way things are. But the story may come in two parts. It may be that one part of the story says which physical way things are, $\emptyset_1$, makes some psychological statement true, and the other part of the story, the part that tells the context, says which different physical way things are, $\emptyset_2$, makes it the case that it is $\emptyset_1$ that makes the psychological statement true. What will be *a priori* accessible is that $\emptyset_1$ and $\emptyset_2$ together make the psychological statement true.[26]

*Australian National University*

NOTES

[1] I mean metaphysics in the sense of speculative cosmology rather than analytic ontology; see D. C. Williams (1966).

[2] Of course part of the reason that Armstrong spends so much of the book doing conceptual analysis is that he thinks (reasonably) that the relevant empirical results are all but in.

[3] I suspect however that Sterelny means something different by 'conceptual truths' and 'conceptual analysis' from what we will be meaning by it. I return to this matter later.

[4] This line of thought is very clearly presented but not endorsed in Michael Devitt (1984), ch. six 'Why do we need Truth?'.

[5] Physicalists who think that there are irreducibly *de se* truths — for instance, about what a person thinks — will of course include the appropriate physical specification of them in their favoured list. I am indebted here to Richard Holton and David Lewis.

[6] What follows is one version of a familiar story, see, e.g., Terence Horgan (1982) and David Lewis (1983). I am much indebted to their discussions.

[7] As far as I can see, it does not matter for what follows what view precisely is taken of non-actual possible worlds: perhaps they are concrete entities of the same ontological type as our world, as David Lewis (1986) holds; perhaps they are abstract entities as Robert Stalnaker (1976) holds; perhaps they are structured universals as Peter Forrest (1986) holds; perhaps they are nothing at all but talk of 'them is understandable in terms of combinations of properties and relations, as D. M. Armstrong (1989) holds.

[8] I am indebted here to André Gallois and David Braddon-Mitchell.

[9] I am indebted to Rae Langton and David Lewis in what follows.

[10] For a survey of various views about singular thought, see Simon Blackburn (1984).

[11] See Lewis (1986).

[12] A traditional objection to idealism has been that no congery of actual and hypothetical facts about sense impressions, no matter how complex and detailed, *entails* that there is a table in the next room; and so idealists must embrace eliminativism about tables. A contemporary criticism of this traditional objection is that it is confusing the ontological thesis of idealism with the analytical thesis of phenomenalism. If we are right, the traditional objection is very much to the point.

[13] I am indebted here to a discussion with Gilbert Harman.

[14] The possibility of this objection was forcibly drawn to my attention by Peter Railton, Lloyd Humberstone and Michaelis Michael. I am not sure where they stand on it however. The terms in which I characterise it are closest to those of Peter Forrest (1992).

[15] I am taking for granted the view that the term 'water' is a rigid designator of (as we discovered) $H_2O$. I am in fact agnostic on the question of whether 'water' as used by the person in the street is a rigid designator. But clearly it is a rigid designator in the mouths and from the pens of many philosophers alive today, and for the purposes of the discussion here we will give it that meaning. It is clear in any case that 'water' *could* have been a rigid designator.

[16] Thus Stephen Yablo (1992) distinguishes fixing from conceptual entailment, see esp. pp. 253-4. What he means by conceptual entailment is, I take it, the same as what we mean by *a priori* entailment below.

[17] Saul Kripke (1980), p. 125, says that 'statements representing scientific discoveries about what this stuff *is* . . . are . . . necessary truths in the strictest possible sense'. This suggests that he does not hold the view in question, though he does not, as far as I know, address the matter explicitly.

[18] I am indebted here to a discussion with John O'Leary-Hawthorne.

[19] At the end of the preface to *Naming and Necessity*, Oxford, Basil Blackwell, 1980, Kripke insists that his concern is with sentences, not propositions; see pp. 20-21.

[20] This kind of theory in its philosophically sophisticated articulations is best known through the work of David Lewis, see, e.g., *On the Plurality of Worlds* (1986), and Robert Stalnaker (1984). But it would, I think, be wrong to regard the folk theory as being as controversial as these articulations. The folk theory is, it seems to me, a commonplace. The Sports section of any newspaper is full of speculations about possible outcomes conveyed by sentences that discriminate among the outcomes in a way we grasp because we understand the sentences. There is of course a major problem about what to say concerning mathematical statements within this framework, but our concern will be with statements about relatively mundane items like water, and with entailments between statements putatively representing the way things are as a matter of empirical fact.

[21] I take it that the account which follows is a thumbnail sketch of the approach naturally suggested by the two-dimensional modal logic treatment of the necessary *a posteriori*, as in for instance Robert Stalnaker (1978), and Martin Davies and Lloyd Humberstone (1980). They should not be held responsible for my way of putting matters.

[22] The example is a variant on one discussed by Stalnaker (1978).

[23] It might be objected that I do know who is being spoken of: I know something unique about the person being spoken of, namely that he is being spoken of and is designated by a certain utterance of the pronoun 'I'. But this is 'Cambridge' knowing who. It is no more knowing who than

Cambridge change is change. I do not know who will win the next election but I know something unique (and important) about that person, namely, that he or she will be the person who will win the most votes.

[24] Thus, there is *a* sense in which understanding 'There is water hereabouts' requires knowing truth-conditions, for it requires knowing a whole lot of things of the form 'If the context is thus and so, then the sentence is true in such and such conditions'. What is not required for understanding is knowledge of the truth-conditions of the proposition expressed. I am indebted here to David Lewis and to Pavel Tichy (1983).

[25] Other views about how the word 'water' gets to pick out $H_2O$ would require appropriately different versions of (2). The view I have used by way of illustration is *roughly* that of Davies and Humberstone, (1980). If you prefer a causal-historical theory, you would have to replace (2) by something like 'It is $H_2O$ that was the (right kind of) causal origin of our use of the word "water"'. Of course, any view about how 'water' gets to pick out what it does will be controversial. But it is incredible that there is *no* story to tell. It is not a bit of magic that 'water' picks out what it does pick out.

[26] Versions of this paper were read at Monash University and the University of New South Wales. I am much indebted to the ensuing discussions.

REFERENCES

Armstrong, D.M. (1968), *A Materialist Theory of the Mind* (London, Routledge & Kegan Paul).
Armstrong, D.M. (1989), *A Combinatorial Theory of Possibility* (Cambridge: Cambridge University Press).
Blackburn, S. (1984), *Spreading the Word* (Oxford: Oxford University Press).
Davies, M.K. & Humberstone, I. L. (1980), 'Two Notions of Necessity', *Philosophical Studies* 38, pp.1-30.
Devitt, M. (1984), *Realism and Truth* (Oxford: Basil Blackwell).
Forrest, P. (1986), 'Ways Worlds Could Be', *Australasian Journal of Philosophy*, 64, pp. 15-24.
Forrest, P. (1992), 'Universal and Universalizability', *Australasian Journal of Philosophy*, 70, pp. 93-98.
Hick, J. (1964), 'The Doctrine of the Resurrection of the Body Reconsidered', in Antony Flew (ed.), *Body, Mind and Death* (New York: MacMillan).
Horgan, T. (1982), Supervenience and Microphysics', *Pacific Philosophical Quarterly*, 63, pp.29-43.
Kripke, S. (1980), *Naming and Necessity* (Oxford, Basil Blackwell).
Lewis, D. (1983), 'New Work for a Theory of Universals', *Australasian Journal of Philosophy*, 61, pp. 343-377.
Lewis, D. (1986), *On the Plurality of Worlds* (Oxford, Basil Blackwell).
Putnam, H. (1975), 'The meaning of "Meaning"' in his *Mind, Language and Reality* (Cambridge, Cambridge University Press), pp. 215 - 271.
Sterelny, K. (1990), *The Representational Theory of Mind* (Oxford: Blackwell).
Stalnaker, R. (1976),'Possible Worlds', *Nous*, 10, pp. 65-75.
Stalnaker, R. (1978), 'Assertion' in P. Cole (ed.), *Syntax and Semantics*, Vol.9 (New York: Academic Press), pp. 315 - 332.
Stalnaker, R. (1984), *Inquiry* (Mass., MIT Press).
Tichy, P. (1983), 'Kripke on Necessity A Posteriori', *Philosophical Studies*, 43, pp. 225 - 241.
Williams, D.C. (1966), *Principles of Empirical Realism* (Springfield, Ill.: Thomas).
Yablo, S. (1992), 'Mental Causation', *Philosophical Review*, 101, 2, April, pp. 245 - 280.

GILBERT HARMAN

# DOUBTS ABOUT CONCEPTUAL ANALYSIS

In these brief remarks, I want to indicate why conceptual analyses by philosophers are unlikely to deliver those sorts of a priori connections that Jackson argues areneeded for armchair metaphysics (Jackson, 'Armchair Metaphysics', this volume).

Paradigm cases of the a priori include basic principles of logic, definitions, and other "axioms" into the truth of which we seem (I say 'seem' because I recognize that this may be an illusion) to have intuitive insight. In such cases we seem to be able to tell that something is true just by considering the matter clearly. In addition, we seem to be able to tell a priori that when certain things are true, certain things are also true. There are patterns of implication that we seem to be able to recognize a priori, modus ponens for example. So, paradigm cases of the a priori include axioms that can be immediately recognized as obviously true and anything that follows from these axioms via one or more steps of obvious implication.

My first point is that philosophical analysis does not often yield a priori results of this paradigmatic sort.

## *Philosophical Analysis: What's Going On?*

It is true that philosophical analyses of 'know,' 'good,' 'refers,' etc. are often presented as accounts of meaning. They are confirmed not by empirical testing but, as Jackson points out, by thought experiments. An analysis is accepted because it cannot be imagined false. So, it may seem that successful analyses aims to provide analytic a priori truths.

But, there is an obvious immediate difference between these cases and paradigm cases of the a priori. We do not normally have the sort of direct intuitive insight into the truth of philosophical analyses that we may seem to have into basic principles of logic. Nor can philosophical analyses ordinarily be demonstrated as following by obvious principles from obvious axioms.

Typically, attempts at philosophical analysis proceed by the formulation of one or more tentative analyses and then the consideration of test cases. If exactly one of the proposed analyses does not conflict with "intuitions" about any test cases, it is taken to be at least tentatively confirmed. Further research then uncovers new test cases in which intuitions conflict with the

*Michaelis Michael and John O'Leary-Hawthorne (eds.), Philosophy in Mind, pp.43-48.*

analysis. The analysis is then modified or replaced by a completely different one, which is in turn tested against imagined cases, and so on.

There is an inductive component to the acceptance of any philosophical analysis that is defended in this way. From the fact that an analysis conflicts with no test cases considered so far, one concludes that the analysis gives the right results for all possible test cases. Compare this with inductive reasoning that might be offered in support of Fermat's Last Theorem: from the fact that no counter examples have been found so far, one concludes that the theorem holds for all natural numbers.

Even in the judgment that an analysis fits a particular case there is an inductive element. When we speak of 'intuitions about cases,' we refer to the fact that, given a description of a possible situation, we find that we are more or less strongly inclined to make a particular judgment about the situation on the basis of that description. Notoriously, such a judgment can depend on exactly what description of the case is given, other assumptions we may be making, etc. It happens more often that some philosophers like to admit that we change our mind about what our intuitions are about a given case, especially if we discover that others have different intuitions.

We have some data, namely, that people P have actually made judgments J about cases C as described by D. Given the data, we infer that P and others would make the same judgments J about the same cases C on other occasions. This is clearly an inductive and fallible inference, because it is a type of inference that is very often mistaken! An analysis is defended in the way one defends any inductive hypothesis, namely, as offering the simplest most plausible explanation of certain data.

### *Analytic-Synthetic*

Philosophical analyses must be understood as involving hypotheses about how we, the people in question, use certain terms. But that is not to say that philosophical analysis must accept a distinction between substantive theoretical assumptions and conceptual truths.

When Jackson speaks of 'extremists' who reject 'conceptual analysis,' I take it that he is referring to those who reject the analytic-synthetic distinction. In my view, the distinction was conclusively undermined at least thirty years ago. I am surprised that this fact has not been universally appreciated.

Why do people reject 'conceptual analysis'? Jackson says they 'are forgetting about biased samples. True, the well known and much discussed examples of putative analyses in the philosophy journals are highly controversial but that is why they are much discussed in the philosophy journals ...'

But this simply misrepresents the opposition. When Quine, Putnam, Winograd, and a host of others raised objections to the analytic-synthetic distinction, they did not mention controversial philosophical analyses. When problems were raised about particular conceptual claims, they were problems about the examples that had been offered as seemingly clear cases of a priori truth—the principles of Euclidean geometry, the law of excluded middle, 'cats are animals,' 'unmarried adult male humans are bachelors,' 'women are female,' and 'red is a color.' Physics leads to the rejection of Euclidean geometry and at least considers rejecting the law of excluded middle (Quine, 1936 and 1953). We can imagine discovering that cats are not animals but are radio controlled robots from Mars. (Putnam, 1962). Speakers do not consider the Pope a bachelor (Winograd and Flores, 1986). People will not apply the term 'bachelor' to a man who lives with the same woman over a long enough period of time even if they are not married. Society pages in newspapers will identify as eligible 'bachelors' men who are in the process of being divorced but are still married. The Olympic Committee may have rejected certain women as sufficiently female on the basis of their chromosomes. (Robert Schwartz pointed this out to me many years ago.) Just as a certain flavor is really detected by smell rather than taste, we can imagine that the color red might be detected aurally rather than by sight (Harman, 1967 and 1973).

The heart of the objection to analyticity was not simply that there are problems with the usual paradigms of analytic truth but that, whatever we are to say about these paradigms, the analytic-synthetic distinction rests on substantive assumptions that turn out to be false.

I do not have the space for a full scale history of the philosophical battle that led many years ago to the abandonment of the analytic-synthetic distinction. But, for the sake of readers somehow not familiar with this history, I will use the remainder of my remarks to sketch some of the main points.

The basic idea behind the analytic-synthetic distinction was that certain propositions could be true solely by virtue of what is meant by the words used to express them and could be known to be true simply by knowing the meaning of these words. These propositions were the analytic truths; all other truths were synthetic. The analytic truths were supposed to be a subset, proper or improper, or the truths knowable a priori. A priori knowledge was supposed to be knowledge that was justified without appeal to experiential evidence. The relevant notion of justification required a 'foundations' account—knowledge of P might depend on knowledge of Q, and so forth, with foundations that were either a priori or derived from immediate conscious experience.

When foundationalism was discredited, so was a priori knowledge (and vice versa).

*The Alternative to a Foundations Approach*

Foundationalism went away once it was realized that one's beliefs are not structured via justification relations. Beliefs do not need justification unless there is a specific challenge to them (Peirce, 1955). Some beliefs may be more "central" than others—more theoretical, more taken for granted (in a way that needs more explication than I can provide). These more central beliefs will seem "obvious," because it is hard to take seriously revising them, but they are not therefore guaranteed to be true and there will normally be circumstances in which such previously obvious beliefs will be revised.

There is no sharp, principled distinction between changing what one means and changing what one believes. We can, of course, consider how to translate between someone's language before and after a given change in view. If the best translation is the homophonic translation, we say there has been a change in doctrine; if some other (nonhomophonic) translation is better, we say there has been a change in meaning. What we say about this depends on context and our purposes of the moment.

*Definitions*

There are various kinds of definitions. Definitions in dictionaries try to capture ordinary meaning. Sometimes definitions are offered in the course of discussion in order to fix meanings. Sometimes this is a matter of making definite a term that is already in use, for example, defining a meter in a certain way. Sometimes this is a matter of setting out a subject in a rigorous way, where there might be several different ways to proceed, e.g. defining numbers in terms of sets. Sometimes a new term is introduced for certain purposes.

No matter what sort of definition is in question, no long term epistemological status for the definition is guaranteed. As time goes by, we may as easily modify a definition as change our belief in something else. And whether we will speak of change in meaning is not determined by whether we changed what was called a "definition"

What about the short term? Here definitions do have a privileged status, but it is a status that is shared by anything being assumed. If we agree that we are going to make certain assumptions, then for the time being we do not challenge those assumptions.

*A Rough Serviceable Commonsense Distinction?*

Grice and Strawson (1965) suggested that the analytic-synthetic distinction was an ordinary one, or that ordinary people could easily be brought to make it and understand it; whether or not we had an acceptable theoretical account of the distinction was another matter, they said.

That was clearly wrong. Ordinary people did not make the distinction and it turned out to be extremely hard to teach students the distinction, which they confused with all sorts of other distinctions.

It turned out that someone could be taught to make the analytic-synthetic distinction only by being taught a rather substantial theory, a theory including such principles as that meaning can make something true and that knowledge of meaning can give knowledge of truth.

I have elsewhere compared the analytic-synthetic distinction with the rough ordinary distinction made in historic Salem, Massachusetts, between witches and other women (Harman, 1973). The fact that people in Salem distinguished witches from other women did not show there was a real distinction. They were able to make the distinction they made only because they accepted a false theory along with some purported examples of witches. Similarly, people can make the analytic-synthetic distinction only if they accept the false analytic-synthetic theory along with purported examples of analytic truth.

This is the point at which it is relevant that many of the supposed analytic truths mentioned by philosophers are either false or easily imagined false—'bachelors are always unmarried,' 'cats are animals,' etc.

*Final Remark*

The fact that once paradigmatic analytic truths are false or easily imagined false shows that we do not have a priori insight into even what seem to be the clearest cases. This reinforces my main objections to armchair metaphysics of the sort defended by Jackson. The analyses that philosophers come up with do not appear to provide a priori connections of the required sort. Prospects for armchair metaphysics are therefore dim. I agree with Jackson that this makes it unlikely that there is a coherent version of physicalism.

*Princeton University*

## REFERENCES

Grice, H.P. and Strawson, P.F. (1965), 'In defense of a dogma', *Philosophical Review* 65.
Harman, G. (1967), 'Quine on meaning and existence, I: The death of meaning', *Review of Metaphysics* 21, pp. 124-151.
Harman, G. (1973), *Thought* (Princeton, NJ: Princeton University Press).
Peirce, C.S. (1955), 'Some consequences of four incapacities', in Justis Buchler (ed.) *Philosophical Writings of Peirce* (New York: Dover), pp. 228-229.
Putnam, H. 'It Ain't Necessarily So', *Journal of Philosophy* 59, pp. 658-671.
Quine, W.V. (1936), 'Truth by Convention', in O. H. Lee, (ed.) *Philosophical Essays for A. N. Whitehead* (New York: Longman's).
Quine, W. V. (1953), 'Two dogmas of empiricism' in *From a Logical Point of View* (Cambridge, Massachusetts: Harvard University Press).
Winograd, T. and Flores, C. F. *Understanding Computers and Cognition: Language and Being*, (Norwood, N.J.: Ablex Pub. Corp.).

ANDRÉ GALLOIS

# DEFLATIONARY SELF KNOWLEDGE

As a number of philosophers have observed, our knowledge of what is passing through our own minds appears to be quite different to our knowledge of other things. I do not, it seems, need to accumulate evidence in order to know what psychological states I am in.[1] Without relying on evidence I am able to effortlessly attribute to myself beliefs, desires, intentions, hopes, fears, and a host of other psychological states. The distinctive knowledge we have of our own psychological states is sometimes labeled privileged access. If 'privileged access' means knowledge that is not evidentially based we certainly seem to have privileged access to some psychological states including those exhibiting intentionality.[2] Nevertheless, some have questioned whether we do enjoy privileged access to our intentional states. One reason for doing so derives from the findings of psychologists. The time honored thesis that we have privileged access to our own psychological states is threatened by such findings (see especially Nesbitt and Wilson, 1977). Moreover, there is threat to privileged access from a different direction. It comes from a philosophical thesis commonly referred to as externalism. Externalism is the view that the content of an intentional state such as a belief is fixed by the environment external to the believer. Some externalists deny that externalism conflicts with privileged access. Donald Davidson and Tyler Burge in particular have developed what might be called a minimalist account of privileged access which, they would claim, reconciles privileged access with externalism. An interesting feature of their account is that it suggests a defense of privileged access in the light of the psychological findings mentioned above. However, that is a topic for another paper. Here I will confine myself to assessing the Davidson Burge account of privileged access, and its implications for the relationship between privileged access and externalism.

## *I. Why The Appearance of Conflict?*

Why should externalism seem to threaten privileged access? Here is what Andrew Woodfield has to say:

*Michaelis Michael and John O'Leary-Hawthorne (eds.), Philosophy in Mind, pp.49-63.*

> One doctrine that will need to be dropped is the Cartesian view that the subject is the best authority on what he or she is thinking. The doctrine has been attacked before, by behaviorists for example, but the present line of attack is more subtle. According to a *de re* theorist, the subject can have full conscious access to the internal subjective aspects of a thought while remaining ignorant about which thought it is. This is because a *de re* thought also has an *external* aspect which consists in its being related to a specific object. Because the external relation is not determined subjectively, the subject is not authoritative about that (1982, p. vii).

Subtle or not, this line of attack against privileged access is less than clear. One way of taking it is as follows. Consider my thought that there is some water nearby. Call knowledge that is not based on evidence non-observational. I have privileged access to a proposition only if it is possible for me to have non-observational knowledge of it.[3] I cannot have non-observational knowledge of the proposition that there is water somewhere. According to some versions of externalism a necessary condition for my thinking that water is nearby is that I have the concept of water, and a necessary condition for my having that concept is that there is water somewhere. So, I cannot have non-observational knowledge of the proposition that I think there is water nearby. Hence, I lack privileged access to that proposition.

As Tyler Burge (1988, pp. 649-663) points out, it does not follow from my having privileged access to some proposition that I have privileged access to every necessary condition for its truth. Moreover, it will not repair the argument to add that I know that a necessary condition for me thinking that there is water nearby is that there is water somewhere. We would have to, at least, add that I can non-observationally know that the existence of water is a necessary condition for my thinking that there is some nearby.

A case can be made for adding the needed premiss. Michael Mckinsey (1991) maintains that if externalism is true then it can be known a priori that the existence of external objects is required for certain thoughts to be entertained. Mckinsey does not spell out what he means by a priori knowledge. However, I take it that anything that is known a priori is known non-observationally. Consequently, if Mckinsey is right, premiss (2) of the following argument will be true:

(A)

(1) It is possible for me to know non-observationally that I think that water is nearby.

(2) A necessary condition for my thinking that water is nearby is that there is water somewhere.

(3) It is possible for me to know non-observationally that there is water somewhere.

At first sight (3) is highly implausible. Surely I need to rely on evidence in order to establish that water exists. It seems that if (2) is true, we will have to reject (1), or deny that (A) is valid.

Why believe (2)? (2) follows from a philosophical thesis I have labeled externalism. If externalism can be known to be true then it is the sort of thesis that can be known a priori to be true. An externalist should be reluctant to concede that externalism cannot be known to be true. Hence, it is a bad idea for an externalist to attempt to avoid a conflict between externalism and privileged access by rejecting (2).

Why not deny that (A) is valid? The possibilities mentioned in (1) and (2) appear to be compossible. If (1) and (2) are both true there is a possible situation in which someone non-observationally knows that she thinks that there is water nearby, and non-observationally knows that a necessary condition for her having that thought is that there is water somewhere. Could such an individual fail to non-observationally know that there is water somewhere? Only if the operator 'X non-observationally knows that p' does not obey epistemic closure. That is, only if we are not entitled to infer X non-observationally knows that q from X non-observationally knows that p and X non-observationally knows that if p then q.

It will be profitable to postpone discussion of this argument for a conflict between externalism and privileged access. Light will be thrown on it by first examining the views of Donald Davidson and Tyler Burge on this issue.

## *II. Why There Might Be No Conflict?*

Donald Davidson has explored the relationship between privileged access and externalism in a number of papers (most notably Davidson 1984, 1986, 1991). He pursues three themes in them. First, it does not follow from the contents of thoughts being determined by factors external to a thinker that thoughts are other than in the head. Second, entertaining a thought is not a matter of grasping a proposition whose content has to be completely known in order to know that one has that thought. Third, that the same external factors determine both the content of a thought and the content of the thought that one has that thought. Davidson takes these three considerations to alleviate any tension between privileged access and externalism. Let us see why.

Davidson is surely right that the external determination of the contents of thought does not imply that thoughts are located outside the head. As he points out something clearly located on the skin counts as a sunburn only if it is suitably related to something outside the skin. However, what bearing this point has on privileged access is not entirely clear. Its relevance would be clear if it could be shown that privileged access goes together with spatial proximity. So far as I can see there is no reason to believe that it does.

As we have seen Davidson also disavows the view that in entertaining a thought an individual completely grasps the content of a proposition, or something playing the same role as a proposition, which is the object of that thought. He writes:

> I am not now concerned with such(now largely disavowed) objects of the mind as sense-data, but with their judgmental cousins, the supposed objects of the propositional attitudes, whether thought of as propositions, tokens of propositions, representations, or fragments of 'mentalese'. The central idea I wish to attack is that these are entities that the mind can 'entertain', 'grasp', 'have before it', or 'be acquainted with' (1986, p. 454).

and continues:

> The basic difficulty is simple: if to have a thought is to have an object 'before the mind', and the identity of the object determines what the thought is, then it must always be possible to be mistaken about what one is thinking (1986, p. 455).

Davidson's argument in these passages is echoed by Tyler Burge who says:

> If one begins by thinking of one's thoughts as objects like physical objects, except that one cannot misperceive or have illusions about them, then to explicate authoritative self knowledge, one makes one of two moves. Either one adds further capacities for ruling out the possible sources of misrepresentation or illusion in empirical perception, or one postulates objects of knowledge whose very nature is such that they cannot be misconstrued or misperceived (1988, P. 660).

In these passages Davidson and Burge are offering a diagnosis of our temptation to think that externalism precludes privileged access. It goes as follows. For the sake of a label call the thesis that thought involves having complete knowledge or grasp of a proposition the Cartesian thesis. We are tempted to adopt the Cartesian thesis. If the Cartesian thesis is true then externalism does preclude privileged access. Why? Because the Cartesian thesis implies that one has complete knowledge of the content of any proposition entertained. However, externalism implies that I may have incomplete knowledge of the content of some proposition I entertain. For example, I may fail to know that the content of the proposition that water is nearby is determined by

H20 even though I am entertaining that proposition. Hence, the Cartesian thesis must be rejected if privileged access is to be reconciled with externalism.

For at least three reasons I am inclined to regard this as a misdiagnosis of our reluctance to combine externalism with privileged access. Davidson compares the Cartesian thesis with other allegedly discredited views such as a sense datum theory of perception. Let us look more closely at that comparison.

Some sense datum theorists hold that a sense datum must appear to have any property that it has. It is not inevitable, but let us take this to mean that if a sense datum one is aware of has a property one knows that it does. No sense datum theorist could take this to apply to all properties of sense data. Suppose I am currently experiencing a red and round sense datum. There will be innumerable properties of that sense datum that I may be unaware that it has: for example, being similar to the last sense datum had by Jack Smart when he inspected his favorite cricket ball. Moreover, there will be intrinsic properties of my sense data that I need not be aware of: for example, being a sense datum.

Call a property of a sense datum which I must be aware of if I have it a self intimating property. In addition, call a property of a proposition that I must be aware of it having if I entertain it self intimating. As we have seen a sense datum theorist will not allow that every property is self intimating. Likewise, an advocate of the Cartesian thesis will not allow that every property of a proposition is self intimating. So, what properties of propositions are self intimating? It is open to us to say: just the property of having a certain content. An advocate of the Cartesian thesis who adopts this position will have to concede that if I have a thought that p then I have privileged access to my having a thought with the content that p: for example, if I have the thought that there is water nearby then I have privileged access to my having a thought with that content. However, Davidson and Burge would not deny that I have privileged access to the content of my thoughts. Indeed, they insist that my having privileged access to the content of my thoughts is compatible with externalism. So, it is not at all clear why a proponent of the Cartesian thesis should deny privileged access given the truth of externalism.

The second reason for being puzzled by Davidson's anxiety to reject the Cartesian thesis in order to protect privileged access is this. Again it is useful to compare the Cartesian thesis with a sense datum theory. Davidson has in mind sense datum theorists who want sense data to do epistemological duty. I enjoy a certainty about my sense data that I do not enjoy about the external world. Now compare a proposition that I am not supposed to have privileged access to, say the proposition that there is water nearby, with one that I am supposed to have privileged access to, say the proposition that I think that there is water nearby. It would make sense to invoke the Cartesian thesis to explain why we are supposed to have privileged access to the latter proposition if the Cartesian thesis had the following implication. When I think that there is water nearby I have a less firm grasp on the proposition I am entertaining than when I think that I think that there is water nearby. Of course, the Cartesian thesis has no such implication. I have an equal grasp on the proposition I am entertaining when I have a thought about the external world as I do when I have a thought about one of my own thoughts.

Finally, in the passages quoted from Davidson 'object' appears to be ambiguous. Consider a demonstrative thought about a cat: that cat is black. In the first passage Davidson appears to mean by 'object' the proposition I am entertaining when I think that that cat is black. In the second he appears to mean the cat itself. Davidson gives no reason for denying that the following could all be true. I completely grasp the content of the proposition I am entertaining when I think that that cat is black. A necessary condition for my entertaining that proposition is that the cat exists. I lack complete knowledge of the cat. Hence, I fail to see why the thesis that in thinking we grasp the content of propositions should be an embarrassment for an externalist who wishes to defend privileged access.

The most important of Davidson's arguments for reconciling privileged access with self knowledge appears to me to be the third. At any rate it is the one I will focus on. Characteristically Davidson introduces the argument by considering what is involved in interpreting a speaker. It goes like this. If I know that you assent to a sentence, and I know what meaning you assign to it, then I know what you believe. Someone else might express information about the meaning I attach to a sentence in the following way:

(4) By 'Water is wet' AG means that water is wet.

For someone else (4) is an empirical hypothesis that could be mistaken. For example, I might be an inhabitant of Twin Earth. (4) will be false if my interpreter is an inhabitant of Earth. Now compare (4) with:

(5) By 'Water is wet' I mean that water is wet.

I cannot be mistaken about (5). I know a priori that (5) is true. Why? Because the sentence mentioned on the right hand side of (5) is bound to synonymous with the same sentence used on the left. Moreover, the same external factors determine the content of the sentence mentioned on the right and used on the left. Since I know a priori that (5) is true, if I know that I assent to 'Water is wet', I know that I believe that water is wet.

In 'Epistemology Externalized' Davidson (1991) makes the same point somewhat differently:

> An interpreter must discover, or correctly assume on the basis of indirect evidence, what the external factors are that determine the content of another's thought; but since these factors determine both the contents of the thought and the contents of the thought one believes one has, there is no room for error about the contents of one's own thoughts of the sort that can arise with respect to the thoughts of others.

In (5) the content of the sentence mentioned on the left must match the content of the same sentence used on the right. Likewise, the content of the second order thought that I have the thought that p must match the content of the thought that p.

In 'Individualism and Self Knowledge' Tyler Burge (1988) takes up this last point. He says of what he calls a reflexive judgment that it 'simply inherits the content of the first order thought', and adds,

> in the case of cogito like judgments, the object, or subject matter, of one's thoughts is not contingently related to the thoughts one thinks about it. The thoughts are self- referential and self verifying. An error based on a gap between one's thoughts and the subject matter is simply not possible in these cases. When I judge: I am thinking that writing requires concentration, the cognitive content that I am making a judgment about is self referentially fixed by the judgment itself (1988, p. 658).

How do these points about the match between the content of a sentence mentioned and a sentence used, or between and higher and a lower order judgment, help to reconcile externalism and privileged access? At first sight calling attention to the a priori status of (5) does not explain privileged access. If I know that (5) is true, it does not follow that I know what I believe. In addition I have to know that I assent to 'Water is wet'. How do I know that? By 'assenting to 'Water is wet"' one of two things might be meant. Either having a disposition to produce assenting behaviour, or believing that 'Water is wet' is true. If 'assenting' means being disposed to produce assenting behaviour then I lack privileged access to my assenting to 'Water is wet'. On the other hand, if 'assenting to "Water is wet"' means believing that 'Water is wet' is true then it remains to be explained how I have privileged access to that belief.

Tyler Burge identifies a type of judgment which he claims constitutes basic self knowledge. I am making a judgment which is an example of basic self knowledge whenever I judge that I am entertaining a certain thought. Burge gives the example of judging that I am entertaining the thought that writing requires concentration. Burge takes such a judgment to be infallible in the familiar sense that if one's judgment that p is infallible then one cannot falsely judge that p. Why, in Burge's view, are judgments that amount to basic self knowledge infallible? Presumably for the following reason. To entertain the thought that p is to entertain a thought whose content is, at least in part, that p. Hence, if I think that I am entertaining the thought that p I am automatically entertaining the thought that p.

What is the relevance of the infallibility of certain thoughts to the relationship between privileged access and externalism? Two problems with this component of Burge's treatment of privileged access have been noted by Paul Boghossian (1989). Boghossian points out that only a limited class of judgments about ones psychological states will be infallible for the reason that Burge has in mind. Burge gives no reason to suppose that my judgment that I want some water to be nearby is infallible. Nevertheless, there is as much reason to suppose that I have privileged access to my desires as there is to suppose that I have privileged access to my entertainings.

In addition, Boghossian points out, my knowledge of my psychological states can be affected by shifts in my attention. However, the infallibility of infallible judgments is not affected by attention.

Apart from these problems there is a third. Consider the choice between a reliabilist and an internalist conception of knowledge. On an internalist conception of knowledge it is hard to see how the infallibility of certain judgments explains our privileged access to them. Given internalism about knowledge, and that I judge that I think that writing requires concentration, how do I know that my judgment is correct? Presumably I have to recognize that my judgment about what I think is infallible, and deduce from my making that judgment that I think that writing requires concentration. So, how do I know that I judge that I think that writing requires concentration?

The infallibility of judgments of basic self knowledge is more obviously relevant to understanding privileged access if we assume a reliabilist view of knowledge. If they are infallible judgments of basic self knowledge are, when made, bound to be true. Hence, the infallibility of such judgments guarantees their truth conduciveness. However, only the most implausible version of reliabilism implies that it is a sufficient condition for a belief amounting to knowledge that there is an invariable connection between holding a belief of that type and it being true. So, even if we assume a reliabilist account of knowledge and the infallibility of judgments of basic self knowledge, it remains to be explained how we know such judgments to be true.

Paul Boghossian characterizes the accounts of self knowledge offered by Davidson and Burge as deflationary. As he puts it, on those accounts self knowledge is not a cognitive achievement. What this means is just that self knowledge is non-observational. However, there is another sense in which the Davidson Burge view of self knowledge is deflationary. In 'Three Varieties of Knowledge' Davidson (1991) puts it like this: 'Knowledge of the contents of our own minds must, in most cases, be trivial.'

In what sense is knowledge of our own minds supposed, for the most part, to be trivial? Light is thrown on this question when we consider an example discussed by Burge and Boghossian. Suppose that over last few years on a number of occasions I have been transported to Twin Earth and back again without being able to tell which planet I am on. When I am on Twin Earth the thought I express by uttering 'Water is wet' is one utilizing the concept of twater. Call that the twater thought. When I am on Earth the thought I express by uttering the same sentence is one utilizing the concept of water. Call that the water thought.

When I am on Earth do I know what I am thinking when I think that water is wet? Davidson and Burge would insist that I do. The reason, most clearly emphasized by Burge, derives from the necessary coincidence in the content of a thought attributing a thought to oneself with the content of the thought attributed. That coincidence precludes a certain type of error.

Consider my current thought that there is a wall in front of me. Had there been no wall, but, instead, a cardboard imitation I would have entertained the same thought. However, my thought that there is a wall in front of me would have been false. Now suppose that I am on Earth, and in fact have the water thought when I think that water is wet. In addition, I think that I am having that thought. Suppose I had been on Twin Earth instead. In that case, would I mistakenly have entertained the same second order thought that I think that water is wet? I would not. The content of the Twin Earth thought I would express by uttering 'I think that water is wet' is determined by the same factors that determine the content of the thought I would express by uttering 'Water is wet'. I would not have mistakenly entertained the same second order thought. Instead, I would have correctly entertained a different second order thought.

In what sense is the account of self knowledge that is prompted by these considerations deflationary? According to Burge, whether I am on Earth or Twin Earth, I know that I think that water is wet. However, if I am subject to being periodically transported from Earth to Twin Earth and back again I may know that I am entertaining one of a pair of thoughts without knowing which one. In some sense, it seems, I may know that I think that water is wet without knowing what I think.

Boghossian takes the deflationary character of the Davidson Burge account of self knowledge to constitute an objection to it. He considers the example of an individual S who is being switched from Earth to Twin Earth and back again, and invokes the following principle: if S knows that p at t1, and if (at some later time) t2, S remembers everything S knew at t1, then S knows that p at t2. Boghossian envisages S at t2 discovering that he has been subject to being switched between Earth and Twin Earth prior to t2. However, at t2 S does not know which planet he was on at t1. He remarks:

> Now, let us ask: why does S not know today whether yesterday's thought was a *water* thought or a *twater* thought? The platitude insists that there are only two possible explanations: either S has forgotten or he *never* knew. But surely memory failure is not to the point. In discussing the epistemology of relationally individuated content, we ought to be able to exclude memory failure by stipulation...The only explanation, I venture to suggest, for why S will not know tomorrow what he is said to know today, is not that he has forgotten but that he never knew (1989, p. 23).

Boghossian's argument appears to be this. Since S does not know today whether he was having a water or a water thought yesterday, today S does not know what he thought yesterday. However, today S can remember everything that he knew yesterday. So, yesterday S did not know what he thought then.

This is unconvincing. Do I know what object is in front of me? Without specifying the relevant contrast that question is unanswerable. I know what object is in front of me insofar as I know that it is a table rather than a chair. I may not know what object is in front of me insofar as I may not know that it is a table from the next room rather than a table from somewhere else. In the Earth Twin Earth switching case do I know what I think? Again it all depends on what contrast is relevant. I know what I think insofar as I know that I think that water is wet rather than that grass is purple. I may not know what I think insofar as I may not know that I am entertaining a water as opposed to a twater thought. It would not be unreasonable for Burge and Davidson to insist that when it comes to knowledge of the contents of ones thoughts only the first type of contrast is relevant.

### *III. Two Problems*

There are remaining problems with the deflationary account of self knowledge that I have attributed to Davidson and Burge. I will conclude by discussing two.

I have represented Davidson and Burge as advancing the following argument. The content of my second order belief that I think that water is wet is determined by the same external factors as the content of my belief that water is wet. I may fail to know what those external factors are. Indeed, they may systematically vary. That does not matter. Whatever those external factors are, and, in consequence, whatever first order belief I have, I am guaranteed to have an appropriately matching second order belief. Whatever my first order belief is I am guaranteed to have a true second order belief about its content.

One problem with this argument is that it is too strong. Suppose I have the belief that I believe that water is wet. There are three types of error that that belief is subject to. First, it may lack an object altogether. There may be no first order state that my second order belief corresponds to. Second, there may be a first order state that my second order belief corresponds to, but the wrong first order state. I may hope rather than believe that water is wet. Finally, there may be a mismatch between the content of the first order and the content of the second order state.

Davidson and Burge address only the third type of error. However, exclusion of that type of error appears insufficient to ensure privileged access. Consider my belief that there is water somewhere. I have no privileged access to there being water somewhere. Nevertheless, in the Earth Twin Earth switching case if my first order belief that there is water somewhere has an object at all it is bound to have the right object. If I am transported to Twin Earth for a sufficient period of time I would not there be mistakenly entertaining the same thought that I am entertaining on Earth when I say that water is wet. Instead, I would be correctly entertaining a different thought.

There is another route to the same conclusion which takes us back to Mckinsey's argument given at the beginning of the paper. You recall that argument goes as follows. If externalism can be known to be true then it can be known to be truc non-observationally. An externalist such as Davidson or Burge will be loathe to deny that externalism can be known to be true. So, an externalist should concede that externalism can be known non-observationally to be true. We are supposed to have non-observational knowledge of our thoughts including those whose content is externally determined including the thought that water is wet. It follows from externalism that a necessary condition for my thinking that water is wet is that there is water somewhere. Hence, if I have non-observational knowledge of externalism, and privileged access to my thoughts, then I have non-observational knowledge that there is water somewhere.

So far as I can see there are just four responses to this argument. The first is one that we have canvassed already. It is to question the validity of the argument by questioning whether 'X knows non-observationally that' obeys epistemic deductive closure. This may be the right response, but it needs defense.

The second response is to resort to the familiar distinction between wide and narrow content. We distinguish between the wide and narrow content of the thought that water is wet. On Earth and Twin Earth the thoughts I express by means of 'Water is wet' have the same narrow content, but differing wide content. What I have privileged access to is having a thought with the narrow content of the thought that water is wet. I do not have privileged access to having a thought with the wide content of the thought that water is wet. However, I cannot deduce from externalism that if I have a thought with the narrow content of the thought that water is wet then water exists.

The difficulty with this reply is that it, in effect, gives up attempting to reconcile externalism with privileged access. All I have privileged access to is having a thought with a certain narrow content. I have no privileged access to having a thought whose content is externally determined.

The third response is the boldest. Perhaps I do know non-observationally that water exists. What does it take to enable me to have non-observational knowledge of something? Roughly, I am able to non-observationally know that p if possessing the concepts associated with p enables me to come to know that p. Suppose I have the concept of water, and know non-observationally that externalism is true. Why should I not come to know simply on the basis of having the concept of water and knowing the truth of externalism that there is some water out there?

This last reply has its attractions. For one thing, if it is successful it constitutes a reply to the first argument for the conclusion that the Davidson Burge view of privileged access is too strong in that it implies one can have privileged access to such propositions as that water exists. Nevertheless, I do not think that it works. Consider a world which I will call the vat world. In the vat world shortly after birth nine out of every ten individuals are made into brains in vats without having any contact with water however indirect. An evatted individual lacks the concept of water, and does not entertain the same thought as a normally embodied individual when she says to herself water is wet. Suppose I am one of the fortunate minority who escape envatting, and have the thought that water is wet. Surely, in the vat world I do not know that water is wet. Nevertheless, on the Burge Davidson view I do have privileged access to my thought that water is wet. Had I been envatted I would not have mistakenly entertained the thought that I think that water is wet. I would correctly have entertained a different second order thought. Hence, if I possess non-observational knowledge of externalism I should be able to deduce, and, so, come to know, that water is wet.

There is one further reply to Mckinsey's argument I would like to consider.[4] So far I have assumed that if one can have non-observational knowledge of the truth of externalism then one can non-observationally know that the existence of water is a necessary condition for entertaining the thought that water is wet. Let us see whether that is so. Suppose I am entertaining the thought that some cats are lovable. Since cats are a natural kind it is plausible to be an externalist about the content of that thought. Moreover, it is natural to predict that an externalist will endorse the following modal claim:

(6) Necessarily (someone entertains the thought that some cats are lov able —> cats exist).

In fact an externalist need only be committed to the following weaker claim:

(7) Cats exist —> Necessarily (someone entertains the thought that cats are lovable —> cats exist).

To see the relevance of the difference between (6) and (7) consider my thought that unicorns are friendly. If I am an externalist about the content of that thought am I committed to supposing that I cannot entertain it unless unicorns exist? I am not. All I am committed to is the following. If unicorns exist, I cannot entertain the thought that they are friendly unless they exist. That is, all I am committed to is;

(8) Unicorns exist —> Necessarily (someone entertains the thought that unicorns are friendly —> unicorns exist).

Even if I am quite certain that I do entertain the thought that unicorns are friendly (8) leaves it open whether they exist.

Suppose that I know a priori, and thus non-observationally, that externalism is true. If the foregoing is correct I am not entitled to deduce from externalism that I cannot entertain the thought that water is wet unless water exists. The most I am able to deduce from externalism is:

(9) Water exists —> Necessarily ( I think that water is wet —> water exists).

Allow that I non-observationally know that I think that water is wet, and that (9) is true. It does not follow that I non-observationally know that water exists. In order to know, in the light of (9), that my thinking that water is wet implies that water exists I have to establish that water exists. Insisting that empirical observation is required to know that water exists is compatible with conceding that I have non-observational knowledge of (9) and the content of my thoughts.

Perhaps the Davidson Burge view of privileged access can be defended against the type of argument advanced by Mckinsey. That argument does not show that if the Davidson Burge view is correct, I have privileged access to facts about the external world. Nevertheless, as we have seen, there is another argument to the same conclusion. Suppose that all it takes to have privileged access to one's psychological states is a guarantee that, in certain circumstances, a change in a psychological state will be tracked by one's belief that one is in that state. If so, it seems that we have privileged access to facts about the external world. After all, in the same circumstances, changes in the external world will be tracked in much the same way by one's beliefs about the external world. I conclude that the Davidson Burge view of self knowledge is inadequate. Privileged access requires more than their deflationary conception suggests. It remains to be seen whether privileged access can be reconciled with externalism.

*University of Queensland*

## NOTES

[1] Notoriously this is denied by Gilbert Ryle. Ryle (1949) claims that our knowledge of our own psychological states differs in degree rather than kind from our knowledge of the psychological states of others. Few have found Ryle's discussion of self knowledge in chapter of the *Concept of Mind* at all convincing.

[2] Some would argue that we enjoy privileged access to some psychological states in virtue of having the evidence afforded by introspection. However, this view is plausible only for states such as pain which satisfy the following condition. In Nagel's sense there is something that it is like to be in them. Here I am only concerned with intentional states such as belief that do not satisfy the condition in question.

[3] Andrew Woodfield's concern is whether we are especially authoritative in self ascribing psychological states. Conceivably we may not be especially authoratitive about our psychologicalstates even if we have non observational knowledge of them. In her book *Intention* Elizabeth Anscombe (1957) argues that we have non observational knowledge of the position of our limbs. Even if she is right, it does not follow that we are especially authoritative about the position of our limbs.

We need to distinguish privileged access from special authority. However, that distinction will make no difference to the arguments I am about to discuss.

[4] I am indebted to John O'Leary Hawthorne for pointing out to me how this reply might be developed.

## REFERENCES

Anscombe, E. (1957), *Intention* (Oxford).

Boghossian, P. (1989), 'Content and Self Knowledge', *Philosophical Topics*, vol. xvii, no 1, Spring 1989.

Burge, T. (1988), 'Individuation and Self Knowledge', *The Journal of Philosophy*, 85,1 (November), pp. 649-663.

Davidson, D. (1984), 'First Person Authority', *Dialectica*, Vol 38, No 2-3, pp. 101-111.

Davidson, D. (1986), 'Knowing One's Own Mind', *Proceedings of the American Philosophical Association* pp. 441-458.
Davidson, D. 'What is Present to the Mind' in *The Mind of Donald Davidson*, Johannes Brandl (ed.), (Amsterdam, Rodopi), pp. 3-18.
Davidson, D. (1991), 'Epistemology Externalized'
Davidson (1991), 'Three Varieties of Knowledge', in *A.J.Ayer Memorial Essays* ed. A. Phillips Griffiths (Cambridge).
McKinsey, M. (1991), 'Anti-Individualism and Privileged Access', *Analysis*, 51.1 (January), pp. 9-16.
Nesbitt, R.E. and Wilson, T.D. (1977), 'Telling more Than We Can Know: Verbal Reports on Mental Processes', *Psychological Review* LXXXIV, 3 (May), pp. 231-59.
Ryle, G. (1949), *Concept of Mind* (Hutchinson).
Woodfield, A, ed. (1982), *Thought and Object* (Oxford: Oxford University Press).

ANNETTE C. BAIER

# HOW TO GET TO KNOW ONE'S OWN MIND: SOME SIMPLE WAYS*

One's own mind should be contrasted, one would think, with others' minds, just as one's own body contrasts with other people's bodies. But in our philosophical tradition we have this odd phrase 'other minds,' which recurs in discussions of our access to facts about minds. In John Stuart Mill's presentation[1] of the infamous argument from analogy (although he does not characterize the argument as one employing analogy), we find 'other sentient creatures,' and 'other human beings,' not 'other minds'. F. H. Bradley has 'other selves' and 'foreign selves,' and Bertrand Russell has 'other people's minds.' 'Other minds' is found in A. J. Ayer, C. I. Lewis, C. D. Broad, John Wisdom, J.L. Austin, Alvin Plantinga, and a host of others.[2] My guess is that 'other minds,' like Bradley's 'other selves' is a secularization of George Berkeley's phrase 'other spirits,' used when he presents what, as far as I know, is the first version of the infamous argument. He is mainly concerned to present a vindication of that really other super-spirit, the Berkeleyan God, but along the way grants that we know not just ourselves as active spirits, making and unmaking ideas at our pleasure, but, less immediately, 'other spirits' who are 'human agents'. We know of them, Berkeley says 'by their operations, or the ideas of them, excited in us. I perceive several motions, changes, and combinations of ideas that inform me there are certain particular agents, like myself, which accompany them and concur in their production.'[3] Presumably for Berkeley, as before him for Descartes,[4] speech is a very important case of such a recognizably spirit-expressing operation, but he would have problems distinguishing the easily intelligible speech of his fellow-Britishers from the less easily decoded language of the 'author of nature,' seen or heard as much in ideas which get interpreted as human speech or other expressive human behavior as those which get interpreted as animal, vegetable or mineral changes.

Thomas Reid, to whom Mill is responding when he presents the infamous argument, wrote that he found no principle in Berkeley's system 'which affords me even probable ground to conclude that there are other intelligent beings, like myself, in the relations of brother, friend, or fellow citizen. I am left alone, as the only creature of God in the universe . . . '.[5] Reid either did not read the passage of Berkeley which I have quoted, or was unimpressed by its claims. He finds these 'egoistic' implications an 'uncomfortable consequence of the theory of Berkeley,' but Berkeley, despite his 'egoism,' is treated gently

*Michaelis Michael and John O'Leary-Hawthorne (eds.), Philosophy in Mind, pp.65-82.*

by Reid compared with the treatment given to Hume. This doubtless was because Reid was at one with Beattie in judging Berkeley to have been a man of conspicuous virtue, who had striven valiantly to overturn what Reid terms the 'fortress of atheism' which sheltered more materialist and less virtuous philosophers.

It is quite understandable that Berkeley should identify human agents with 'spirits,' and so contrast himself as a spirit with 'other spirits'. But later thinkers who shared neither his theology nor his idealism speak very oddly when they contrast their own minds with 'other minds,' rather than with others' minds. Austin (whose 'How to talk, some simple ways' my title imitates), rightly remarks, 'misnaming is not a trivial or a laughing matter.' (1961, p. 51, n.2). This is not exactly a misnaming, and may be a matter best laughed at, but whatever sort of mis-speaking it is, it is no trivial mismove. Did all these writers, like Berkeley, assume that they *were* their minds? Surely Broad, Ayer, and company took living bodies more seriously than that would suggest. Wisdom and Austin use the phrase in implicit shudder quotes, which is what it deserves. In 'Other Minds,' Austin demotes Mill's and Russell's supposedly 'deep' problem of our epistemological warrant for attributing any mental life at all to any others and of attributing particular states of mind to particular others at particular times to an addendum to a discussion of the folk epistemology of bird watching, making it seem just as silly to keep asking 'how do you know' of the person who claims to know that another feels angry as it would be to try to create a sceptical problem for the bird watchers who are identifying bitterns, gold finches and gold crests in the agreed ways for making such identifications. He also dismisses the basis of the solipsist's worry in one brief final note: '*Of course* I don't introspect Tom's feelings (we should be in a pretty predicament if I did)' (1961, p. 83). This witty paper really did, I think, subvert a whole, if admittedly reassuring short,[6] tradition in the philosophy of mind, and did so several years before the publication of Wittgenstein's *Philosophical Investigations.* ('Other Minds' dates from 1946 — it may show the indirect influence of Wittgenstein through the acknowledged influence of John Wisdom's *Other Minds*, but its method, namely that of laughing a silly pompous philosophical position out of court, is quite different from Wittgenstein's, or indeed from Strawson's later exploration of such issues in *Individuals*. By then, the high seriousness of the Wittgensteinian manner, the turning of philosophy into a species of religious devotions, had come to hold sway, whereas Austin came very close to actually doing what Wittgenstein himself allowed as a possibility, albeit one he certainly did not realize, of writing a philosophy book which took the form of a series of jokes.) Witty moral psychology is taking quite a while to make any sort of comeback, after its partial eclipse under the sombre Wittgensteinian mantle of quasi-religious solemnity. John Passmore once complained of the tediousness of the aesthetics of the time, and I think there is ground for some current complaint about the humorlessness of that part of the philosophy of mind that is seen as having any link with ethics. Dan Dennett has certainly brought humor to his treatment of the philosophy of mind, but even he turns very solemn in papers like 'Conditions of Personhood' where his philosophy of mind is brought into relation to matters of moral importance. Hume's plea for a morality that has room for gaiety and frolic has, in the subsequent tradition in ethics and moral psychology, fallen on pretty deaf ears.

C. D. Broad should be given credit for introducing at least an occasional note of salutary ridicule into the philosophical discussion of our cognitive access to the minds of our fellow persons, in his Tanner Lectures of 1923, long before Wisdom and Austin. Before giving his fairly solemn and very complicated revised versions of Mill's argument (one version takes a page and a half to state, the other two and a bit pages), as a way of answering any sceptical doubts that might arise about minds other than one's own, Broad distinguishes the proper role he takes this argument to have, to vindicate the natural convictions we all have that we are not the sole exemplars of human nature and human mentality, from its improper use to give an account of how we came by our solipsism-banishing belief in human company. He writes 'the notion that, as a baby, I began by looking in a mirror when I felt cross, noting my facial expression at the time, observing a similar expression from time to time on the face of my mother or nurse, and then arguing by analogy that these external bodies are probably animated by minds like my own, is too silly to need refutation.' (1937, p. 324). This sort of use of ridicule as a philosophical tool, one might add, was condoned even by Reid, that relatively solemn critic of the wittier Hume. Reid ridiculed Hume's views throughout his *Essays on the Intellectual and Active Powers of Man* and at one point defended his own methods by claiming 'Nature has furnished us with . . . [ridicule] to *expose absurdity*, as with . . . [argument] to *refute error*. Both are well fitted to their respective offices, and are equally friendly to truth when properly used.'[7] Since ridicule 'cuts with as keen an edge as argument,' and Reid aimed to cut the Humean view to shreds, he supplemented his argument if not with wit, at least with plenty ridicule. I do not know if Broad read Reid as well as Mill's reply to Reid, which he surely had read, but he certainly takes himself to be warranted to supplement lengthy argument with passing ridicule of absurdities. Another nice example of his humor is found at the beginning of his chapter on 'Mind's

Knowledge of Other Minds' (an extraordinary title, where 'mind' is either a sort of abstract particular, getting to know from what it is abstracted, and sorting its concrete base into one introspected mind plus other 'extraspected' ones, or an abstract particular getting to know other related abstract particulars, a sort of 'Life meets Extra-terrestrial Life'). Broad's ironic explanation of why the supposed 'problem' of knowledge of an 'external world' had received more philosophical attention than that of 'other minds,' runs like this: 'We should be doing too much credit to human consistency if we ascribed this to the fact that all convinced Solipsists have kept silence and refused to waste their words on the empty air . . . I think that the real explanation is that certain strong emotions are bound up with the belief in other minds, and that no very strong emotions are bound up with the belief in matter. The position of a philosopher with no one to lecture to, and no hope of an audience, would be so tragic that the human mind naturally shrinks from contemplating such a possibility' (1937, pp. 317-18).

Broad's sense of the tragic predicament of the solipsist philosopher is an ironic intellectualized variant of Reid's sense of the emotional discomfort of those of Berkeley's followers who found their relatives and friends converted into 'parcels of ideas,' and of the moral predicament of Fichte, finding himself saddled with a moral law demanding that he show respect for the rights of others to whose real personhood he had no access from theoretical reason. He was driven to make the reality of other right-holders a sort of extra postulate of practical reason, by a transcendental argument from the reality of his duties to the reality of those mentioned in them. For sheer comic value, Fichte's argument for other minds may take the prize.[8]

Reid noted that ridicule is not well tolerated on religious topics. 'If the notion of sanctity is annexed to an object, it is no longer a laughable matter.'[9] Since morality was so long allied with religion, some remnants of protective sanctity still cling to it. And self knowledge is one of those topics where the dead seriousness thought appropriate to ethics has infected the treatment of the topic in epistemology and philosophy of mind. Since 'γνῶθι σεαυτον' is a hallowed commandment, the modes of self knowledge have been taken to have the character of serious moral exercises — if not the severe self examination of conscience, at least the serious inward gaze of introspection. Descartes deliberately uses the religious genre of 'meditations' as the vehicle for his attempt to 'understand this "I" that necessarily exists,' and until Wisdom and Austin took their fresh approach, most other philosophers also took the topic of self knowledge to be one to be treated with some solemnity, even when, like Broad, they let knowledge of others' minds occasionally be laughing matters.

Hume of course had introduced a definitely light-hearted tone when he asked religiously-inclined self-seekers, in particular Samuel Clarke, just how they conceived of the conjoining of an indivisible spiritual thinking self with one of its own impressions or imagistic representations of spatially extended body. 'Is the indivisible subject . . . on the left or on the right hand of the perception?' (T. 40), echoing the very challenge that he had earlier put into such 'theologians'' mouths, when he has made them ask if the indivisible thought were on the right or the left of the thinker's divisible body, 'on the surface or in the middle? On the back or the foreside of it?' (T. 234). He himself claimed that when he tried to use their preferred method of 'intimate entry' by introspection he failed to find anything whatever answering to the rationalists' conception of a spiritual perceiver, distinct from its own perceptions. This famous finding or nonfinding of Hume's led Reid at least to exempt him from the charge of 'egoism,' which he had levelled at Berkeley. But he has other charges.[10]

The same supposedly sceptical Hume, whose system, Reid charged 'does not even leave him a *self*,' in Book Two of the *Treatise* cheerfully announces that 'our consciousness gives us so lively a conception of our own person that 'tis not possible that any thing can in this particular go beyond it' (T. 317).[11] So vivacious does Hume take this 'idea or rather impression of ourselves' to be that he explains our ability to transform mere ideas of other people's joys and sorrows into sympathetic fellow feelings by supposing that the surplus liveliness of our self perception spills over into our ideas of others' experiences, thereby enlivening them from mere ideas of feelings into fellow feelings. For us to have an *impression* of ourselves, not just impressions of pride enlivening our mental life, we must have become considerably more bodily than we seemed to be in 'Of Personal Identity', so that sense and kinaesthetic impressions of ourselves now play a role in our self-perception. Our perceived relations to others, relations of resemblance, contiguity and causal blood ties, are taken by Hume to 'convey the impression or consciousness of our own person to the idea of the sentiments or passions of others, and makes us conceive of them in the strongest and most lively manner' (T. 318).

This extraordinary thesis of Hume's neither repeats nor implies Berkeley's infamous argument by analogy for knowledge of other people (which some think they find elsewhere in Hume — in particular in *Treatise* Book Three, at T. 576). Hume here in Book Two simply seems to assume that we reasonably believe, for example, that another person really is suffering when we hear them moaning. 'Tis obvious,' he says, 'that nature has preserv'd a great resemblance among human creatures, and that we never remark any passion or principle in others, of which, in some degree, we may not find a parallel in ourselves.' (T. 318) He finds it obvious not only that he is not the only intelligent being, but that he resembles the others so closely that nothing human is alien to him — nor much that is animal either. The sequence he gives here is that we 'remark' others' passions and then look for parallel ones in ourselves. It is almost as if, for getting to know minds, *we* become the 'other,' they the main case. What he is explaining in this passage is not our firm belief in the reality of other's feelings, which he takes as unproblematic, but rather our natural sympathy, our tendency to come to 'catch' another's joy or suffering, as if by infection. As to how and with what warrant we come to be so sure that others are indeed suffering when they moan, or are delighted when they laugh and jump as if for joy, in Book Two (as indeed in Book One also), he ignores these particular 'abstruse' questions, and simply says that we do take ourselves to be members of a species who have the same range of feelings and express them in the same ways. 'The minds of men are mirrors to one another,' (T. 365). The 'rays of passions' that are mirrored are, of course, not in any way mysterious, since they are rays of *expressed* passions. As Darwin was later to emphasize, we depend vitally on natural expression for communication and coordination of emotions. And even when we do not need to know how others feel, we often do know, and fairly often come to share their feelings. 'A chearful countenance,' Hume writes, 'infuses a sensible complacency and serenity into my mind; as an angry or sorrowful one throws a sudden damp upon me' (T. 317). This sympathetic spread of feelings is said to depend upon our sense of similarity to those with whom we sympathize. This sense shows both in the fact that we expect to have feelings parallel to those that we 'remark' in others, and also in our being naturally guided by 'a kind of *presensation*; which tells us what will operate on others, by what we feel immediately in ourselves' (T. 332). There is symmetry between oneself and others as far as cognitive access to minds is concerned. Only later, in Book Three, does he do anything at all to transform our natural convictions about what others are feeling at particular times into the conclusions of 'inferences' that proceed from sense perception of the typical causes or effects of some passion, in another's case, to a belief in the other's feeling that passion.

But even there Hume does not make such an inference an inductive generalization from one's own case to that of others, nor into an argument from analogy. Hume had discussed inductive arguments from analogy in the Book One section 'Of the probability of causes'. In such reasoning, he wrote, we do have a fully constant conjunction for some class of cases, which is then extended to a case or cases which to some degree resemble the first set, but only imperfectly. 'An experiment loses its force when it is transferred to instances, which are not exactly resembling.' (T. 142) By this account, our taking peacocks to feel pride when they strut would be such an argument from analogy, but scarcely our taking fellow human persons to feel pride when they strut, since Hume never suggests that we each see ourself as something only a bit like our fellows as far as our nature goes. It is *human* nature, not his own nature, that he treats in his famous work, even if his first reviewers did think he fell into 'egotisms'. Only very briefly, in the Conclusion of Book One, did he contemplate the possibility that he might be unrepresentative of humanity, and even there it was not scepticism about other minds that he entertained, but fear that he might himself be 'some strange uncouth monster'. Even at the height of his despairing doubt he still can ask 'Whose anger must I dread?'. No suspicion that he might be the only one capable of real anger ever seems to have entered his mind, even at the depth of his doubting mood. The inference at T. 576 is presented by Hume as a standard inference, a 'proof,' not a probability, and not as that imperfect sort of inference that he calls argument from analogy. (Mill follows him here, claiming that the argument conforms to 'the legitimate rules of experimental enquiry. The process is exactly parallel to that by which Newton proved that the force which keeps the planets in their orbits is identical to that by which an apple falls to the ground.'[12])

In Mill's version of the inductive generalization in question, we generalize from our own case to that of others, taken to be closely similar in anatomy, powers of speech, etc. Similarity of mind is inferred from similarity of body and behavior. In the Book Three passage, Hume *begins* rather than ends with the assertion that 'the minds of all men are similar in their feelings and operations'. The inference to belief in another's particular state of mind is taken as an instance of the already granted truth that no one can 'be actuated by any affection of which all others are not, in some degree, susceptible' (T. 576). This shared susceptibility facilitates the sympathetic sharing of particular emotions. 'As in strings equally wound up, the motion of one communicates itself to the rest; so all the affections readily pass from one person to another and beget correspondent movements in every human creature' (ibid.). Hume treats the perceived causes or effects of another's passion as the cue that activates in us the general certainty that others feel as we do, for the same sorts of reasons, and a specific belief as to what someone is now feeling, leading naturally to the sharing of that passion. The 'signs' of others' feelings are not treated as giving us premisses for some inference needed to establish that others ever feel as we do. He does say that we 'infer' others' specific passions, but he also says that the inference is 'immediate'. 'When I see the effects of passion in the voice and gesture of any person, my mind immediately passes from these effects to their causes' (ibid.). We immediately anticipate the pain of the patient undergoing 'any of the more terrible operations of surgery,' even if we ourselves have been so far spared the surgeon's knife. The causal inferences involved here do not seem to depend on constant conjunctions that the inferrer need herself have verified in her own case — we are taken by Hume to have a predisposition to believe that we are not unique in our reactions, and that our conspecifics' reactions to, say, surgery without anaesthetic are the best possible indication of what our own reactions would be. As he presents the matter, the signs of anxiety in the surgery patient as the instruments are prepared and the irons heated have a great affect on the sympathetic spectators' mind, exciting pity and terror. The generalization is as much from the way the patient reacts to the way one expects one would oneself react, as vice versa.

We might in theory try to take Hume's remarks about the inferential nature of our 'discovery' of what another feels on a given occasion, from the 'effects' that feeling has in how the person looks, sounds, and behaves, to mean that the spectator's inference runs like this: 'I would feel terror if those instruments and irons were being prepared for me. The patient is pale, and shuddering, which is how I would be if dreading the surgery, so I suppose that he really feels terror.' From the causes — sight of the preparations, and the effects — anxious behavior, one infers the passion of terror. But if one asks how one knows that one would oneself pale if in terror, one soon realizes that the inference is not from purely first personal 'discovery' of the typical terror causal chain, all the way from terror-causing expectations through feeling to paling and gibbering, for one's knowledge that one would pale, or does pale, seems as extrapolation-dependent as the certainty that others do really feel. For, to repeat Broad's point, how many of us have been offered a mirror when we pale with fright? Yet we have no reasonable doubt that we do, like everyone else, pale when frightened, flush when angry, and so on. Even Descartes, supposed believer in our privileged access to our own mental states, had no sceptical doubts on that score, indeed he thought he knew all sorts of interesting facts about the natural expression and physiology of human emotions, such as that love speeds the digestion (*Passions of the Soul*, Arts. 97 & 107). To verify that one does oneself exhibit the typical face, voice, and gestures of the sad person, when one is sad, or of the frightened person, when one is frightened, or of the angry person, when one is angry, one depends upon what others, or what mirrors or cameras or tape recorders, tell one about the expression of such emotions. (To realize the effects of love on one's digestion, one depends on others a little less.) One surely knows from one's own case that feelings tend towards their own expression, but just how that expression appears to others one learns only with others' help.

It is indeed from such other people who do take one's expressive face to express what such a face normally does that one learns the words and concepts we have for discriminating different emotions, moods, and attitudes, even the difference between being awake and being asleep and dreaming.[13] So one very simple way to get to know one's own mind, at least as far as its emotional states go, is to return to this original source of knowledge about such matters, and to ask others to share their knowledge of one's state with one. I think that it is pretty obvious to all of us that we are not especially good at recognizing our own emotional states for what they are, whereas our friends, or even our enemies, have a much more reliable access to this dimension of our minds, or should I here say hearts, than we do ourselves. Of course they, especially our enemies, may refuse to share their knowledge with us, since knowledge is power, and their ability to manipulate us for their own ends will be greater the less self knowledge we have.

It might be conceded that one simple way to know what emotions are eating one up is to ask a well-disposed but frank companion to read one's face for one. But, it might be said, emotions are not everything, and my title spoke of 'knowing one's own mind,' which refers more properly to one's intentions than to one's passions. Against this objection I would argue firstly that intentions are not so separate from ruling passions, and that even those aspects of them that are not passion-determined may be as clear or clearer to a well placed observer as to the agent herself. So the simple ways already mentioned, namely asking a friend what we appear to be up to, or having one's behavior audio and videotaped so that one can get an approximation to a privileged outsider's viewpoint on it, still hold good. We do, surely, pretty confidently predict our close acquaintances' future decisions, whereas our own seem to be veiled from us.

What we do often know better than others about our own intentions is the answer we will give when asked how we came to be doing what we are doing, and what it is that appeals to us in the goal we are aiming at — though even this is something that an old and perceptive friend can often anticipate, both when our answers are self deceptive and when they are sincere. The passing thoughts and images that deck out our decision making and our intention-formation, that are strewn along our paths to the carrying out of our intentions and adorn our subsequent satisfaction or regret, are what are safest from public scrutiny, at least by current techniques. But who would say that our automatic access to our own stream of consciousness is enough to enable us to know our own mind? The command 'know thyself' is not obeyed simply by avoiding deep anaesthetics and blows on the head.[14] Locke, talking about reflection, took it at least to require some special attention.[15] Reid, commenting on Locke, emphasized that reflection was difficult, and took training and practice.[16] One reason why it was difficult, he thought, was what Hume had called the 'inconceivable rapidity' (T. 252) of our changes of perception. But even if we could survey a stretch of our own stream of consciousness in slow motion, giving our full attention to it, we still might not find our thoughts self-characterizing. Some interpretation of the drift of our fantasies, the direction of our thoughts, needs to be offered before we can even pretend to self knowledge. What is needed for self-knowledge is not just consciousness, but raised consciousness. And for that, most of us need a little help from our friends and enemies.

Knowledge is contrasted both with error and with ignorance. It is easy to be in error about one's own resentments, motives and character traits, and ignorance about one's deepest wishes is presumably what many people pay their psychiatrists large sums to have remedied. Why do we grant others, as we surely do often grant them, the right to correct us in our self descriptions, and to teach us how to read our own motives? Not just because we cannot see our own give-away faces, nor even hear our own expressive voices in quite the way others do, but because one thing that most of us do know about our own motives is that we frequently have motives for deceiving ourselves about ourselves, and that self deception is not easily unmasked by the self deceiver, all on her own. Donald Davidson calls self deceit 'an anomalous and borderline phenomenon' (1987, p. 441) Either he must be a lot freer of it than most of us, or he is subject, as we all tend to be, to meta-self-deception. Of course we do often have motive for avoiding the truth not just about our own but also about our friends' motives and characters, so that there can be conspiracies of mutually aided self deception. Still, for most people their self conception is more jealously guarded from threat than their conceptions of even their best friends. Hume, discussing the relations between pride and love wrote, 'The passage is smooth and open from the consideration of any person related to us, to that of ourself, of whom we are every moment conscious. But when the affections are once directed to ourself, the fancy passes not with the same facility from that object to any other person, how closely soever connected to us.' (T. 340). Hume is not here denying the force of love, nor of sympathy, but simply noting the special importance to us of our self assessment, of the centrality of self-directed passions among the full array of human passions. There is an important asymmetry not so much in our knowledge of ourselves and our fellows as in our concern about what is known about ourselves and others.

To be capable of the sort of pride, humility, and resentment of insult that Hume attributes to us, it is not enough that we tend like peacocks to strut, like beaten dogs to skulk and whimper, and like offended cats to spit and strike — we also need to have a self description, a sort of self written testimonial to ourselves secretly stored away, ever sensitive to challenge and on the watch for endorsement. In discussing the pride of animals, Hume emphasizes both the fact that the higher animals do seem to seek and take pride in our approbation of them, and also the fact that their pride is taken solely 'in the body, and can never be plac'd in the mind' (T. 326). This, he writes, is because they have 'little or no sense of virtue or vice,' so no conception of their own character traits. We do have a conception of our own Humean virtues and vices, that is to say our dispositions and abilities or disabilities, and it is this which we guard so jealously, which motivates most of our self deceit, and so throws us back on frank talk from friends or enemies for any hope of becoming less deceived about ourselves.

Hume, a little cynically, lists another's willingness to flatter our vanity among the chief causes of love (T. 347-349). Since 'nothing more readily produces kindness or affection to any person than his approbation of our conduct and character' (T. 346), then those we love as friends will have to be very careful about the 'service' they render us in the way of disabusing us of our more optimistic self interpretations. If 'a good office is agreeable chiefly because it flatters our vanity' (T. 349), then the office of frank critic is not likely to be perceived as a good one. Hume notes both our tendency to make friends with those who will not perform this thankless task too often, and also our tendency to discount the assessments of those whom we do not ourselves respect. Since we do not easily admit that we respect flatterers, then we often have some difficulty finding people who can fulfill one of the friend's main roles, that of 'seconding' our evaluations (including our self evaluations), while at the same time avoiding appearing too openly to be willing to 'flatter our vanity'. 'Tho fame in general be agreeable, yet we receive a much greater satisfaction from the approbation of those whom we ourselves esteem and approve of, than those, whom we hate and despise. In like manner, we are principally mortified with the contempt of persons, upon whose judgment we set some value . . . ' (T. 321). This sets the scene for self deception about oneself to need to be allied with some self deception about the friends who nourish and support one's self delusions. Maybe the best solution is to look to one's enemies, not one's friends, to keep one honest, while making some allowance for their special bias. Strangers, although free of bias, will not do, since they tend to be polite, and in any case they have no reason to give particular notice to the way we are, but enemies do tend to keep track of us, and to be more willing to violate norms of politeness. Nor is their bias likely to be any greater than our own, so it can serve nicely to correct it.

There is a wonderful example of the plight of the person who is left to his own devices for knowing his own mind in the novel, *The Margin*, by André Pieyre de Mandiargues (1969), interestingly discussed by Adrian Piper in her paper 'Pseudorationality' (1988). The forty-two year old hero or antihero, Sigismond, is during the time of the novel's narrative artificially isolated from his normal companions, since he is on a brief visit to Barcelona, where he has never been before, and knows no one. In his three days there, the only person he gets at all to know is a pretty young Castilian prostitute, Juanita. She speaks no French, he little Spanish, so he does not disclose much of himself or his preoccupations to her. He is virtually on his own, as far as knowing his own mind goes. His is a good case, for our purposes, for what he clearly needs some help with includes knowing what he believes, knowing what he feels, and knowing what he intends. He has travelled from his home in France, leaving his wife and little son there, to do some business in Barcelona for his cousin. On the way he bought some condoms, so that he could try the pleasures of the city, of which his cousin had given him graphic accounts. He is portrayed as a fairly self deprecating man, happy in his marriage to the younger and more vivacious Sergine, whose frequent mockery of his more stolid ways he treats as loving mockery. Then on his arrival he receives a letter from home, from his old servant. He opens it and reads enough to know that it reports some woman's suicide by throwing herself from a tower on his property. He reseals the letter, puts it on a table in his hotel room, builds a little shrine around it and engages in a sort of ritual at this shrine, regularly caressing both it and the phallic bottle, in the shape of the Colon tower, with which he has weighted it down. (After receiving the letter, he had made for the Colon tower, ascended to the top of it, and considered how it would feel to throw oneself from it.) He adds a black tie to the outfit he is wearing and for nearly three days lives in what he terms 'a bubble', explores the city's red light district, sleeping each evening with Juanita. He thinks repeatedly of his wife, remembering times together, imagining what her reactions would have been to the sights he is seeing and the Catalan food he is eating, even imagining how she would look as one of the 'sea of prostitutes' his cousin promised him he would move in, in Barcelona, and that he does move in. He thinks of Sergine's body as he looks at Juanita's body, and he thinks of her in what is ambiguous between the present and the past tense. But when he sees just the gift to take (to have taken?) back to her, he does not buy it. He does not bother to use the condoms he has prepared himself with, nor to wash after making love to Juanita. He does not bother to change his clothing. Eventually he reads the letter properly, learns that it is indeed his wife who has killed herself, after their little son's accidental death by drowning in a pond in their garden, while she was reading nearby. He then drives out of the city, pulls over into a quarry, and shoots himself through the heart.

Sigismond's state of mind is described by the author, near the end, thus. 'He feels no surprise, upon reflection, for he knows that without having exact knowledge, he had apprehended the disaster in its irredeemable totality, and he prides himself on having been capable of sidetracking it for two days before it burst upon his consciousness at a moment of his own choosing'.[17] Sidetracking apprehension of disaster and choosing the time to bring it onto the main track is a very interesting thing to be proud of. The novel makes it quite clear that Sigismond has, since the letter arrived, been bent on suicide, that his final fling is a desperate one. He was not really self deceived, just unnaturally self controlled. He delayed full 'exact' knowledge, full acknowledgement, for three days. Why? That is the more difficult question, and it raises the question of his motives. Is his suicide occasioned, as Piper believes,[18] by the destruction of his idea of himself as centre of his wife's life? Or simply by despair at the prospect of life without wife and son? Or did it take disgust at his own chosen style of mourning — namely going ahead with his eagerly awaited binge in Barcelona — to firm up his suicidal intentions? These questions are not answered by the novel, and human motives can remain enigmas even to omniscient observing or creating novelists. But any room-mate or companion could have corrected Sigismond's impression that he had really sidetracked his apprehension of disaster. To make a fetish of a letter is not to sidetrack it — on the contrary, it is to put it in a central place. A simple question to Sigismond as to why he was wearing a black tie, and not bothering with normal hygiene, could have ended any pretence that he was having his planned good time, as distinct from revving up to shooting himself. (Indeed Juanita's question on their last meeting, when she notes his disregard for hygiene, and asks 'Are you ill?' may be taken to be what breaks Sigismond's bubble.) He both knew his own belief, despair, and suicidal intentions well enough to act intelligently in the light of them, and yet he was not wrong in thinking that he did choose the moment to let this knowledge, this despair, this intention, 'burst on his consciousness'. He performed an elaborate act of postponing, out of somewhat murky motives. Determination not to be cheated out of his fun weekend? Uncertainty as to what his response to his wife's suicide should or would be? Need for time for the news to sink in? We can understand St. Augustine's prayer 'Lord, let me be pure, but not yet,' but the prayer 'let my empty life end, but not while I am having such a good time,' is a little harder to make sense of. And was it because he could, or because he could not, drown his grief in dissipation, that his suicidal drift turned into purposeful action? The novel is to be recommend for the disturbing doubts it leaves us with, as much as for its fine writing and its less ambiguous psychological insights.

Sigismond does, through the letter, get insight into how others, in particular his wife, must have seen, overlooked, or ignored him. But his ability to delay full acknowledgement of the impact of the news of her suicide, to keep his bubble unbroken, was a function of his virtual isolation in a foreign city, moving among strangers. Another case of delayed self knowledge through isolation, and one closer to the philosophical nerve, is that of Descartes in his *Meditations*, and it is a bit like Sigismond's in that pride is taken in the very act of putting aside what is granted to have a legitimate demand on his attention. In Descartes' case, the isolation is broken once copies of what he wrote are circulated by Mersenne to those who wrote the *Objections*, and they were quick to challenge Descartes' version, in the text, of what he had done there, intentionally and according to plan. Descartes in his *Replies* of course makes a very good self defense, but the *Replies* are as interesting as they are to us as much for the *new* information they give us about how the Cartesian story goes as for their mere repetition and clarification of the narrative or argument line of the original text.

The expressed intentions of Descartes' meditator, at the end of the first *Meditation*, include this: 'I think it will be a good plan to turn my will in completely the opposite direction, and deceive myself, by pretending for a time that these former opinions are utterly false and imaginary. I shall do this until the weight of preconceived opinion is counterbalanced, and the distorting force of habit no longer prevents me from perceiving things correctly. In the meantime, I know that no danger or error will result from my plan, and that I cannot possibly go too far in my distrustful attitude. This is because the task in hand does not involve action, but merely the acquisition of knowledge'.[19]

Descartes' announced plan is to treat the uncertain as if it were the false for a limited time, namely until his earlier prejudices are counteracted. But how long exactly is this? He does not tell us when he judges himself cured of prejudice,[20] so can revert to treating the uncertain as uncertain, the probable as probable, instead of treating, or pretending to treat, all such claims as false. Did he forget to? Did he think that we the readers should see when was the time for him to end the controlled self deception? Or did he deliberately confuse the shift from pseudo-negative claims to seriously meant negative claims?

Mersenne, in the *Second Set of Objections* wrote that the 'vigorous rejection of the images of all bodies as delusive was not something you actually and really carried through but merely a fiction of the mind enabling you to draw the conclusion that you were exclusively a thinking thing. We point this out in case you should perhaps suppose that it is possible to go on to draw the conclusion that you are in fact nothing more than a mind, or a thinking thing.[21] Descartes in reply points out that it was not until the essence of body had been shown to be intellectually comprehensible spatial extension, in the *Fifth*, and the essences or natures of body and of mind compared, in the *Sixth*, that the real distinction of himself as a thinker from any body was inferred. But Mersenne's implicit gentle charge, that what began as a 'fiction' had not been clearly concluded as a fiction, still stands unanswered.[22]

Arnauld in the *Fourth Set of Objections*, Gassendi in the *Fifth Set of Objections*, and Bourdin in the *Seventh*, all join Mersenne in querying the move from 'I can know I am a thinking thing without yet knowing whether I am an extended thing, that is, while pretend-denying that I have a body,' to 'I know I am a thinking and unextended thing,' which is what is asserted in the *Sixth Meditation* proof of the real distinctness of mind and body. Gassendi and Bourdin are less gentle than Mersenne. Gassendi suggests that Descartes' policy amounts to adopting a new prejudice, rather than relinquishing old prejudices, and says that critics will accuse Descartes of artifice and sleight of hand. Now if Descartes in fact makes the move from 'I could be and was certain I was a thinker while still uncertain that I had a body, that is while I was pretend-denying that I had a body (as a counter-ploy to any powerful beings who might be trying to deceive me into uncritical belief that I did have a body)' to 'I am certain that I am a thinker who does not need (even if he in fact has) a body in order to be a thinker,' then he has indeed either taken himself in, has done what he said he intended to do but deceived himself more thoroughly than he intended, or else there is intentional 'sleight of hand'. For he will be guilty of confusing his pretend denials, those of the *Second Meditation*, with serious denials, those needed for the reasoning of at least the *Fifth* and *Sixth* to succeed, and to have done this by blurring the transition from fake to real denials. He has forgotten or covered up the fact that his denials are untrustworthy, or rather that some of them need decoding.

It took the watchful eye of Descartes' friendly and less friendly critics to raise the question of when the pretence, or controlled temporary self deception, came to an end, and whether the cogency of the *Sixth Meditation* demonstration of the real distinction of soul and body depended on forgetting to put an end to it. Descartes of course indignantly rejected such a suggestion, and gave a restatement of the notorious demonstration that is supposed to save it. But even if his defence works, it took Mersenne and the others to get him to attend explicitly to the question of how consistently he had carried out his announced plans, of how many changes of mind he had undergone along the way in the supposed six days of mental labor. Our friends and enemies often play this indispensable role of reminding us nor merely of our earlier promises to them, but of our earlier expressed intentions, which appear from our behavior to have been either dropped or conveniently forgotten. And just as often we throw a retrospective appearance of coherence over our doings, reconstructing certain past moves to make them look as if we did not swerve too wildly nor lose control of our benign self deceptions, as if we *did* know our own mind throughout. But we usually convince our friendly and less friendly critics of this no more successfully than Descartes has convinced his critics of the lucidity of his intentional thought moves throughout the *Meditations*. Taking heed that we may be mistaken about what we intended and intend, what we believed and believe, what we wanted and want, is a vital step in the improvement of our knowledge of our own minds. We will have taken it only when charges of self-misunderstanding are, if not welcomed, at least respectfully entertained. This is as true of our intentions in our philosophical writing as of any others. I am pretty sure what Descartes was up to, but what am I up to?

*University of Pittsburgh*

## NOTES

* A version of this paper was given as the Irving Thalberg Memorial Lecture, at the University of Illinois, Chicago Circle, in September 1992, another version as the Keynote Address at the Northern New England Philosophical Society, University of New Hampshire, October, 1992. I am grateful to the many people who have helped me better to know what I think about this topic.

1 J. S. Mill, *Examination of Sir William Hamilton's Philosophy,* Ch. XII.

2 James Conant, spurred into action by my offensive suggestion that the 'problem of other minds' may have been a straw problem invented by Wittgenstein, as material for dissolution, drew my attention to the discussions of 'the problem of other minds' to be found in Mach, Carnap, Schlick, and C. I. Lewis. I restricted my list to those who wrote in English, and used 'other', rather than 'others'.

3 George Berkeley, *Principles of Human Knowledge,* A145. See also A148. I discussed Berkeley's problematic account of the 'concurrence' between divine and human agents in Baier (1977).

4 In Descartes' *Discourse on the Method,* Part Five, the criteria given for distinguishing real thinkers from automata include versatile verbal response to what is said to the thinker. Descartes does not offer these criteria to answer any sceptical worry about 'other minds,' but rather to emphasize the non-mechanical nature of thought. His passing reference in the *Second Medita-*

*tion* to the possibility that what he takes to be men passing outside his window may for all he knows be automata in hats and coats makes a point about seeing, and interpreting what we see, following on his discussion of the wax. It is not intended to present grounds for doubt on new matters, to add to the *First Meditation* doubts. (Descartes' meditator seems to care very little whether or not he is the only finite mind.)

[5] Thomas Reid, *Essays on the Intellectual Powers of Man*, II, X, XIII. I am grateful to Myles Burnyeat for drawing my attention to this passage.

[6] Myles Burnyeat has drawn my attention to the report of Diogenes Laertius (IX, 69-70) that Theodosius said that sceptics should not call themselves Pyrrhonians, since they did not know Pyrrho's mind's movements. But as Theodosius is reported as using the first person plural as confidently as Hume did, he clearly did not doubt that Pyrrho and others had minds, and ones that did move and have dispositions.

[7] Thomas Reid, *Essays on the Intellectual Powers of Man*, VI, IV, XVIII.

[8] See Fichte, *Sämmliche Werke*, vol. III, 39. I am grateful to Jerry Schneewind for pointing me towards this philosophical treasure-trove.

[9] Reid, op.cit., XI, IV, XIX.

[10] One of Reid's loaded 'Questions for Examination,' placed at the end of his presentation of Hume's views in Essay II, reads: 'Hume exceeds the Egoists in disbelief and extravagance?'

[11] T = *A Treatise of Human Nature*, David Hume, edited by L. A. Selby-Bigge and P. H. Nidditch (Oxford: Clarendon Press, 1978).

[12] Mill, ibid.

[13] I discuss this in 'Cartesian Persons,' in Baier (1986).

[14] Elizabeth Anscombe (1981, p. 36) cites William James' account of poor Baldy who, when he came to after a blow on the head received when he fell out of a carriage, asked 'Did anyone fall out?' When told 'Yes, Baldy did,' he replied 'Poor Baldy' (William James, 1901, p. 273 n.). Anscombe takes this story to show that one can be conscious without being self conscious, and so it does. So my point needs to be put by saying that neither consciousness nor self consciousness ensure self-knowledge.

[15] I am indebted to David Finkelstein for drawing my attention to Locke's words in *Essay* II, 1, A7, that unless one turns one's thoughts towards the operation of one's mind, and considers them attentively, one will not have clear and distinct ideas of reflection.

[16] Reid, op.cit., I, VI, I.

[17] André Pieyre de Mandiargues, *The Margin*, op.cit., p. 206.

[18] I discuss Piper's interpretation more thoroughly in 'The Vital but Dangerous Art of Ignoring: Selective Attention and Self Deception,' given at a conference on 'The Self and Deception' at the East-West Center, Honolulu, August, 1992, and forthcoming in *Philosophy East and West*.

[19] *The Philosophical Writings of Descartes*, vol. II, translated by John Cottingham, Robert Stoothoff, Dugald Murdoch, Cambridge University Press, 1984, p. 15. (This volume will henceforth be referred to as C.S.M.) Adam and Tannery, vol. VII, p. 22 (henceforth A.T.).

[20] He does, at the beginning of the *Fourth Meditation*, claim now to have no trouble attending to intelligible as distinct from sensory things, but not all the old beliefs made doubtful in the *First* were sense-based, so it is not clear that all his former prejudices are yet counter-balanced. His earlier assumptions about the scope and freedom of his will are yet to be corrected.

[21] C.S.M., 87, A.T., 122.

[22] I discuss what trouble this makes for Descartes' argument in the *Fourth Meditation*, and for his various claims in the *Meditations* about deceit and self deceit, in 'The Vital but Dangerous Art of Ignoring: Selective Attention and Self Deception'.

## REFERENCES

Anscombe, E. (1981), 'The First Person' in *Metaphysics and the Philosophy of Mind: Collected Philosophical Papers*, vol. II (Minneapolis: University of Minnesota Press, 1981).

Austin, J.L. (1961), *Philosophical Papers* (Oxford).

Baier, A. (1977), 'The Intentionality of Intentions,' *Review of Metaphysics* (March).

Baier, A. (1986), *Postures of the Mind* (University of Minnesota Press, 1985 & Methuen Press).

Baier, A. (forthcoming), 'The Vital but Dangerous Art of Ignoring: Selective Attention and Self Deception,' *Philosophy East and West*.

Berkeley, G. *Principles of Human Knowledge*.

Broad, C.D. (1937),*The Mind and Its Place in Nature* (Routledge & Kegan Paul).

Davidson D. (1987), 'Knowing One's Own Mind,' *APA Proceedings and Addresses*, 60.

Descartes, R. *Meditations on First Philosophy*.

Descartes, R. *Discourse on the Method*.

Descartes, R. (1984) *The Philosophical Writings of Descartes*, vol. II, trans. J. Cottingham, R. Stoothoff, D. Murdoch (Cambridge University Press).

Hume, D. *A Treatise of Human Nature*, L. A. Selby-Bigge and P. H. Nidditch, eds. (Oxford: Clarendon Press, 1978).

Fichte, *Sämmliche Werke*.

James, W. (1901), *Principles of Psychology* 11 (London).

Locke, J. *Essay on Human Understanding*.

de Mandiargues, A.P. (1969), *The Margin*, translated from the French by R. Howard (New York: Grove Press). (Winner in 1967 of the Prix Goncourt.)

Mill, J.S. *Examination of Sir William Hamilton's Philosophy*.

Piper, A. (1988), 'Pseudorationality', in *Perspectives on Self Deception*, ed. B. P. McLaughlin and A. O. Rorty (University of California Press), pp. 297-323.

Reid, T. *Essays on the Intellectual Powers of Man*.

HUW PRICE

# PSYCHOLOGY IN PERSPECTIVE

If recent literature is to be our guide, the main place of philosophy in the study of the mind would seem to be to determine the place of psychology in the study of the world. One distinctive kind of answer to this question begins by noting the central role of intentionality in psychology, and goes on to argue that this sets psychology apart from the natural sciences. Sometimes to be thus set apart is to be exiled, or rejected, but more often it is a protective move, intended to show that psychology is properly insulated from the reductionist demands of natural science. I am interested here in the general issue as to how this move to insulate intentional psychology should best be characterised—how to make sense of the idea that there can be a legitimate enterprise of this kind. I shall concentrate on what is perhaps the best known version of such a view, that of Daniel Dennett. I think that my conclusions apply to other versions as well, but Dennett provides a particularly accessible example.

Dennett argues that intentional psychology is to be understood as a product of a particular way of viewing certain natural systems or structures; the view from what he calls *the intentional stance*. He has often compared the recognition of intentional states to the recognition of patterns (see most recently Dennett 1991). He tells us that just as a pattern is only salient from a particular perspective—that of the pattern recogniser—so beliefs and desires are only salient from the standpoint of the intentional stance. This is a *perspectival* view of intentional psychology, then.

But perspectivalism comes in more than one flavour. In this paper I want to suggest that Dennett's perspectivalism is too bland, and to recommend a richer though less fashionable version. Among the advantages of this richer blend is the fact that it would provide Dennett with the best prospect of an answer to those of his critics who accuse him of wavering between realism and instrumentalism (the issue which motivates Dennett's return to the pattern recognition analogy in 'Real Patterns'— he says for example that 'it is this loose but unbreakable link to ... perspectives ... that makes "pattern" an attractive term to someone perched between instrumentalism and industrial-strength realism', 1991, p. 32).

This more strongly flavoured perspectivalism has much in common with that of Ryle, and hence ultimately of Wittgenstein. In my view Dennett is not the only one in the philosophy of mind who would benefit from a rather selective infusion of Rylean and Wittgensteinean themes. Imbibed in the cor-

*Michaelis Michael and John O'Leary-Hawthorne (eds.), Philosophy in Mind, pp.83-98.*

rect manner, they provide the most promising approach to the perennial problem of explicating the dependence of psychology on the physical sciences—that is, to reconciling psychology with naturalism. This may seem implausible. Contemporary Wittgensteinians are not friends of naturalism, by and large. But these contemporary writers have failed to notice the naturalistic character of Wittgensteinian reflections on language. 'What does this part of language do for us; what is its function?' To ask such questions is to adopt an *explanatory* stance towards an aspect of one's own linguistic practice. The stance here is thoroughly naturalistic — that of a biologist, in effect. And yet it should come naturally to a Wittgensteinian, for of course one of the Master's main themes is that philosophy needs to pay close attention to the uses of language, in order to avoid its characteristic traps.[1] (Later I shall argue that it is the naturalism of this explanatory standpoint that takes the mystery out of a non-reductive defence of intentional psychology.)

The first task is to distinguish three different kinds of perspectivalism: the rather bland kind which Dennett seems to have in mind, an intermediate strain, and lastly the richer variety that I want to recommend.

*Three kinds of perspectivalism*

*1. Epistemic perspectivalism.* Many aspects of reality can only be seen from particular viewpoints. There are many trivial examples. The eyes of the Mona Lisa may follow you around the room, but they won't follow you into the next room, so if you want to see them you'd better stand in the right part of the building. For a less trivial example, suppose we identify colours with underlying physical properties. Redness is then perspectival in the epistemological sense. The property exists independently of any observer, and may be characterised by someone who lacks colour vision; but its instances are normally accessible only to an observer whose visual apparatus is attuned to physical properties of just the right kind. Much the same is true if redness is thought of as a disposition to affect normally sighted observers in a particular way. In this case the characterisation refers to normal observers—to be red is to be such as to evoke a particular response in a normal observer under suitable conditions—but one needn't *be* a normal observer to make sense of that characterisation. (One may grasp the concept without possessing normal colour vision.) But again, only a normally sighted observer occupies the perspective needed to *detect* such a disposition, at least in a direct way.

In these cases the perspectivalism is *epistemic*. The states of affairs in question exist independently of whether the perspectives in question are ever occupied; but are only directly detectable when and in so far as the perspectives are occupied.

*2. Conceptual perspectivalism.* A stronger form of perspectivalism is obtained if we deny that colour concepts can be grasped by someone who lacks normal (or near normal) colour vision. Thus it might be held that although redness itself is a physical property, just as the reductionist contends, the *concept* red is not the *concept* of that physical property. Rather the sense of the term 'red' is one which can only be grasped by someone who experiences colour vision 'from the inside'. A creature without normal human colour vision could in principle refer to the same elements of reality, but not by means of the human system of colour concepts. (This form of perspectivalism is a natural partner to the common view that at least some of the conceptual joints at which we carve the world are not absolute, but reflect our own interests and responses. Note that it is not incompatible with metaphysical realism. In one possible version, for example, it is the view that properties are simply classes, all of which are equally real, even though only a small number are picked out in any given system of concepts.)

*3. Functional perspectivalism.* Now think of a projectivist view of colour — the view that colour discourse projects our subjective colour experience. A creature without such experience wouldn't have a use for an element of language with this function. The basic idea here is a simple one: the tools a creature has a use for depend on its circumstances. Dolphins have no use for bicycles, and telepaths none for mobile phones or fax machines. If different parts of language serve different functions then presumably the same sort of thing may be true in linguistic cases. (Omniscient beings would have little use for the interrogative mood.) Similarly, if colour discourse serves the function of expressing and projecting the normally sighted observer's visual perspective, then unless a creature occupies that perspective, it lacks what it takes to make use of that discourse; the function of the discourse is one which simply has no place in such a creature's life. In such cases the dependence on perspective is not merely epistemic or conceptual—I call it *functional* perspectivalism.

Projectivism is normally an irrealist position—projectivists are typically either non-factualists or eliminativists. We might expect the same to be true of functional perspectivalism more generally. On the face of it, the view seems to involve a denial of the independent reality of whatever falls under the concepts in question. Later I want to argue that this impression is misleading. But for the moment we have three kinds of perspectivalism, epistemic, conceptual and functional.

Where does Dennett's position fit in? The fact that functional perspectivalism is prima facie irrealist gives us some reason for taking Dennett's perspectivalism to be of a non-functional sort, for Dennett claims to be a realist (of sorts) about intentional states.[2] More importantly, the pattern recognition analogy seems best read in epistemic (or perhaps conceptual) terms. Dennett himself tells us that 'other creatures with different sense organs, or different interests, might readily perceive patterns that were just imperceptible to us. The patterns would be *there* all along, but just invisible to *us*.' (1991, p. 34, Dennett's italics). This certainly sounds like epistemic perspectivalism. It is true that even a projectivist will allow that colour vision, moral sensibility, or whatever, involves what might be thought of as a pattern recognition mechanism — something in us that responds to particular conditions in the world. But the projectivist will deny that the discourse concerned—about colours, moral value, or whatever — should be understood as *referring* to the relevant physical patterns. Its function is to express our responses to these patterns, which is a different matter.

However, I think the most striking way to exhibit the epistemic character of Dennett's perspectivalism is to contrast it with that of a philosopher whose perspectivalism about the mental is clearly non-epistemic, namely Ryle. The most crucial difference between Dennett and Ryle is simply this: Ryle is unashamedly a linguistic philosopher, whereas Dennett belongs to the generation for whom language became so to speak "unmentionable", at least in the context of philosophical discussions of other topics. Accordingly, Ryle attempts to trace the Cartesian myth to a philosophical misunderstanding about the role of the *language* of psychological ascription. His central instrument is the notion of a category mistake—the Wittgensteinian idea that as theoreticians we may be misled by the grammatical form of an utterance, and hence misunderstand its logical function. Like Dennett, Ryle thinks that psychological language is distinctive; unlike Dennett, however, he thinks that its distinctiveness lies in its function, and not merely in the aspect from which it regards its subject matter. For Ryle failure to adopt a perspective is not simply like failure to stand in the right viewing position; it is failure to adopt a form of speech, failure to put language to use in a particular way.[3]

A number of aspects of Ryle's view fall flat for contemporary readers. One such aspect is the linguistic approach itself, of course, though here the fault may lie as much with the prejudices of the contemporary reader as with

Ryle. But even modern readers not embarrassed to discuss philosophical problems in linguistic terms are likely to find Ryle's account of what is distinctive about talk of the mental a little quaint. Ryle tells us that mental talk is talk of behavioural dispositions. To modern ears this seems severally inadequate: first, because we are now so accustomed to dispositions in philosophy that it is hard to take seriously the claim that dispositional talk has any special logical status;[4] second, because the dispositional brush seems hopelessly broad to capture what is distinctive about the mental; and third, because even when the focus is narrowed to *behavioural* dispositions the result is a position whose drawbacks have since been pointed out by a generation or so of functionalists.

All the same, there might well be room for a Dennettian version of Ryle's perspectivalism—a view which would take over Ryle's Wittgensteinian idea that beneath their surface similarities, different parts of language serve different functions, while incorporating Dennett's account of the function of the intentional stance. Thus we might abandon the behaviourism,[5] but keep the notion of a category mistake, and the associated diagnosis of what was wrong with the Cartesian tradition.[6]

### *Ryle, Quine and existence*

Could Dennett justifiably resist this move to recast his views in Rylean form? An obvious thought is that he would see it as having unwelcome consequences concerning the existence of beliefs and desires—as leading to irrealism in some form. Dennett says he favours a 'mild realism', weaker than 'Davidson's regular strength realism', but stronger than 'Rorty's milder-than-mild irrealism, according to which the pattern is *only* in the eyes of the beholders' (1991, p. 30).

But is Ryle necessarily an irrealist? What does the issue of the existence of beliefs and desires look like from a Rylean viewpoint? At one point Ryle says

> It is perfectly proper to say, in one logical tone of voice, that there exist minds, and to say, in another logical tone of voice, that there exist bodies. But these expressions do not indicate two different species of existence, for 'existence' is not a generic word like 'coloured' or 'sexed'. They indicate two different senses of 'exist', somewhat as 'rising' has different senses in 'the tide is rising', 'hopes are rising' and 'the average age of death is rising'. A man would be thought to be making a poor joke who said that three things are now rising, namely the tide, hopes and the average age of death. It would be just as good or bad a joke to say that there exist prime numbers and Wednesdays and public opinions and navies; or that there exist both minds and bodies (1949, p. 24).

This might seem to give Dennett grounds for resisting Ryleanisation, namely that Ryle's appeal to 'two different senses of "exist"' falls victim to Quinean lessons concerning existence. As Quine himself puts it:

> There are philosophers who stoutly maintain that 'true' said of logical or mathematical laws and 'true' said of weather predictions or suspects' confessions are two uses of an ambiguous term 'true'. There are philosophers who stoutly maintain that 'exists' said of numbers, classes and the like and 'exists' said of material objects are two uses of an ambiguous term 'exists'. What mainly baffles me is the stoutness of their maintenance. What can they possibly count as evidence? Why not view 'true' as unambiguous but very general, and recognize the difference between true logical laws and true confessions as a difference merely between logical laws and confessions? And correspondingly for existence? (Quine, 1960, §27).

Dennett elsewhere 'Quines qualia', and in 'Real Patterns' endorses Arthur Fine's rather Quinean attitude to ontological issues; so this invocation of Quine seems a move that might well appeal to him.

However, if (as Quine tells us) matters of ontology amount to little more than matters of quantification, then this objection to Ryle looks rather beside the point. Ryle would presumably not deny that we should quantify over prime numbers, days of the week and dispositions. The respect in which he takes the categories to differ must be some other one. Indeed, Ryle might say that in denying that there are 'two species of existence' he is agreeing with Quine. Ryle simply needs to say that what we are doing in saying that beliefs exist is not what we are doing in saying that tables exist—but that this difference rests on a difference in talk about tables and talk about beliefs, rather than on any difference in the notions of existence involved. So far this is exactly what Quine would have us say. The difference is that whereas Quine's formulation might lead us to focus on the issue of the difference between tables and beliefs *per se*, Ryle's functional orientation—his attention to the question as to what a linguistic category *does*—will instead lead us to focus on the difference between the function of the *talk* of beliefs and the function of the *talk* of tables.

This functional linguistic perspective does not appear to violate any hallowed Quinean doctrine. For one thing the issue as to the function of a particular part of language is a scientific one. It is the anthropologist who asks 'What does this linguistic construction do for these people?' (Or even more basically, the biologist who asks 'What does this behaviour do for these creatures?'.) Moreover, Quine is in no position to insist that there is a common semantic function for the various indicative discourses in question, such as that of being *descriptive*, or *fact-stating*. For these are meaning categories, to which Quine's irrealism about meaning would seem to prevent him appealing.

It is no use Quine's saying for example that the discourses in question are alike in aiming at truth; for the Rylean may agree, but counter that in view of Quine's own minimalism about truth, this concession is quite compatible with the claim that they serve different functional roles. Truth, like existence as quantification, belongs to what Wittgenstein refers to as the 'common clothing', which overlays and masks the interesting differences between discourses.

To sum up: a position that combined Ryle's explanatory linguistic stance with Dennett's account of what psychological discourse is *for*, wouldn't fall fowl of Quinean ontological minimalism (nor apparently of the Finean version on which Dennett himself claims to rely). On the contrary, the minimalist themes we find in both Quine and Wittgenstein appear to cut away the ground from someone who would argue that functional perspectivalism is necessarily irrealist. So we can enjoy this rich blend of perspectivalism without endangering our realist figures. Of course this sounds too good to be true, and in one sense it is; as I've argued elsewhere (Price, 1992), the trick turns on the fact that minimalism also undermines the traditional (rather anorexic) conception of what a good realist figure looks like. That conception depends on the assumption that there is a single unifying semantic function properly served by indicative discourse—that of 'stating the facts', or 'describing the world'. Given such a function, realism concerning a discourse can then be cast as the view that the discourse succeeds in performing this function; i.e., that the discourse in question is genuinely descriptive (as opposed say to *expressive*), and that the descriptions it offers are not uniformly false, in virtue of the non-existence of that to which they claim to refer. But semantic and ontological minimalism undercut this conception at its foundation; for they entail that our notions of truth, factuality, existence and the like are simply too insubstantial to bear the metaphysical weight.

This is not the place to go over that argument in detail. However, I do want to try to give some impression of what sort of stance this functional perspectivalism will involve. In particular I want to distinguish it from two other positions advocated by contemporary Wittgensteinians. And I also want to argue, as promised, that it embodies the most plausible approach to the old problem of making reasonable sense of the intuition that the mental depends on the physical.

*Three ways to be Wittgensteinian*

A prominent theme in Wittgenstein's later work is that the surface uniformities of language hide an underlying diversity. Common syntactic forms and structures mask distinctions among a variety of different discourses or language-games. Wittgenstein combines this linguistic pluralism with a kind of quietism, saying that it is not the job of philosophy to try to remove this diversity. Hence Wittgenstein's anti-reductionism: the view that it is not the job of philosophy to try to reduce one discourse to another.

These themes have often been deployed in defence of varieties of non-reductionist realism. Simon Blackburn calls the resulting position 'dismissive neutralism' (1984, ch. 5). Crucially, Wittgenstein is thus seen as denying that there is any external standpoint from which it makes sense to criticise a discourse for failing to correspond with the world, or to see it as reducible to some other discourse. As Blackburn elsewhere points out ('The Message of Wittgenstein', typescript, p. 1), this is a prominent theme in late twentieth century philosophy, taken up perhaps most notably by Rorty, who finds it not only in Wittgenstein, but also in Sellars, Quine and Davidson.

Blackburn goes on to argue that the view is wrongly attributed to Wittgenstein. In particular, he argues that is wrong to take Wittgenstein as 'a writer bent on showing that reference to an objective reality is something that is aimed at by all indicative sentences, all of which express propositions, in the same way' ('The Message of Wittgenstein', typescript, p. 2). On the contrary, Blackburn makes a persuasive case to find in Wittgenstein something closer to his own *quasi-realist* position: the view that indicative sentences divide into those that genuinely express propositions (and have genuine truth values), and those that only appear to do so (and hence whose apparent truth values the quasi-realist needs to explain).

Our present concern is not to locate Wittgenstein on the philosophical map, but to sketch some elements of the map itself. The crucial point illustrated by the possibility of a quasi-realist reading is that Wittgensteinian non-reductionism is not incompatible with the viewpoint which asks, 'What is this discourse *for*; what function does it serve?' (and hence 'Why does it take indicative form?') Questions of this kind come naturally to the quasi-realist. *And they are not excluded by the neutralist's insistence that there is no Archimedean standpoint from which to compare a discourse to an external reality.* As I have emphasised earlier, the standpoint here is not Archimedean but simply scientific—it is that of the biologist, in effect, who wants to know what function this element of linguistic behaviour serves in the lives of the creatures in question. This scientific perspective does not deny anything that the neutralist says; it simply promises to say more. It promises to tell us about an aspect of the plurality that the neutralist simply remains mute about, namely the issues as to what different discourses are *doing*, and as to why there should be more than one (or even one, for that matter).

Blackburn's neutralist opponents might reply that this scientific perspective may yield no basis for the quasi-realist's distinction between those discourses that require the quasi-realist treatment—i.e. those with respect to which it is appropriate to ask 'Why is truth used here?'—and those have their truth values by birthright, as it were. It saying this they might of course claim Wittgenstein's authority; they might argue, contra Blackburn, that Wittgenstein thinks that there can be no distinction between the genuinely factual (or proposition-expressing) uses of language and the rest of the indicatives. Two responses will then be in order. First, in so far as it is an empirical matter, we shall just have to wait and see. But second, and more important, even if both points are conceded to the neutralists, the effect is not to take us back to their cosy quietism. For even if there is no factual–non-factual distinction, we can still ask the quasi-realist's scientific questions about the functions of different bits of language. The only thing that changes is that the quasi-realism now goes all the way down.

In summary then, the linguistic stance admits at least three possible positions: the neutralists' quietist realism, quasi-realism and the position I call 'explanatory pluralism'. Whatever Wittgenstein's own views on the matter, there seems to be no conceivable reason for favouring the first position over the latter two. (The explanatory issues that divide them are not ours to reject; we may choose not to engage with them, but that is a different matter.) The choice between quasi-realism and explanatory pluralism is more difficult, and really turns on the availability of a substantial distinction between genuinely factual uses of language and those which the quasi-realist wants to regard as merely quasi-factual. I have argued elsewhere that such a distinction isn't available (Price, 1989), but I don't want to try to press the point here.

However, it is worth noting that discourse pluralism is especially well placed to defend a position of Dennett's kind against the charge that it offers an instrumentalist account of the mental. For the instrumentalist too relies on a distinction between full-bloodedly descriptive uses of language and those uses that call for an instrumentalist reading. In rejecting such a distinction the discourse pluralist thus rejects a fundamental presupposition of instrumentalism. This may seem a rather pyrrhic victory. After all, doesn't the discourse pluralist defeat the instrumentalist by being anti-realist about everything? Again, I don't think so, for the reason mentioned at the end of the previous section. As I have argued elsewhere (Price, 1992), discourse pluralism supports minimal realism, and that in effect this is all the realism that anyone is entitled to, once we take on board the lessons of Quinean and Wittgensteinian quietism about truth and ontology. (In other words, I think that quietist minimal realists are right about realism, and only mistaken in ignoring the further explanatory issues.) This richer functional brand of perspectivalism thus holds out the prospect of a much more powerful response to instrumentalism than Dennett's epistemic variety. It does so, in effect, by taking up the question as to what is at stake in the issue between realists and anti-realists in such a case, and hence arguing that there is no valid basis for the instrumentalist challenge. Dennett's mistake—a mistake that flows from his failure to acknowledge the linguistic and functional dimension to an adequate perspectivalism—is to take uncritically on board too much of his opponents' conception of what is at issue between realists and anti-realists about intentional psychology.

Dennett's mistake is a very pervasive one, of course. There is a vast Flatland in contemporary philosophy, a land whose inhabitants simply fail to notice the dimension of linguistic complexity which is addressed from the explanatory perspective. Unable to direct their gaze in the required direction, they simply fail to ask the question, 'What does this part of language *do*?' Even Wittgensteinians, who were once taught to ask questions of precisely this kind, have tended to lose the ability. Having rightly taken Wittgenstein's minimalism to provide a reason for not asking one sort of question, they wrongly take it to provide a reason for asking no questions at all—or else simply fail to notice that there is another question to be asked.

I want to emphasise again that the questions being overlooked here are of an empirical and naturalistic nature. Though in practice they may arise (or fail to arise!) in a philosophical context, their content is scientific. They belong to the study of language users as natural creatures. Two corollaries follow from this. The first is that the explanatory stance that takes such questions seriously is not to be dismissed by armchair philosophy, or not at any rate by any armchair philosophy that claims to respect the scientific perspective. To illustrate the second, let us turn to the issue as to how intentional psychology may be reconciled with naturalism.

*The dependence of psychology on physics*

How, if at all, are we to preserve psychology in the face of the naturalist intuition that in the final analysis, everything depends on the physics? Of all the distinctively philosophical questions about the study of the mind, this one is perhaps the choicest. A familiar range of answers are currently on offer in the philosophical literature. At the two extreme ends of the range, for example, we find on the one hand the eliminativist answer, that psychology—or folk psychology, anyway—has had its day, and shouldn't be defended; and on the other various anti-scientist positions which reject the naturalist intuition itself, claiming that psychology is not really under attack. (Some of the latter draw on the Wittgensteinian themes we have described above.) In between these extremes there is a group of compatibilist positions, which seek to save both psychology and naturalism. I want to suggest that explanatory pluralism not only has the credentials to join this group, but promises to become its most successful member.

Let us begin with three desiderata, one stemming from the desire to save psychology, one from the desire to accord priority to the physical, and one perhaps from more general philosophical sources.

1. We should respect Occam, avoiding (unmotivated?) ontological profligacy.
2. We should respect and account for the intuitive priority of the natural realm—i.e., justify the intuition that in some sense, everything else depends on physics.
3. We should respect psychological descriptions and explanations, both in principle, and (by and large, at any rate) in practice.

It is important to keep in mind that although we are here concerned with the relation between psychology and physics, similar issues arise elsewhere. Indeed, they arise both on the scientific hierarchy (e.g. concerning the relation between biology and physics) and elsewhere in philosophy (e.g. with respect to the status of moral or modal discourse). I don't want to prejudge the issue as to the relation between these two sorts of cases, or, in so far as they are distinct, that as to in which camp intentional psychology should properly be located. All the same, if our treatment of the psychological case is not to seem ad hoc, it ought to meet a fourth desideratum along these lines:

4. If possible, the solution we adopt in the psychological case should be one which will generalise to other cases in which we are confronted with the task of reconciling modes of description other than those of physics with naturalist intuitions.

I shall come back to this requirement at the end of the paper.

Addressing the issue as to how we might defend higher-level explanation (including causal explanation) in the face of naturalism, Philip Pettit (1992) notes in a recent paper that there are three main compatibilist options currently on offer: the identity theory, the supervenience view, and Pettit's own preferred position, the Jackson-Pettit program explanation model. Of these, the identity theory has long been popular, particularly in the so called *token identity* version. It has obvious attractions, most prominent of which are that it is absolutely non-profligate in ontological terms, and yields a straightforward account of the priority of physics: higher-order states are identified with physical states, but not in general vice versa. But it also has defects. Some of these are identified by Jackson and Pettit, who argue in particular that the identity theory gives implausible results concerning higher-order causal connections. Other defects have recently been pressed by Mark Johnston (1992) and in unpublished work by David Braddon-Mitchell and myself. In sum, I think it is fair to say that the causal reductionism that drives these theories turns out to be doubtfully compatible with requirement 3.

Jackson and Pettit have argued that the supervenience model also gives counterintuitive results concerning higher-order causal connections, and thus fails requirement 3. I want to suggest that it also does poorly on 1 and 2. For one thing, the ontological proliferation permitted by the supervenience model appears to be unconstrained. The model doesn't tell us where the multiplicity of levels comes from—it appears simply as a brute fact. Secondly, it is far from clear that the supervenience model provides a sufficiently substantial account of the priority of the physical. That is, the supervenience relation itself appears to be left as a brute fact about the world, not explained by the model.[7] So the supervenience model *describes* the dependence of the higher on the lower—gives its logical form—but does not *explain* it.

I think the same is true of the program explanation model. Jackson and Pettit explicate higher level causal relevance in terms of the idea that higher level properties may "program" for the presence of causally efficacious lower level properties. Pettit describes the model as follows:

Suppose that there is no question about the causal relevance of properties at a given level L to the occurrence of an event E, of a given type. According to the program model, a property P at some other level will be causally relevant to that event just in case two conditions are fulfilled.

1. The realisation of P *makes it more probable* than it would be otherwise—at the limit, it may ensure—that there are certain properties realised at level L which are apt to be causally relevant to the occurrence of such an E-type event.

2. Properties of the probabilified kind are in fact causally relevant to the occurrence of E. (1992, pp. 10-11 in ts., my italics).

It seems to me that the program model faces a dilemma with respect to the relevant notion of probabilification (of lower level properties by higher level properties). On the one hand we might read this notion in a very thin frequentist, or associationist, sense. The claim will then be that the higher level property simply *varies in parallel* with the presence of a lower level property of the required kind. But this will leave the view vulnerable to the objection I just raised to the supervenience view, viz. that at best it is *describing* a regularity in the world, whereas what we want from account of the priority of the physical is an *explanation* of this regularity. A graphic way to make the point is to note that even a Cartesian event or property dualism will meet the condition, so long as the dualist allows that mental events run in parallel with physical events in the brain. In other words the condition might be met in a world in which psychology simply did not depend on physics in the way in which we take it to do so in fact.

The program model might of course reply that it has some stronger or more substantial notion of probabilification in mind. The difficulty is that the metaphysics needed to support a more substantial probabilistic connection between levels is likely to bring with it an independent account of the dependence of higher levels on physics; after all, what it has to exclude is precisely the case in which we have association without the right sort of dependence. So there is a dilemma here: the crucial notion of probabilification must be neither too weak nor too strong, but it not clear that we have been given any acceptable middle course.

How does an explanatory pluralism avoid these difficulties? The solution turns on the fact that the explanatory stance is itself naturalistic. In explaining a discourse our aim is to say what function it serves for the natural creatures who employ it, what responses, capacities or other features of these creatures it depends on, and so on. To take one of the simpler examples, consider colour. We take colour properties to supervene on a physical basis. On occasion we might take colour to be causally relevant—we say that it was the redness of the flag that enraged the bull, for example—and we might account for this in program terms. Why aren't these features of colours mysterious? Because both are readily explicable in terms of a naturalistic account of colour vision. It is in virtue of the fact that colour vision involves an interaction between physical states in the world and physical states in the visual system that so long as the latter is held fixed (and is reasonably deterministic), a difference in the colour judgements it outputs requires a difference in the physical state it inputs. (This is similar to a point long made by Blackburn in support of moral projectivism: Blackburn (1971, 1984 and 1985) argues that only the projectivist can account for moral supervenience, since for the realist it is simply a mystery. Note that in both cases the force of the point is explanatory, not justificatory or normative; we are explaining why the practices of making colour judgements or value judgements normally exhibit supervenience, not demonstrating within colour or value discourse that they do so. Indeed, it seems that to attempt such a demonstration should properly be seen as involving a category mistake, on this view.)

Similarly, it is in virtue of the fact that colour vision involves an interaction between physical states in the world and physical states in the visual system that the physical circumstances which normally produce the output 'That's red' are typically of the sort whose effect is to enrage bulls. In both cases it is the naturalistic explanatory standpoint that demystifies the bare logical relationship. Leave aside this illuminating standpoint and nothing stands in the way of interpreting supervenience or the program relation in terms of a brute parallel dualism about colour.

I want to emphasise that this explanatory standpoint is compatible with taking either supervenience or the program model to provide the appropriate *description* of the dependency of higher levels on lower levels. However, it is the explanatory background that demystifies the bare logical dependency, that explains the formal priority of physics. Thus explanatory pluralism provides the "cement" that prevents multiple discourses from floating free of one another. And the cement is only visible when we take the explanatory "step backwards"—when we disengage ourselves from the discourses in order to ask what they do, what features of speakers and the world they depend on, and so on. At the same time the explanatory pluralist acquires an answer to the identity theorist's original worry, namely that anything except the identity theory involves ontological profligacy. The plurality of discourses is not arbitrary and unconstrained; each is accounted for from the explanatory standpoint. (The plurality is one of linguistic functions or forms of behaviour.) Again, the supervenience view and the program model appear to have nothing to say here.

Thus in explaining dependence of the psychological on the physical functionalist perspectivalism does better than the three main alternatives, namely the identity theory, the supervenience view, and the Jackson-Pettit program explanation model. The crucial points are (i) that the perspectivalism provides the respects in which discourses differ, thus countering the charge of ontological profligacy; (ii) that the functional perspective is itself naturalistic, which underpins the felt priority of physical explanation; and (iii) that the connection between discourses only emerges from the explanatory viewpoint adopted by the functional perspectivalist—these connections remain mysterious for supervenience views and for the program explanation model.

Finally, let us return briefly to the issues raised in my formulation of the fourth desideratum at the beginning of this section, namely that the solution adopted in the psychological case should generalise to other cases in which we face the task of reconciling a level or mode of description other than that of physics with our naturalist sympathies. I noted that there seem to be two sorts of cases to be dealt with, those which arise in the scientific hierarchy and those which arise elsewhere in philosophy. This distinction seems to me to be an important one. There seems to be an important difference between the relation between say chemistry and physics, on the one hand, and that between colour talk or modal talk and physics on the other. Roughly, the former distinction depends on levels of structure in the world, the latter distinctions more on functional differentiations in the sorts of things we humans do with language. Of course it doesn't follow that these categories are sharp. The cases mentioned might simply lie at the opposite ends of a continuum. But assuming for the moment that the distinction is a sharp one, the question arises as to where intentional psychology lies—With science? Or with other projects?

I think it is the latter alternative that should appeal to Dennett, and others in contemporary philosophy who want to argue for the irreducibility of intentionality. Of course, it is widely believed that to detach psychology from the scientific hierarchy is to consign it to the anti-realist wilderness. What I want to stress in closing is that this is simply not so. To read a functional perspectivalism about the mental as an irrealist view is to commit a category mistake, to judge psychological discourse by a standard which ought never to have been applied to it. As Ryle saw well, the place of philosophy in the study of the mind is to point this out.[8]

*University of Sydney*

NOTES

[1] Perhaps the reason this stance is so unpopular is that its naturalism offends Wittgensteinians, while its linguistic character offends post-Wittgensteinians. If so, then it is time for the two sides to adopt the Sprat family's approach to conflict-resolution: hard-nosed post-Wittgensteinians should embrace the stance for its lean naturalism, and Wittgensteinians should embrace it for its rich linguisticism.

[2] It is true that I want to suggest that functional perspectivalism is not necessarily irrealist after all, but this is a rather unorthodox view, and Dennett certainly does not offer any argument to this effect.

[3] Why is this not simply conceptual perspectivalism? Because of Ryle's emphasis on logical function. A conceptual perspectivalist need not deny that the perspective-dependent concepts are all employed for the same single descriptive purpose.

[4] This doesn't necessarily mean that Ryle is wrong, of course. It may simply mean that in ignoring the functions of language, contemporary philosophy has lost a sensitivity that earlier philosophers possessed.

[5] Note that this would simply by-pass some of the major criticisms of Ryle, such as that of Fodor (1975), who accuses Ryle of being a functional reductionist. It is doubtful whether it is fair to term Ryle a reductionist of any sort—the point is nicely brought out in terms of a contrast between analysis and explanation—but even if he were, in ditching his behaviourism we would ditch what was problematic about his reduction base.

[6] Dennett is not the only contemporary philosopher of mind whose view of intentional psychology might be given this Rylean flavour, of course—Davidson, McDowell, Pettit are others—but for present purposes let us continue to use Dennett as our example.

[7] A similar point against the supervenience approach has been made by Stephen Schiffer (1987), pp. 153-4. As Schiffer puts it, 'Supervenience is just epiphenomenalism without causation' (p. 154).

[8] There is a powerful and somewhat Rylean card to be played against anyone who would exile psychology on these grounds, namely to point out that according to the philosophical view in question, where psychology goes modality goes also. To exile psychology would be to exile not only Ryle's dispositions, but also probability, causality, laws, and so on—and let science try to survive without that! This card needs to be backed by a defence of the pragmatist treatment of modality on which it relies, of course, which is a task for another time. All the same, one of the ironies of contemporary philosophy of mind is that hard-nosed physicalists make great use of appeals to causality, and in particular to the intuition that all causality is ultimately physical causality. Yet it is surely a conceivable view—indeed it is a view held, as well as conceived, both by Russell and by Ayer—that a *really* hard-nosed physicalism would regard causation itself as an anthropocentric adornment to the pure physical description of the bare four-dimensional lattice of the world. So contemporary physicalists are apt to build their case on foundations which are doubtfully compatible with their own exacting standards.

## REFERENCES


Blackburn, S. (1971), 'Moral Realism' in Casey, ed., *Morality and Moral Reasoning* (London: Methuen), pp. 101-24.

Blackburn, S. (1984), *Spreading the Word* (Oxford: Oxford University Press).

Blackburn, S. (1985), 'Supervenience Revisited' in Hacking, ed., *Exercises in Analysis: Essays by Students of Casimir Lewy* (Cambridge: Cambridge University Press, 1985), pp. 47-67.

Blackburn, S. (unpublished), 'The Message of Wittgenstein'.

Dennett, D. (1991), 'Real Patterns', *Journal of Philosophy*, 88, pp. 27-51.

Fodor, J. (1975), *The Language of Thought* (New York: Cravell).

Johnston, M. (1992), 'Constitution is not Identity', *Mind*, 101, pp. 89-105.

Quine, W.V. (1960), *Word & Object* (Cambridge, Mass.: MIT Press).

Pettit, P. (1992), 'The Nature of Naturalism', *Proc of the Arist Soc, Supp. Vol. 63.*

Price, H. (1989), *Facts and the Function of Truth* (Basil Blackwell: New York).

Price, H. (1992), 'Metaphysical pluralism', *Journal of Philosophy*, 89, pp. 387-409.

Ryle, G. (1949), *Concept of Mind* (Peregrine Books).

Schiffer, S. (1987), *Remnants of Meaning* (Cambridge, Mass.: MIT Press).

KARL-OTTO APEL

# CAN THE PHILOSOPHY OF LANGUAGE PROVIDE THE KEY TO THE FOUNDATIONS OF ETHICS?

(TRANSLATED BY LISABETH DÜRING*)

## *I Exposition of the Thesis*

The central question which I want to take up here is this: Is it possible to identify an internal relation between language and ethics, between the language we speak and the object of ethical inquiry? This is the fundamental issue that I want to address. But my question needs to be formulated even more specifically, and I'd like to put it in the following terms: Could it be that the hope of a rational justificaton of ethics will be fulfilled with the help of a proper philosophical conception of language?

At first blush this project, which the title of my presentation signals, will strike most philosophers as unreasonable, if not bizarre. Even for those who have heard of communicative ethics the project continues to sound implausible. Why should this be so? Before I can go on to give an affirmative answer to my two initial questions, I must first deal with this initial scepticism.

## *II The traditional view of the relationship between speech and moral intentions as one-way traffic: Some reflections on the possible shortcomings of this view.*

In the philosophical tradition from Aristotle to Kant and more recently Husserl, it has been generally supposed that we can talk of an internal relation between ethics and practical reasoning or the will. But the idea of a relationship between ethics and speech remains foreign. This is so because for most of that tradition language operates merely as an external signifier enabling the expression of rational or volitional intentions. Language on this account translates the ideas or intentional states of the mind just as the conventional written signifiers of a language translate the phonemes of what is spoken. This sort of picture, dominant in most philosophical treatments, was classically represented by Aristotle in 'On Interpretation' (I, 161, I).

*Michaelis Michael and John O'Leary-Hawthorne (eds.), Philosophy in Mind, pp.99-117.*
*Kluwer Academic Publishers, 1994.*

Language, under this rubric, is made up of conventionally differentiated signs and sign-systems which have meaning only insofar as they refer back to and rely on subjective states, on conditions of mental life identical in every speaking subject. Such a position has recently received vigorous defence from John Searle in his book *Intentionality* (1983).[1] For ethics, the implication of this line of argument is clear. If ethics is to have a foundation, this must be located in the intentional states of the conscious subject. That is so because it is essential to the ethical that it derive from the inner form of subjective intentions and choices: not, as would be the case for law, from the external forms of human conduct. This distinction is one Kant has made particularly clear.

Insofar as philosophy ever imagined a connection between language and subjective morality, the moral problem posed by discourse was interpreted as one of 'sincerity', meaning the genuine expression of inner feeling or conviction. Language, the reasoning went, must presuppose the sincerity of those who use it: a subjective, moral state is hence the precondition of all meaningful use of words. This is true because if the opposite were the case, and lying were made the universal principle of discourse, verbal communication would be impossible.

Wittgenstein acknowledged this point. Language, when it is in working order, acts as a certain guarantor of the intentions being expressed in it. For even communicative acts composed solely of lies only function if they presuppose a norm of language used correctly. Lying is, in a sense, parasitic on the truthful representation of intentions.

Something similar can be found the relationship between persuasion in the sense of being talked into something (i.e., persuaded by suggestions that may be false or deceitful) and persuasion in the sense of being convinced (i.e., persuaded by arguments or reasons). For if deceitful persuasion is to be effective, it must above all create the illusion that it is rational arguments which are doing the work. Thus, in a certain sense, the art of deceitful persuasion depends more or less parasitically on the art of persuasion through rational argument. And in the same fashion the effectiveness of lying depends parasitically on the sincere expression of intentions or states of mind, which are feigned or dissembled.

This last allusion, implying a deeper connection between speech and morality, however raises a new and rather different point. No longer can we say with certainty that the only moral terms which are relevant derive from the expression of intentions, translated into what is essentially an ethically neutral medium. For as we saw, verbal communication must in its very function assume that the convictions of the single individual who speaks are sincere. Yet that on its own is no sufficient guarantee. There is a way in which verbal communication lacks any solid ground unless it can also assume that those who enter into the discourse do so honestly. Does this mean that those inner-subjective intentions, the individual's convictions or states of mind, which we did accept as morally relevant, can on their own have no significance unless we are able to assume that inter-subjective communication between persons is possible and reliable? The point can be taken even further: not only is the possibility of such communication the condition for intentional sincerity or truthfulness, but the very possibility that meanings could have intersubjective validity can be seen to rely on this same condition. This brings my argument to its crux: It is possible to ask whether we found here a criterion for the normative validity of that human interaction we call communication? Could we take the case of undistorted coming-to-understanding between human beings as a normative standard applicable to individual subjective morality? (cf. Apel 1991A). Indeed once we grant this, won't it entail that the ideal of the autonomous conscience, if it fails to acknowledge the normative condition of intersubjective understanding and communicability, will risk degenerating into a state of narcissistic wilfulness?[2]

### *III. How the methodological solipsism of modern philosophy caused the idea of a 'Foundation' for Morality to break down and how the 'Turn' to speech pragmatics saved the day*

However I don't want to respond to this question immediately. Before doing so I will first try to point out some of the difficulties which have been caused by the traditional conception of speech and communication as being somehow external to rationality itself. This conception has been responsible for the one-sided view that verbal communication is secondary to subjective intentions and the morality based on them. Now such a view is far from uncommon: it has dominated the philosophical history of modernity. If its assumptions are not challenged, it seems inevitable that, in the case of morality, we will see the capitulation of all rational validity claims that could be intersubjectively binding.

But is this going too far? What has the aporetic situation of ethics I have just described to do with how philosophy evaluates the relationship between reason and language, reason and verbal communication?

I would answer as follows: as the traditional grounding of morality and law in the Christian religion lost its legitimacy, a certain conception of philosophical reason, which had acquired with modernity an autonomous position, became increasingly important as the foundation for the philosophical discourses of ethics and law. I will mark out here only a few of the significant stages in this process: the names of Hobbes and Kant stand for the first two stages, and the third can probably be best designated as the period in the first half of this century dominated by the seeming alternatives of a 'value-free' scientism or a purely private morality.

Thomas Hobbes was the first to seek a rational foundation for morality and the state of law whose necessity and even possibility can only be established throught the exercise of fredom and reason. However what Hobbes understood by 'reason' was what we would call instead instrumental or that strategic rationality based on the logical consideration of calculating self-interest. And what he understood by 'freedom' we would rather call the freedom of inclination, the unmotivated act of choice which alone determines my exercise of self-interest.[3]

So very early in the piece, indeed at the dawn of modernity, we glimpse in Hobbes' presuppositions a central axiom of modernity: the turn to methodological individualism and solipsism . On its terms it is not possible to see the relevance of language and communication for the normative conditions of morality. For the fiction of a merely strategic state of nature as the primordial context of relations between human beings, which for Hobbes is foundational to his system, can on no account be made compatible with the need for an intersubjective understanding that is linguistic. I will return to this point later.

In the first instance, the 'state of nature', that essential concept of Hobbes' argument, is meant to represent human society before the establishment of the state, and before the concentration of power in the form of the state. As Hobbes conceives it, the state of nature is composed purely of strategic interactions between individuals, who exist in a 'war of each against all'. Hence, and this is the second stage in his argument, the so-called 'natural laws' of morality, on which the state of law rests, can only be understood as laws of strategic rationality governed by self-interest.[4] Thirdly, the transition from this catastrophic state of nature to a state of law - involving the voluntary renunciation of the individual's rights to everything and their transference to the sovereign -is also a function of the strategic exercise of calculation. This in particular is true for the third law of nature. For this law commands, indeed obliges as a rational imperative, that one keep to the social covenant, because to do so lies in the interests of all and also in the interest of the individual.[5]

Now in practice are these assumptions of Thomas Hobbes capable of providing a rational grounding for morality and justice? Is such a foundation sufficient to *oblige* the natural human being - remembering that in Hobbes, man in the state of nature is a 'wolf'- to keep his covenants as a member of civil society, and above all to keep to that covenant which institutes the commonwealth and makes it possible to sustain a state of law?

Again and again attempts have been made to find a demonstration for morality and law which would presuppose only that kind of causal and functional explanation which Hobbes seems to provide: that is to say, one that would require only the involvement of strategic rationality, the individual's calculation of what is advantageous to self-interest. The most recent version applies the conceptual scheme of game theory to the problem of the welfare economy.[6] However it has to be said that the results of games theory are morally indifferent, if not negative: If in fact the only rational motive to be taken into account is calculated self-interest, then the most rational procedure for the individual would be to enter into contracts while all the while planning not to fulfil them whenever a suitable occasion arises, and when there are no sanctions to be feared. This in effect would mean that whoever enters into such contracts - including even the social contract itself - entertaining this secret reservation, will profit by them over and above his fellows. He exploits the good faith of the others to his own advantage. Thus in a sense he relies parasitically on their honest intention to abide by the contract. Ultimately this shows that, given Hobbes' assumptions, the behaviour of the free rider is the most rational.

In view of this situation many contemporary thinkers, especially those who like Hobbes (and Martin Luther) equated practical reason with instrumental and strategic rationality, were led to the conclusion that morality could not be left to reason for its justification, but instead had to be based on feelings and affect, even on those 'instinctual residues' which man shares with the beasts (Cf. Lorenz, 1965). Indeed what happened was that even this early criticism of the modern notion of rationality brought into doubt the assumptions of methodological solipsism and the meaning of Hobbes' 'fiction', the state of nature where only strategic motives apply. Alternative models have been sought in those co-operative bonds between members of a community which functioned in traditional cultures prior to european modernity and which still continued elsewhere, associations inspired by a sense of 'group solidarity'. And it was hoped that they could be preserved or revived, possibly at the level of a modern, global discourse community of the human race.

I don't want to deny the partial justice of these alternative positions. Nor am I contesting the fact that such ideals can be considered as already embracing, at least implicitly, the notion of an intrinsic link between communication and ethics. But I cannot myself be satisfied by them, because for me what remains missing is anything like a rational foundation of the ethical. Such a foundation is for us at this time indispensable, whether or not we are considering philosophy's own requirement of rationality. In support of this claim I want only at this stage to make these few points:

The pre-modern notion of ethics as based on feelings and sentiments, or even on instincts, applied to the self-contained moral life of particular groups. It was a morality for insiders. Its fundamental values were loyalty, love, piety, protection and readiness to offer support. These were in effect universalised in the discourse of the great religious systems. But we cannot help noticing that modern mass society precludes such relationships and the forms of behaviour based on them: in multinational economies and international politics relations between individuals are detached, anonymous; because of the powers of technology to collapse distance, the relations mediated by modern means of communication are rarely localised or intimate. Under these conditions what happens to those virtues and dispositional behaviours necessary to group solidarity? It is hard to see where you could find anything that might convincingly inspire them. It would be expecting too much of the old 'virtue morality' and its codes if they were asked to apply to the regulation of political-social relations beyond the boundaries of face to face communities. Such divergent analysts as the ethnologist Konrad Lorenz (1973) and the economist Friedrich Hayek (1987) treat this reservation as conclusive.

To take one example, Hayek believes that for the international industrial complex there can be only one appropriate moral injunction which is rational: Contracts must be fulfilled and business done honestly. But even this requirement, as we saw with Hobbes, first needs to be given a rational justification. And above and beyond this framework, there is the issue of a global standard of social justice that, for instance, could apply to relations between the First and the Third Worlds. While this is an objective which Hayek abandons on ideological grounds, it seems to me both to need and admit of such a rational grounding. What it doesn't need, and certainly would be less than useless to this end, is any further appeal to archaic group-moralities with their pre-rational codes of virtue.

Finally - last but not least - we are now facing for the first time the demands for a planetary macroethics concerned with collective responsibility for the future consequences, immediate and less immediate, of humanity's collective activities: by this I mean the activities of science, technology, industry and politics.[7] There is no way that such demands can be coherently met by a pre-rational morality of sentiment or group loyalty. What is needed in the way of rational foundation both here and in the case of an internally applicable code of social justice, is a concept of universal validity based on the solidarity of the human race as a whole. But how can this even be conceivable as long as reason is pictured on the lines of instrumental-strategic rationality, which is to say calculated self-interest?

Now I want to turn to the second distinct stage in modern philosophy's project of justifying morality through reason. This one bears the name of Immanuel Kant.

Kant's principle of universalisability is the first attempt to formulate a criterion by which all maxims of moral conduct can be tested. When Rousseau proposed the axiom of the 'general will' as sufficient to account for the legitimacy of the state, he had in mind the case of the individual nation-state, where legal institutions are created on republican principles. Kant went further. For him, the political implications of the principle of universalisability transcend the particular formations which we know as nation-states. They require instead that politics be seen in terms of a just, international ('cosmopolitan') order guaranteeing peace and the rule of law.[8] (In my view Kant was right to believe that national and international rules of law must be developed at the same time, that in fact neither can be effective unless their mutual interdependence is recognised.)

As we saw with Hobbes' derivation of the state of morality and law, it is again human reason and freedom which are called upon as solely responsible for this principle of universality, which Kant called the categorical imperative. However the Kantian version of these fundamental ideas is totally different from that of Hobbes, and this distinction is not very often recognised. While it is true that Kant was perfectly aware of the existence of instrumental or strategic rationality (which he referred to as the basis of the 'hypothetical imperative' and the 'rules of prudence'), for him this was not to be confused with practical reason. Practical reason rather is the use of reason according to certain principles, pre-eminent among them that universal moral law through which reason determines the will in the direction of justice. Such a rational determination of the will must a priori transcend any empirical instance of self-interest. Furthermore freedom, which Kant describes as the autonomous, as opposed to heteronomous, determination of the will, can be sharply distinguished from the freedom of choice or spontaneous inclination, where self-interest may indeed figure as a motive. Freedom is in fact the power of human reason to legislate to itself: as Kant puts it, reason is free when it is related in an apriori way to the universal moral law and thereby precludes all external or empirical content.

Yet nonetheless Kant holds in common with Hobbes the presupposition of a basic methdological solipsism and individualism. The intentional and substantive applications of the moral law, i.e., the conditions under which the law acquires a content, are envisaged in a particular way. The moral law, when applied, can ignore those discursive relations between the very individuals for whom the law is prescribed, and who can in their turn determine that law in practice. (But indeed such a statement is fully in harmony with Kant's convictions, for, to take a different case, when he discusses epistemology, he argues that when we want to confirm something as true, the agreement of other knowing subjects is only relevant as a 'subjective criterion': doubtless of importance on psychological grounds, to correct the individual's tendency towards error, but not much more.)[9] What is foreign to Kant's thought is any conception of communicative consensus as a 'transcendental-hermeneutic' condition for the intersubjective validity of linguistic representations of the world : such a conception might arguably provide an 'objective criterion' of truth, even if only a provisional and fallible one.

As it stands, has the principle of universalisability been successful? Has it enabled Kant to offer a rational foundation for what we are looking for, that post-traditional macroethics for the human race which I have already postulated?

To answer this I think we need initally turn to the main objections to Kantian ethics. If there is a problem about the rationality of the ethical it would appear that the Kantian approach must share the responsibility. But the positions assumed by the opponents of Kant were even less equipped to establish any convincing 'grounding' of morality. Together they have managed to bring the enterprise of philosophical ethics to a critical moment at the beginning of the twentieth century. From that time on there can be no doubt about that capitulation of reason which I have already had occasion to announce.

1. One of the most important objections to Kant, and one which has not lost its currency today, was put by Hegel. Kant's moral law, according to Hegel, was purely formal. Its claim to universal validity was based on abstracting from all empirical objectives or contents of the will. Against this, Hegel proposed his idea of 'substantive ethical life', the product of an 'objective spirit' active in the historical world. From the perspective of this 'objective spirit' the very notion of an 'autonomy of conscience', which Kant had made fundamental to the categorical imperative, begins to look like that same merely subjec-

tive inclination or free choice which Kant thought he had also repudiated. The grounds Hegel gave for his polemic were that such appeal to individual freedom of conscience can only work under conditions that encourage the individual's narcisstic turn inward; this narcissitic preoccupation with the self must discourage the individual's willingness to enter into a serious dialogue about real principles of conduct, even his readiness to uphold them. Indeed, such a state of affairs, as Hegel puts it, amounts to 'trampling underfoot the roots of humanity.'[10]

At this point it is clear that Hegel has in mind something like the necessity for subjective intentions to be mediated through a discursive community. What Hegel needed was a way of defining his category , 'the ethical subject of a substantive ethical life'. This term, which he designates as the 'concrete universal' must be understood as referring to a particular community, indeed ideally the nation-state. But to maintain this category he was obliged to abandon in practice the Kantian ideal of a universalist and cosmopolitan ethic. Such a sacrifice was in fact deeply problematic for Hegel's systematic conception; in effect it contradicts his own aim of reaching a totalised state of complete mediation, subsuming everything partial and particular. However Hegel could always avert these undesired conclusions by invoking his historico-teleological idea of the progress of the human race towards freedom, which involves a necessary succession from one national culture to another. But if we reject the consolations of these metaphysics of history, then the Hegelian objective ethics will unavoidably lead to a realm where individual conscience (which for Kant and for Christianity is truly only oriented towards the universal as such) is subordinated, constrained by the fact that it can only represent one particular position in the conflict between the different rights-claims of the nations.

As far as I can tell, any attempt to revive the Hegelian category of 'ethical substance' (including the neo-Hegelianism of the English-speaking 'communitarians')[11] must finally reproduce precisely the same strengths and weaknesses of the Hegelian 'concrete universal'. Even if the principle of universalisability were, (for the sake of argument), made subject to the conditions of a particular historical community and tradition, I cannot see how the new, 'concrete' category would be a preservation of the old 'formal' one. The hope that 'ethical life' could be an 'Aufhebung' of abstract morality, the desire, in other words, to see a Kantian ethic of justice transformed into the terms of ethical substance with its 'concrete-universality', is doomed to frustration (cf. Apel 1988).

2. Yet Hegel's accusation of formalism does strike at a weakness of the Kantian position. Kant's conception doesn't provide for the way in which its

major premise - that the choice of moral maxim be based on considerations of universalisability - must be transmitted through a given ethical practice and its own self-understanding. Moreover,the general thrust of Hegel's objections here is reinforced by the utilitarian critique from its own perspective. Utilitarianism complains that what matters to Kantian ethics is only the form of the 'pure moral will', not the consequences of any course of action. Indeed the claim that the consequences of an action are irrelevant to the moral content of that action is incoherent unless one means by that what Kant means: Kant is only thinking of 'consequences' from the point of view of the agent's personal interest in punishment and reward in the afterlife, or success in this life. If, on the other hand, we include under 'consquences' the effects of a course of action on other people, as for example the way in which a politician is responsible for the life of the community, then the problem looks completely different. In such cases it is extremely implausible to insist, as Kant does, on the priority of those duties derived from the categorical imperative over and above any consideration of possible consequence. But that he was indeed driven to comparable absurdities is clear from the debate with Benjamin Constant over 'whether it can ever be permitted or even necessary to tell a lie',[12] or when he recommends to the 'moral politician' the maxim: 'fiat iustitia, pereat mundus' (let justice prevail, even if the world perish).[13]

At least this has become clear from the problem raised by 'responsibility for consequences': For reasons which are fundamentally metaphysical and stem from the dualistic schema of his philosophy, Kant's ethical program was committed to a formal and methodological solipsism. This led his reasoning to the following aporia. Because the only criterion by which right can be judged is the postulate of universalisability dictated by the categorical imperative, any rival claims posed by a morality of ethical substance or a utilitarian ethic are secondary and depend for their validation on it. Yet the way in which this is meant to happen is not clear. For how can there be any relation? There is no common ground between the principle as norm in its formal purity and the particular practices of an ethical tradition, no common ground between it and the world in which factual consequences flow from actions. Indeed the only figure which Kant proposes as ethical subject is a metaphysical monad, a rational entity without a context: embodying only the intentions of the 'good will' and as such, isolated from all communicative interaction with other people. This 'individual' is in no position to function as mediator.

According to Kant it needs no particular gift or experience of the world to know how to behave morally.[14] But this can only be true when it is a matter of determining what conventional morality requires of individual behaviour, where in any case the presence of a received ethical tradition is taken for granted. It cannot work in situations larger than the individual, in which we must speak of a 'post-conventional' macro-ethic: here social justice must be determined by reference to international standards. Nor does it apply to the ethical questions of responsibility in modern industrial societies where the global consequences of collective human activity transcend individual subjectivity and experience. On the one hand the Kantian universality principle as a norm first requires a connection to socio-cultural reality, to the actual practices and forms of life within an ethical community; on the other hand it must be made to consider pragmatically what are the possible outcomes and corollaries of particular courses of action and rule-governed way of so acting.

But how is this possible? Even to raise such a question brings us closer to the situation for philosophical ethics at the begining of this century. The key figure in this crisis was Max Weber, who mapped this terrain with two key ideas: First was his notion of the 'value neutrality of the sciences',[15] which included even that social science which is commited to 'hermeneutic understanding'. And second with his famous distinction between an 'ethic of conscience' and an 'ethic of responsibility'.[16]

Both of Weber's positions can only properly be understood when he is seen as a moralist, who (nonetheless) no longer believes in the possibility of an intersubjective crietrion of validity. If there is anyone who can adhere to the categorical imperative, says Weber, it would have to be the practitioner of the 'ethics of conviction', who places all responsibility for the consequences of his actions, say in the political sphere, in the hands of God. This categorical imperative as Weber together with the neo-Kantians understands it, turns out to be nothing more than an unconditional norm of value or fundamental principle. On the other hand, the follower of an ethic of responsibility imagines his unconditioned principle of value not on rational grounds, as Kant does, but by orienting his action towards foreseeable consequences, which are value-free in themselves (cf. Schlucter, 1979, p. 192, and Habermas, 1981, p. 380 ff.). For a universally valid principle, such as Kant's prohibition against lying, can not take properly into account the consequences of actions, argues Weber.

In this way Weber(and this is his fundamental premise) reads the situation as saying that, if there are unconditionally valid principles of value, they are a matter of irrational, private, existential choices. This is even true for the moral option whch he prefers, that of consequentialism, and it must also be true for the ultimate evaluation of consequences, an evaluation which can only be established by a rationality in itself exempt from valuation.

At this point we already have the two complemetary aspects of each alternative before us, at least as they have been presented by philosophers in the first half of this century. On the one side is the value-free rationality of the sciences, which is capable of determing which are the consequences we are responsible for. From this camp no rational justification of ultimate moral principles can be expected. Hence it seems that there can be no rational, that is to say, intersubjectively valid, foundation of morality. This is the position of scientism, and logical positivism.

On the other side there is a realm where it is possible to speak of principles of value which are unconditional, but these are limited to irrational, private, existential choices. Such a moral position was already anticipated by Hegel when he argued that Kantianism encourages the 'narcissism of conscience'. Since Kierkegaard and Weber, not to mention Nietzsche, this has been the position of a radical subjectivity, of which the clearest statement in philosophy is the 'freedom' of the existentialists. Here the self-legislating moral subject of Kant has had every trace of content removed, even that of the universalisability principle. (It is an analogous case which we see in early Sartre (1946) when he talks about 'choice of self'; and again, in his later work, after finding his way back to Kant's universal, when he insists that the free individual would choose for mankind.)

In the face of these alternative or rather complementary options - a value-free scientism and an irrational existentialism (cf. Apel 1973 and 1988A) - there arose in our epoch the following paradoxical situation. We could call it the dilemma of ethics in the age of science and technology:

On the one hand we find science together with the technology it has brought about. Thanks to these it is a matter of considerable urgency that we develop, on a solid rational foundation, a macroethics for a post-traditional world. For the effects of our collective action have been magnified in space and time by science and technology operating together. And now the collective responsibility for these consequences returns to us as an ethical problem. (One need only reflect here on the immediate problem posed by the ecological crisis which we cannot properly think about unless we re-examine the problematic relationship between the First and the Third Worlds.)[17]

On the other hand science itself has made us excessively concerned with a concept of rationality based on the value-free rationality of the sciences, and so in effect has made it impossible to conceive of any intersubjectively valid ethical norms grounded in reason alone. From this point of view morality is simply a private matter, as religion has been for quite a while. Even such a minimal moral code as Hayek's conception of the 'honesty in contracts' rule, which was meant to apply to the market economy and the sphere of positive regulations, can not hope to enjoy intersubjective validity. And it becomes hopeless to even imagine such validation for a macroethics of global solidarity, justice and responsibility.

Is there any way out of this paradoxical situation?

It is possible to see some changes: since the first half of the century, that is, since the epoch dominated by the choice between scientism and existentialism (the two complementary systems of western ideology), philosophical ethics has begun to recover from those original fears. Indeed ethics has been enjoying a boom in the last decades. However those contradictory alternatives have not really been overcome. There is still no clear path to the rational reconstruction of a universally valid macroethics. Rather what we have seen most recently is a singular convergence of pragmatism and neo-Aristotelianism, which has argued for the following thesis: Both moral philosophy and concrete ethical practice are only conceivable insofar as they rely on the 'contingent consensus' (as Richard Rorty and Alisdair MacIntyre would hold) of a particular cultural tradition. Thus the claim to validity which they express only holds within certain historical limits.[18] Gadamer (1967, p. 179ff.) borrows from Aristotle a description of the possibility of ethics which he identifies through the slogan: hos dei plus phronesis. I choose to paraphrase that as advocating the prudent, appropriate and flexible way in which the members of a community might apply and reflect on their traditional norms and practices.

But the famous hermeneutic-pragmatic turn seems to have been taken in directions which we can only call relativistic. The responses of Gadamer, Rorty et al. avowedly do nothing to solve the problem of how we are to construct a satisfactory macroethics for the human race in the age of technology and science. So are we to take this as the last word about the situation of moral philosophy in our time?

I think it is necessary to point out at this stage that there is one consequence of the hermeneutic-pragmatic turn which is of great relevance to our problem: In the work of Gadamer, Rorty and also for many post-Wittgensteinians, the theoretical priority of 'methodological individualism or solipsism' has been forcefully challenged. More attention should be paid to the fact that notions like that of 'rule-following' or 'understanding-as' do not make sense when applied to the 'singular individual', in whom we must recognise another version of the autonomous transcendental subject. What they presuppose is the kind of interpretation and mediation which only happens when people share a language and belong to a community of discourse.

Thus it seems that this hermeneutic-pragmatic orientation makes it difficult if not impossible to conceive of a universally valid ethic. And further it also leads to the belief that if there are any intersubjectively valid rules or norms, then they resemble the conventions of meaning within natural languages: they function within particular communities which interpret them according to their distinct and diverse traditions or forms of life. A move which began with Hegel and with those 'communitarians' which appeal to Hegel continues here: not only is Kant's principle of universalisability rejected, on the grounds that it cannot provide a criterion against which the substantial norms of different ethical practices could be tested (cf. Habermas 1984). But the very idea of a universal standard for ethics is abandoned. Is such a position required by the 'linguistic-pragmatic turn'? Does it have to lead us to the belief that there is nothing but specific natural languages and those communities formed around their conventions, forms of life in Wittgenstein's sense?[19]

It is exactly this moment that I maintain is the place to raise in a new way the question about the conditions for universal, intersubjective normativity, which includes the demand for a fundamental ethical principle. What this means is that I have posited that question of a transcendental ground. which Kant had himself posited in the 'Groundwork for the Metaphysics of Morals'. However, the answer he gave there was dropped at the beginning of the 'Critique of Practical Reason', and replaced by the mere stipulation of the 'apodeictic' 'fact of (practical) reason'.[20] In the background of this shift we can detect the difficulty Kant encountered with the logical circularity of his definitions: the moral law was the 'ratio cognoscendi' of freedom, but freedom itself was the 'ratio essendi' of the moral law. Yet the real reason for this difficulty in Kant's framework, the problem that made a transcendental foundation of ethics impossible without invoking metaphysics (cf. Apel 1990A), is in my opinion to be found in the following conditions:

The cartesian cogito is for Kant as for Husserl the primordial point of transcendental reflexion that one cannot go beyond. But if that cogito of the 'transcendental solipsism' (Husserl) is what I presuppose, then there cannot be any reflexive investigation of the moral principles which are already assumed. And with this as my presupposition, it is impossible even to constitute the very *meaning* of the question about a moral principle, about what Kant called the 'moral law'. For a moral law is unlike a law of nature: it cannot be referred back to one of the possible objects of the cogito. If a moral law can be established through transcendental reflexion as a principle already acknowledged, and hence as a non-empirical 'fact of reason', this would need further to be proved and demonstrated reflexively through the transcendental relation of intersubjectivity. It is in its connection to this intersubjective relationship that the normative ethical principle always already acknowledged must be reflexively demonstrated through thought (cf. Kuhlmann, 1990).

Now: for this to be possible, it is clear that what needs to be overcome is the transcendental paradigm of the cogito and the subject-object relation, both of which have dominated the philosophical assumptions of modernity. But I would argue that we do have the means to replace that paradigm with a paradigm which we could call 'transcendental pragmatics'. In place of the transcendental ego and the subject-object relationship, here speech pragmatics proposes as an organising structure the statement 'I argue' and the subject-subject relationship. The ultimate term then for reflexion is not the autarchic 'I think' of the philosophical tradition, which neither requires nor can ground a morality. Rather its ultimate term is that process of argumentation which presupposes as its subject a member of a discursive community which in principle excludes no one.

This subject of argumentation must in actual practice already acknowledge the normative, morally-relevant principles of an ideal discursive community. These principles belong as a matter of necessity to the process of argumentation, a process which cannot take place without the assumption that there are validity claims redeemable via consensus. Such principles include the idea of a truth which would be intersubjectively acknowledged as such, as well as the idea of a subjective truthfulness, and finally might include the normative idea of 'rightness', insofar as such a term can be applied to questions of morality (cf. Habermas, 1981, Bd. I, Chap. III). For we cannot be interested in the redemption of truth-claims through argument on our own. We must extend the same rights and duties, and the same responsibilities, to the other partner in the discourse. The other contributor to the discussion is just as responsible for raising and responding to those problems which can be resolved through argument. This phenomenon is what I would call the 'fact of reason' as transcendental pragmatics sees it. I believe it cannot be disputed without bringing the entire activity of argumentation into a performative contradiction. That fact alone is ultimately grounds for its validity.

Lack of time compels me simply to assert that the procedural principles for a practical discourse are already established by this way of grounding ethics through the reflexive constructions of transcendental pragmatics. By procedural principles I mean those which can govern the resolution of moral conflicts and the formation of particular norms in response to concrete moral and legal situtations. To explain this adequately requires the architechtonic development of a discourse ethic, which the work of the last few years has shown to be very complex indeed (cf. Apel 1988A and 1990A).

I would like to draw attention to a point where the transcendental-pragmatic approach coverges with that of the 'linguistic-hermeneutic turn' of contemporary philosophy, but also to show how the former avoides the latter's particularising and historically-relative consequences.

I'm not saying that discourse ethics, with its foundation in transcendental pragmatics, must ignore or underestimate the insights of contemporary speech pragmatics and hermeneutics. What has been demonstrated by these inquiries is the way in which thought, mediated by langauge, is dependent on the historical background assumptions of a specific socio-cultural form of life. I believe that transcendental pragmatics has learned its lexicon from Collingwood, Wittgenstein, Heidegger and Gadamer and continues to take their work into consideration. Clearly the resolution of conflicts, and the establishment by consensus of concrete contextualised norms at the level of practical discourse cannot simply arise from ultimate grounds in the sense of universally-valid procedural norms of discourse. They need to be understood, and put into practice, in the terms of particular, historically-conditioned traditions, if for no other reason, this would be required by the fact that meaning-validity claims cannot be articulated except through the mediation of language, but also on account of the socio-cultural presuppositions of our truth claims and on account of the abiding norms of the prevailing ethical life. It is from these presuppositions that any problematising of tradition and convention must start.

But this is in no way to admit that practical discourse can only contain contingent, culturally-dependent presuppositions as means to the evolution of consensus. For it is that same practical discourse which we look to for the resolution of conflicts about international and intercultural norms. To argue otherwise, as I take Rorty and Macintyre as doing, simply overlooks the fact that the same argumentative discourse, which they contribute to with their books and conference papers, brings its own normative presuppositions with it. This is what I mean by the procedural presuppositions of a discourse morality, which in a certain sense represents a 'metainstitution' in respect of all contingent social institutions, and already in practice transcends the merely historical presuppositions of any particular preconception of the world and the norms handed down by it.

Nor is such a reflexive transcendence of all historical and regional contingencies through a discourse ethics unreachable or irrelevant at the level of actual practice. For its function is precisely to turn Kant's universalisability principle, which Kant had presented as the criterion for all moral maxims, into a procedural principle for the coming to agreement over validity claims in the 'real world' i.e., in historical and social actuality. But this does not mean that there is an intention to cancel or denounce ethical life in its plurality, insofar as that diversity represents really the different forms in which the 'good life' is being realised. However some of the plurality of ethical life would be limited insofar as we could come to agreements about certain actions and normative regulation of ways of acting, judging them through that knowledge of consequences which is avaiable to us now. What we would be judging them by is their ability to put into play the conditions for a just coexistence and a cooperative interaction of differenet forms of life, assuming equally sharing of responsibility. Already that practical discourse which I am postulating here is functioning in various ways, not least in the publically-accessible forms of those scores of conversations and conferences, in which the normative preconditions for the resolution of humanity's problems is being debated every day.

These discussions and conferences need today to be measured against their own pretensions. That would mean: we should be able to analyse their communicative practices: are they simply practices of power, as Foucault would argue, or are they contributing to the essential task, to bringing about an organisation in which all men would be responsible for the global consequences of our collective activities (our science, technology, economy and politics)?

This is the present mission of a discourse ethic. Such an ethic is only possible, as we saw, if moral philosophy ceases to rely on the idea of the isolated intentions of independent individuals (as Hobbes and Kant did), and begins to understand that the conditions of morality's possibility and its actuality are to be found in language and communication.

*Johann Wolfgang Goethe University, Frankfurt*
*Translator: University of New South Wales*

NOTES


[1] See my critical comments in K.-O.Apel 'is Intentionality more Basic than Linguistic Meaning?' (1991).
[2] Cf. G.W.F.Hegel, *Phänomenologie des Geistes*, in E. Moldenhauer and K.M.Michel, eds.(1973), p.64f. And see also his *Rechtsphilosophie*, para.139.
[3] On this point and the subsequent stages in the argument see K-O Apel (1989).
[4] Cf. *Leviathan*, ed. Iring Fetscher (Neuwied/Berlin: 1966), p.122; see also p.99.
[5] *Leviathan*, chap. 15. p.110; see also p.122.
[6] Cf. O.Höffe (1975), part I. Nor has the book of R.Axelrod (1984) demonstrated to my satisfac-

tion that moral behaviour in all situations of life (that is to say, including those outside such 'play situations', in which moral behaviour is strategically presented as cooperative activity) can only be grounded on egotistic rationality.

[7] See Hans Jonas (1978). Further discussion in K-O.Apel (1987, 1990 and forthcoming).

[8] Cf. I. Kant, 'Idea for a Universal History with a Cosmopolitan Purpose' and 'Perpetual Peace' *Werke* (1968).

[9] Cf. I Kant, *Critique of Pure Reason*, A 820f., B 848f. See also his 'Reflections' nos. 2163, 2171 and 2175, in *Anthropology from a Pragmatic Point of View*, Part 1, § 2, and §3.

[11] Cf. Footnote 4.

[10] Cf. A. MacIntyre (1981 and 1988), and further C.Taylor (1989) and M. Walzer (1983).

[11] Cf. I. Kant, 'Über ein vermeintes recht, aus Menschenliebe zu lügen' in *Werke*, Bd. VIII, p.431-38. (Footnote 12)

[12] Cf. I.Kant, 'Perpetual Peace', Appendix 1, p.431-38. Cf. Footnote 12.

[13] I.Kant, *Foundations of the Metaphysic of Morals*, Akad.,Textausgabe, Bd. IV, p.403.

[14] Cf. Max Weber 'The Objectivity of Social Science Knowledge', in *Archiv für Sozialwissenschaft und Sozialpolitik* 19 (104), p.24-87, and ibid.'The meaning of the Value-neutrality of the social and economic sciences', in *Logos* 7 (1917), p.49-88, and 'Science as a Vocation' in *Gesammelte Aufsätze zur Wissenschaftstheorie* (Tübingen, 3rd ed., 1958), p.493-548. These three texts are translated in Max Weber (1949).

[15] On this see Max Weber (1958).

[16] See also the works cited in Footnote 11.

[17] Cf. my debate with A.MacIntyre and R.Rorty in *Discourse and Responsibility*, p.414ff.

[18] Cf. The contributions to this controversy in D.Böhler, T.Nordenstam and G.Skirbekk, eds. (1986). Further discussions can be found in B.McGuinness, J.Habermas, K.-O.Apel, R.Rorty, C.Taylor, F.Kambartel and A.Wellmer (1991).

[19] Cf. I. Kant, *Critique of Pure Reason*, Akad.Textausgabe, Bd.V., p.46f.

## REFERENCES


Apel, K-O (1973), 'The apriori of the discourse community and the foundations of Ethics', in *Transformation der Philosophie*, Vol. II (Frankfurt a M). English translation:*Towards a Transformation of Philosophy* (London: Routledge & Kegan Paul, 1980).

Apel, K-O. (1987), 'The Problem of a Macroethics of Responsibility to the Future in the Crisis of Technological Civilization: An Attempt to Come to Terms with Hans Jonas' "Principle of Responsibility" ', in *Man and World*, 200, pp.3-40.

Apel, K-O. (1988), 'Kann der postkantische Standpunkt der Moralität noch einmal in substantielle Sittlichkeit 'aufgehoben' werden?', in ibid., *Discourse and Responsibility* (Frankfurt a. Main: Suhrkampf, 1988).

Apel, K-O (1988A), The Conflicts of our Epoch and the Demand for an ethico-political reorientation', in *Discourse and Responsibility*, p.15-41.

Apel, K-O. (1989), 'Normative Ethics and Strategical Rationality: The Philosophical Problem of Political Ethics', in A.Schürman, ed., *The Public Realsm: Essays on Discursive Types in Political Philosophy*, (Albany, SUNY Press), pp.107-132.

Apel, K-O. (1990), 'The Problem of a Universalistic Macroethics of Co-responsibility' in S. Griffioen , ed., *What Right does Ethics Have?* (Amsterdam: VU University Press,), pp.23-40.

Apel, K-O. (1990A), 'Diskursethik als Verantwortungsethik. Eine postmetaphysische Transformation der Ethik Kants', in P.Fornet-Betancourt (ed.), *Ethik und Befreiung*

(Concordia, Monograph Series: Aachen), pp.10-40.
Apel, K-O. (1991),'Is Intentionality more Basic than Linguistic Meaning?' in E. Lepore and R. Van Gulick, eds., *John Searle and his Critics.* (Cambridge, Mass: Basil Blackwell), pp.31-56.
Apel, K-O. (1991A), , 'La foundation pragmatico-transcendentale de l'entente communicationelle illimité', in H. Parret , ed., *La communauté en paroles* (Liege: Mardaga,), p.15-34.
Apel, K-O. (forthcoming), 'The Ecological Crisis as a Problem for Discourse Ethics', in A Öfsti ed., *Ecology and Ethics.*
Aristotle, *On Interpretation.*
Axelrod, R. (1984), *The Evolution of Cooperation* (New York: Basic Books).
Böhler, D.,Nordenstam, T, and Skirbekk, G., eds.(1986), *Die pragmatische Wende.Sprachspielpragmatik oder Transzendentalpragmatik,* (Frankfurt a Main: Suhrkamp).
Gadamer, H-G. (1967),'On the possibility of a Philosophical Ethics', in *Kleine Schriften* (Tübingen,), Bd.1.
Habermas, J. (1981), *The Theory of Communicative Action* (Frankfurt a M., Suhrkamp), Vol I.
Habermas J. (1984), 'Über Moralität und Sittlichkeit - Was macht eine Lebensform 'rational'?' in H. Schnädelbach (ed.), *Rationalität* (Frankfur a. Main: Suhrkamp) p.218-235.
von Hayek, F.A. (19 ), *The Fatal Conceit. Part One: Ethics: The Taming of the Savage* (London: Routledge & Kegan Paul).
Hegel, G.W.F, *Phänomenologie des Geistes,* in E. Moldenhauer and K.M.Michel, eds., *Werke* (Frankfurt a.M., Suhrkamp, 1973).
Hegel, G.W.F, *Rechtsphilosophie.*
Hobbes, *Leviathan,* ed. Iring Fetscher (Neuwied: Berlin, 1966).
Hoffe, O. (1975), *Strategies of Humanity: Towards an Ethic of Public Decision-making.*(Freiburg: Munich),
Jonas, H. (1978), *The Principle of Responsibility* (Frankfurt a. Main: Insel).
Kant, I. 'Idea for a Universal History with a Cosmopolitan Purpose', in *Werke* (1968), Akad.-Textausgabe, Bd. VIII (Berlin: de Gruyter, 1968), pp. 15-32.
Kant, I. 'Perpetual Peace', in *Werke,* pp. 341-386.
Kant, I. *Critique of Pure Reason.*
Kant, I. *Anthropology from a Pragmatic Point of View.*
Kant, I. 'Über ein vermeintes recht, aus Menschenliebe zu lügen' in *Werke,* Bd. VIII, p.431-38.
Kant, I. *Foundations of the Metaphysic of Morals,* Akad.,Textausgabe.
Kuhlmann, W. (1990), 'Solipsismus in Kants praktischer Philosophie und die Diskursethik', in K.-O.Apel/R.Pozzo (eds.), *Zur Rekonstruktion der praktischen Philosophie: Gedenkschrift für K.-H. Ilting* (Stuttgart-Bad Cannstatt: Fromann-Holzboog), p.246-82.
Lorenz, K. (1965), *Animal and Human Behaviour,* 2 volumes (Munich: Piper).
Lorenz, K. (1973), *The Eight Deadly Sins of the Civilised Human Race* (Munich: Piper).
MacIntyre, A. (1981), *After Virtue* (Notre Dame: University of Notre Dame Press).
MacIntyre, A. (1988), *Whose Justice? Which Rationality?* (London: Duckworth).
Sartre, J-P. (1946), *Existentialism is a Humanism* (Paris: Nagel).
Searle, J.R. (1983), *Intentionality* (Cambridge University Press).
Schluchter, W. (1979), *The Development of western Rationalism. An analysis of Max Weber's Social History* (Tübingen: Mohr).
Taylor, C. (1989) *Sources of the Self* (Cambridge, Mass.).
Walzer, M. (1983), *Spheres of Justice* (New York: Basic Books).
Weber, M. (1949), *The Phenomenology of the Social Sciences* (Glencoe/Ill.: Free Press).
Weber, M. (1958), 'Politics as a Vocation', in *Gesammelte politische Schriften,* 2nd ed. (Tübingen), pp.493-548.
Wellmer, A. (1991), *Der Löwe spricht ...und wir können ihn nicht verstehen* (Frankfurt a M.: Suhrkamp)

LEE OVERTON

# UNPRINCIPLED DECISIONS

## *Introduction*

### A question of approach

What approach should be taken in the study of deliberation and practical reasoning? Should the primary concern be to describe and explain a natural process? Or should it be to articulate a normative framework for deliberation? It's tempting to think that the choice is between taking reasoning and deliberative practice as it is, on the one hand, and shaping that practice, on the other. But it's not that simple.

The point of articulating a normative framework is to improve deliberative practice, to make it more rational. Mere description of reasoning seems impotent to provide a normative framework for deliberation. If we assume that the only way to improve deliberation is to develop principles telling us how one ought to deliberate, and if we assume furthermore that a descriptive theory cannot yield principles of that sort, then it looks like a descriptive theory will be impotent to improve deliberation. But is this conclusion right? Instead of arguing with the second assumption—the more commonly trod route—we might instead argue with the first. In addressing it, we set aside the hope of getting normative principles of deliberation out of a theory of reasoning; instead we consider the possibility of an descriptive theory's having a shaping effect on deliberative practice without generating normative principles.

### Self-description, self-reflection and the theory of reasoning

Writing on self-description and self-reflection tends to focus on people's attempts to form adequate descriptions of their beliefs, desires and other attitudes and tendencies; these descriptions are accompanied often by judgments about what sort of persons they are. This sort of reflection, and the accompanying *evaluative* judgments, may have a shaping or organizing effect on these very attitudes and tendencies in various ways. Some writers argue that self-descriptions can have an immediate *constitutive* effect on their object: by describing yourself, you actually articulate who you are. Velleman (1989) suggests this, and Taylor (1985) sometimes pushes this view. Others suggest a

*Michaelis Michael and John O'Leary-Hawthorne (eds.), Philosophy in Mind, pp.119-139.*

less direct effect. In thinking about yourself, you may find that you like or dislike what you discover about yourself, and reactions of this sort may have indirect effects on the sort of individual which you become. This view is familiar from the papers on 'second order desires' by Frankfurt (1988).

Less common is to think of the theory of reasoning and deliberation as having consequences for the kinds of person we become. Nonetheless, if our thought in this area is part of an attempt to understand ourselves, then we will want to employ our settled thought about the nature of reasoning and deliberation, as we would want to employ any bit of self-understanding, in our further decision-making. And when we do, even though this understanding is not explicitly normative, it nevertheless affects the character of our particular deliberations. Comparing this to the two types of shaping effect by self-reflection that I have just mentioned—namely *constitutive* and *evaluative*—what I have in mind is closer to the constitutive idea.

### Reasoning distinguished from deliberation

Take a moment to consider some different ways we might understand the relation between actual reasoning and deliberation, and the theory of them, keeping in mind the question of why actual reasoning and deliberation for the most part conforms pretty closely our best theories of reasoning and deliberation. One sort of explanation says that such theories are roughly true to life because their whole point is to describe how we actually reason and make up our minds. Another says that such theories are roughly true to life because they make explicit the principles to which we are trying to conform in our reasoning, and naturally, once we make them explicit, we try to conform to them even more. The second sort of explanation says that what the first sort takes to be *self*-descriptions are in fact articulations of a normative, rational ideal; they only appear to be self-descriptions precisely because of their status as ideals: naturally we tend to conform in our deliberative practice to our explicit conceptions of how we ought to deliberate.

I'm not happy with either of these sorts of explanation. The problem is that both fail to carefully distinguish the concepts of reasoning and deliberation. They debate the question whether reasoning, or what they take to be the same thing, deliberation, is a process which is under our control. One sort of explanation says we don't control it; we just describe it. The other says we control it by developing normative frameworks for deliberation, and that reasoning correctly is reasoning according to the correct (rational) principles of deliberation.

Here's an understanding of the relation which distinguishes the concepts of deliberation and reasoning. Reasoning is a relatively automatic process whose outcome is not under our control. For example, given some evidence and two or more competing hypotheses explaining the evidence, one or the other of the hypotheses will likely stand out to us as the more sensible or reasonable. The ability to make such discriminations is the basis of our capacity to reason. But it is, to repeat, a relatively automatic process. Deliberation, on the other hand, is a revisable process in which we exploit our capacity to reason. For example, in deliberation, we employ maxims of reflection—rules that say what to consider, what to pay attention to more closely, etc.—in order to set things up in a way that best exploits our capacity to reason. These maxims arise from our theory of how reasoning actually and potentially works. Deliberation is thus something we do naturally, but which we can revise and improve upon.

Thus we have a third story about the relation between actual reasoning and deliberation, and the theory of the two: our theory of reasoning is roughly true to life because it is largely descriptive. And it informs our deliberation, which is guided by maxims of reflection. We conform to our thought about reasoning by adjusting our *deliberative* practice, not our reasoning. Our deliberative practice is guided by maxims or rules of thumb which are in turn based on our understanding of the nature of our capacity to reason. Thus, a theory of reasoning is not only a theory about how we *actually* exploit our capacity to reason, but also a theory of how our capacity to reason can *potentially* be exploited. The maxims of deliberation are actually a part of the theory of reasoning. That is, we explain reasoning in part by saying what we need to pay attention to in deliberation so that we take most complete advantage of our capacity to reason.

Our answer, then, to the question of how an explanatory theory may shape deliberation is this: a descriptive theory of reasoning may yield, instead of normative principles of reasoning, *maxims of reflection*. How we explain our reasoning determines what we explicitly try to exploit about ourselves when we deliberate. There is a kind of success that descriptive theories of reasoning can achieve that is not simply capturing how people actually reason. Namely, it can be successful in more fully capturing our potential as deliberators. Knowing what we can be is in a way a sort of *constitutive* knowledge, since it leads us to approach actual deliberation differently.

Summary

In deliberating, we employ guidelines for deliberation. These guidelines determine the character of our deliberation, and are themselves determined in part by our approach to the study of reasoning. One approach aims at producing a normative framework that tells us what we ought to conclude given certain input; this approach tends to conflate the concepts of reasoning and deliberation, and yields one or another *framework* for rational deliberation. Another approach aims at a descriptive theory of reasoning, clearly separating reasoning from deliberation, and yielding *maxims of reflection* as guidelines for deliberation.

## *Principled and unprincipled deliberation*

I echo Harman (1986) in saying that I understand practical reasoning to be the reasoned revision of intentions. This includes revisions one makes matter-of-factly in one's intentions, without worrying much whether one is making the appropriate revision: one routinely updates and fills in one's intentions in a regular way, depending on normally reliable immediate dispositions to adopt intentions that cohere with and further our plans and desire, and to adopt desires that give substance to our intentions. My special concern, however, is with deliberation, which I take to be a kind of exploitation of our capacity for practical reasoning. I understand deliberation to be the explicit attempt to make up one's mind, guided by an understanding of the nature of reasoning.

There are two aspects, or parts to emphasize, about deliberation: deliberation can be principled, and it can be unprincipled. Not only do we deliberate instrumentally, planning rationally in the pursuit of our desires, but we also deliberate non-instrumentally, searching for desires to substantiate and give meaning to our plans. It is this non-instrumental part of deliberation that is unprincipled. Deliberation can be helped along not only by appealing to a framework for rational decision, but also by trying on different frames of selective attention that characterize various desires. The reason for this is that reasoning works pretty much automatically on what we happen to be attending to. To exploit this, we simply attend to different things to see what the outcome of reasoning will be, given that particular focus of attention. In this sense, deliberation may involve trying on different desires.

### Substantiating plans

Given the usual emphasis on rational decision making, there is a tendency to think of deliberation as a process largely concerned with rational revision of plans and intentions. The sort of deliberative uncertainty most emphasized in the study of deliberation is the uncertainty about what is the rational thing to do. But in deliberation one may also be concerned with finding attitudes that substantiate one's plans and intentions, attitudes that allow one to *identify* with ones plans and intentions. And deliberation often involves an uncertainty of a sort that arises from lack of such substantiation: one's normally reliable dispositions to adopt desires which substantiate one's plans and intentions have come up dry, and although one may have a plan or intention, one is in a sense in a state of confusion for one hasn't the feel for acting on one's plan. Thus, one of the things that one might do in deliberation is search for desires which substantiate one's intentions.

Having formed a plan, obviously one ought to adopt desires that substantiate the plan. Usually one does. But sometimes one is, as we might put it, conatively confused. In such cases, one may form explicit beliefs about what kinds of desires one should adopt. But even though one may form beliefs about what desires (under some description of them) one should adopt, believing that one should adopt desires of a certain sort, on the one hand, and being familiar with and able to adopt these attitudes, on the other, are quite distinct matters. Furthermore, the search for desires may come up short. Thus, one's beliefs about what one ought to do—which plan is better, etc.—and one's beliefs about what desires one ought to have, are often subordinate to what one *can* do in the sense in which ability may be limited by what it is possible for one to desire. For this reason, I suggest that a significant component of the deliberative process may consist in *trying on* different postures of desire, in a search for one that not only fits with one's intentions and plans, but which are also possible for the deliberator to adopt—desires that are, so to speak, desirable.

We often think about what good reasoning is, and explicitly employ our conception of it in the revision of beliefs about what we *ought* to believe or intend. That is, we often lay out our reasons before us, so to speak, and form conclusions about what we ought to believe and intend. These conclusions characteristically, or at least very often, lead to the formation of the beliefs and intentions prescribed.

This psychological fact has led some writers to speak of *this* manner of revising our beliefs and intentions as paradigmatic of the way we revise our beliefs and intentions. That is, we are supposed to lay out the reasons, form a conclusion about what is best to believe or intend, and then believe or intend it. A failure to do the last bit—form the prescribed attitude—is sometimes thought to be a bit of irrationality.

Obviously I think this is a mistake. Thinking about what one ought to believe or intend is reasoning all right, but it is theoretical reasoning about what beliefs we will or would adopt. But this doesn't seem to me to be the only, or even the most important part of the way we make up our minds. There are limits to what we can desire, and there are limits to our creative abilities to articulate our possibilities of desiring. Practical deliberation has a lot to do with probing these limits.

My hypothesis is that the nature of deliberation and decision is such that we cannot possibly capture all of what may qualify as deliberation in a set of principles. And I don't mean that a set of principles would be an idealization not always followed in actual deliberation. I mean that deliberation legitimately may be in part unprincipled.

The uncertainty that a deliberative agent encounters is usually framed as uncertainty about what is best to do. What I mean by suggesting that deliberation has an unprincipled aspect is that deliberators may be pitted against another sort of uncertainty, distinct from the sort that theories of rational decision are designed to handle. A first approximation is to think of this sort of uncertainty as a kind of confusion regarding what attitudes can even be held by the deliberator in his situation. The idea is that one doesn't always enter deliberation with clearly articulated desires. Rather, deliberation is concerned to some degree with articulating one's desires. And the deliberation is going to be in part an exploration of what it is possible for the deliberator to desire.

Practical deliberation is to a great extent instrumental in nature, and we're right to portray much of it as aimed at the attainment of desires, and as doing it in a satisfactorily efficient way. We have desires, and we form intentions in their pursuit. Furthermore, practical deliberation may be characterized in terms of the aims of maintaining coherence among, and of being conservative about making changes in, one's plans and intentions. Once formed, one's plans and intentions seem to acquire a momentum of their own which allows them to survive the desires which first motivated them (Harman, 1986). Also, long range intentions are a normal part of our lives. When one forms new intentions and plans, one is sensitive to their coherence with plans and intentions that one already has. An idealization, or principled characterization of this process will attempt to say what the coherence-making features are, so that we can explicitly try to preserve coherence.

But, as I've indicated, this seems to me to be only part of the story. Although it may appear that the formation of an intention is the conclusion of an argument which proceeds from what one already wants, intends and believes, to the new intention, there is a bit of deciding what one wants that goes on, and hence one is articulating one's premises as well as the conclusion. Planning cannot be the whole story, since we are continually articulating the desires that underlie our plans.

### *Desirability*

I want now to turn for a moment to discussion of *desirability*. How we understand the concept of desirability turns out to be relevant to how we approach the study of practical reasoning.

#### Questions about ends, and questions about means

Moore distinguished two questions he thought were often confused—'What is good?' and 'What ought one to do?' The former is supposed to concern what ends are desirable; the latter, what means to take to promote the desirable ends. The distinction is useful, since it supports a framework within which to articulate questions about value and obligation—namely, the means-end conceptual framework. But the framework is not entirely satisfactory insofar as it represents deliberation as simply a search for the most efficient way to balance and promote a plurality of ends. People point out that there are deontological constraints on acceptable means which have nothing to do with the efficient promotion of desirable ends. Unfortunately, this objection often seems a bit silly, because we tend to speak of the deontological constraints in terms of principles according to which certain sorts of means are not desirable, suggesting that in fact we *can* represent deliberation as teleological. What is not so often recognized is that the distinction presupposes a notion of *desirability* to which there is a credible theoretical alternative. The presupposition is that 'desirable' means roughly the same as 'what ought to be desired'.

#### Is desirability like visibility?

'Questions about ends,' wrote Mill, in *Utilitarianism* (1861) 'are, in other words, questions about what things are desirable.' When you consider that by 'desirable', he seems to have meant 'able to be desired,' the claim is striking. Recall his famous comparison of desirability and visibility:

> The only proof capable of being given that an object is visible, is that people actually see it. The only proof that a sound is audible, is that people actually hear it: and so of the other sources of our experience. In like manner, I apprehend, the sole evidence it is possible to produce that anything is desirable, is that people actually desire it. (ch. 4, 'Of What Sort of Proof the Principle of Utility is Susceptible', p. 44).

Moore, in *Principia Ethica,* dismissed the suggestion of an analogy between the concepts of desirability and visibility; philosophers since have tended to follow his lead. According to Moore, to say of something that it is desirable is to say that it merits—either intrinsically or instrumentally—the appreciation involved in desiring it. Questions about ends, then, on Moore's view, concern what things merit our desire for them. Far from finding an analogy, his response to Mill is to gesture towards a great *disanalogy* between the concepts of visibility and desirability, suggesting that Mill had made ' as naive and artless a use of the naturalistic fallacy as anybody could desire.' Listen to Moore:

> Well, the fallacy in this step is so obvious that it is quite wonderful how Mill failed to see it. The fact is that 'desirable' does not mean 'able to be desired' as 'visible' means 'able to be seen.' The desirable means simply what *ought* to be desired or *deserves* to be desired; just as the detestable means not what can be but what ought to be detested and the damnable what deserves to be damned (p. 67).

But our question isn't really about the use of a word. Moore's response exasperatingly misses the point. The question raised by Mill's striking claim is whether questions about ends are concerned primarily with what it's possible to desire. Since the sticking point for Moore is the inability even to find this question intelligible, the interesting challenge is to explain what Moore apparently fails to see: how questions about ends could concern the limits on our capacity to desire.

## Summary

The standard conceptual framework for thinking about desire and desirability rejects any analogy between *visibility* and *desirability* : ' visible' means roughly that certain beings, under certain conditions, can see it. But when we say that something is desirable, we do not mean that it *can* be desired. We mean, rather, that it merits the appreciation involved in our desiring it. I want to challenge this framework by suggesting that (1) deliberation is largely of ends, and (2) questions about ends concern what it is possible to desire. This means that deliberation is concerned with what one *can* do, rather than with what one *ought* to do, with *possibility* rather than *obligation.*

*Is deliberation of ends?*

Aurel Kolnai (1977) defines a dispute by posing the question whether in deliberating, we choose between means or choose between ends. I want to approach the issue by dividing thought about the nature of deliberation into two camps, one camp holding that deliberation is of means, and the other holding that it is of ends. The notion of 'deliberation of ends' that would satisfy Kolnai seems rather elusive, mainly because the position that I think he wants, and that I'm interested in explaining, does not comfortably fit into either side of the dispute, as the sides are normally construed. Saying why will help to articulate the position I'm after. Basically, the problem is that a decision about ends cannot be construed as *choosing* among ends, and therefore it needs to be explained what deliberation among ends is if it is not an attempt to *choose* between them.

The rough idea

Let's begin with some a rough idea which we'll try to sharpen up. Kolnai uses a passage from Aristotle to get the debate started.

> We deliberate not about Ends, but Means to Ends. No physician, e.g., deliberates whether he will cure, nor orator whether he will persuade, nor statesman whether he will produce a good constitution, nor in fact any man in any other function about his particular end; but having set before them a certain end they look how and through what means it may be accomplished. [*Nichomachean Ethics*, Book III, 112b]

Kolnai thinks that 'if deliberation was not of ends but of means' this would mean that it is 'reducible to rational computation' (p. 57). Wiggins (1987) makes a similar claim. Like Kolnai, he suggests that thinking of deliberation as concerned only with means leads to thinking of practical reasoning as merely a kind of calculation.[1] It leads, as he puts it, to

....that pseudorationalistic irrationalism, insidiously propagated nowadays by technocratic persons, which holds that reason has nothing to do with the ends of human life, its only sphere being the efficient realization of specific goals in whose determination or modification argument plays no substantive part. (p. 227)

These writers gesture in the direction of an insight, but they have not sufficiently articulated the insight. It will take some work to flesh it out. Begin with the point that a deliberator's problem is not the simple one of being faced with an array of means from which to select one in order to achieve a further end. The typical deliberative situation has a different character: the deliberator comes to a situation not only with possible means to choose in order to promote his ends, but also a plurality of ends and concerns which may be put into different arrangements of competition and conspiracy, with each other and even with the means he is considering and already undertaking. His problem is as much one of dealing with possible ends as it is dealing with possible means. As Kolnai puts it,

[It] may be more conveniently said that in deliberating, we confront mutually conspiring and mutually competitive ends with one another than that in deliberating we seek out the available and efficacious means to one determinate end. For the latter activity is a subsidiary requirement of the former, whereas the converse is not true. (p. 50)

One aspect of the deliberative problem, then, it appears, is deciding what one wants, from among competing ends. But how is the agent to decide among ends? Kolnai again:

Do we then, after all, consider the ends between which we have to choose as if they were means to something else?... No, for we feel inclined to pursue them for their own sake, without any logically necessary or previously given reference to further ends... (p. 50)

If we are to truly get away from the 'deliberation is of means' picture, one has to deal with the competing ends without the guidance of a further, or higher, end which the end in point is supposed to serve. Here, Kolnai turns to the notion of a certain 'fittingness' of ends with other ends already present:

[S]uch ends are more strongly weighted as display, in addition to their worth considered in itself, a particular 'fittingness' in respect of important other ends already established and as it were axiomatic. ... [We] compare ends *in the light of further ends* they are likely to promote or to hamper respectively, either in a causal sense or at any rate in the sense of concordance or discordance... (pp. 50-53)

It appears, that in considering how to decide between ends, you are supposed to give greater weight to an end if it coheres to a greater degree with other ends you already have. That is, even granting that one sort of consideration that may favor an end is that end serves another already established one, another sort of favoring consideration is simply that the end 'fits with' as opposed to 'serves' one's other more stable ends. That gives us a principle of choice for ends: choose ends, or rankings of ends, which promote the greatest coherence among ends while being conservative with respect to ends one already has.

But this makes it look like Kolnai may have simply substituted a different normative principle: instead of maximizing the satisfaction of one's desires, one is concerned to maximize the coherence of one's desires. One is concerned, in other words, that one's desires meet certain normative standards. In a footnote, Kolnai makes the following revealing comment:

> What we decide is whether to approve or to disapprove a desire of ours... In an indirect and limited sense, we even choose our desires (p. 51 fn.).

This is suggestive. What it suggests is the possibility of distinguishing two levels of desire. Harry Frankfurt (1988) has pursued this idea. He writes of its being a structural feature of the will of persons that there be both first order and second order desire sets, second order desires having as their objects desires of the first order. At the first level, one's desires compete with one another and deliberation is a rational maximizing game. At the second level, one approves or disapproves of different rankings of first level desires, in effect shaping one's first level attitudes, and therefore the outcomes of first level deliberations.

There are two issues that arise regarding 'hierarchical' stories like this, both of which have been discussed in the literature. The first is that this story of two levels of desire does not really give an alternative to the 'deliberation is of means' story, since it just introduces a higher level of means-end deliberation. We have a maximizing story at both levels. For, unless we suppose that there is a single end we are pursuing at the second level, a single thing that we are trying to maximize, then we have the same problem all over again of how to weight the various second order desires. That is, if we again have to choose between competing second order desires, how do we choose? Which second level desires are the ones which shall win out and determine the arrangement of first level ends? This criticism came up in Watson (1982), where the question was how we are to avoid being wantons at the second level. Frankfurt had introduced the first/second level distinction in order to distinguish persons from wantons: you're a person if you form second order desires (volitions), and a wanton otherwise. What Watson wanted to know was how we keep the second order volitions from being wanton. Either the second order desires are already weighted among themselves, or else they are not. If they are, then we are wantons with respect to these desires; if they are not, then we have the same problem all over again of what deliberation among ends is supposed to amount to.

The second problem has to do with interpreting the notion of taking responsibility for one's own attitudes as a matter of *choosing one's attitudes*: this is a disconcerting interpretation. Surely one doesn't simply choose one's desires, at least not if we understand the expression ' choose' in the ordinary way.

### Choosing one's attitudes

I'll address the second problem first. I do not believe that 'choosing to desire' is the most perspicuous way to put what we are talking about. Kolnai shouldn't have put it that way. It suggests the wrong thing. The point that wants to be made is simply that we are active with respect to our desires. We take responsibility for them by actively shaping them. However, there is a way of thinking about the mind which both pushes us towards the idea of a second level of desire and, at the same time, makes that very idea seem very perplexing. According to this way of thinking, there are two basic kinds of psychological attitude: the so-called ' pro-attitudes' or desire-like attitudes, and the 'cognitive' or belief-like attitudes. These attitudes are distinguished functionally in terms of what is sometimes called 'direction of fit'. Desire-like attitudes are said to have a world-to-mind direction of fit. To have a desire is to lean into the world, attempting to make an impression on it, to make it go your way. Belief-like attitudes, on the other hand, have a mind-to-world direction of fit. They are the results of the world leaning into and making an impression on you. Desire-like attitudes keep on leaning, despite the perceived state of the world, whereas belief-like attitudes change to reflect the perceived state of the world.

This provides a way of distinguishing between what an agent does, i.e., instances of activity, and events that merely happen to him, i.e., instances of passivity. One is said to be active when one pursues desires according to one's beliefs, and passive otherwise. Active, that is, when one leans into the world, and passive when the leaning goes the other way. So, one would be active with respect to one's own beliefs and desires only insofar as one leaned into them, changing them to suit other desires one has.

The problem is that on this way of thinking about activity, there's no room, except in a disconcerting way, for the commonsense notion of making up one's mind as something that one does. It's inadequate because it doesn't allow for the notion of our taking responsibility for our beliefs and desires by *deciding* what to think and do. On this model, the expressions 'deciding to believe' and 'deciding to desire' are disconcerting because they tend to get interpreted as 'choosing to believe' and 'choosing to desire' . That is, it's really the idea that we choose our beliefs and desires that we are uncomfortable with.

Consider the notion of choosing to believe. It is natural to object to it, to say that choosing to believe is something we don't, or at least shouldn't, do. It just *reeks* of irrationality. It's irrational to believe things just because you want them to be true or because you want to believe them. Williams (1973), argues the irrationality of deciding to believe. The flavor of what he says there is largely captured by the following quote.

> We might well think that beliefs were things that we, as it were, found we had (to put it crudely), although we could decide whether to express these or not. In general one feels that this must be on the right track. It fits in with the picture offered by Hume of belief as a passive phenomenon, something that happens to us (p. 147-8).

To suggest that we are active with respect to our own attitudes, especially our beliefs, clashes with this model of the mind that Williams prefers because the model only allows 'active' to be cashed out as ' goal-driven' . But surely our belief-revision *is* an active process, and what we want to do is not to reject the notion that belief revision is an active process, but instead reject one way of unpacking the notion of activity. What's objectionable is the notion of choosing to believe, and identifying deciding with choosing transfers what's objectionable to the notion of deciding what to believe. Deny that deciding is simply choosing, and you lose the reason you had for embracing the *passive* view of belief-revision. Ditto for deciding to desire. That is, if there is such a thing as unprincipled activity, then there is room for the notion of deciding to believe and deciding to desire.

But what is a decision if it is not a choice? This brings us back to the first of the two problems I mentioned, the problem of how to avoid higher and higher levels of choice. For, if decision is not a choice, won't it be arbitrary?

#### Picking, choosing, deciding

Many practical situations are such that we are faced with equally good options, and it is permissible by our own lights to be arbitrary. We just pick. Appealing to deciding to desire may sound like an appeal to picking.

Not so. Suppose you wonder which of several equally attractive Mayan sites you shall visit on your upcoming trip to Mexico. Since they are equally ranked by the standard criteria, you have to come up with a new criterion for choice. If you are arbitrary and just pick, it is this which gets picked—the new, but arbitrary feature of the sites. Here's the way picking goes. Suppose there are two sites. You can name one 'Heads', and the other 'Tails'. Then, you can pick, as the relevant feature, the property of having the same name as the side of the coin which faces up from the ground after you flip it and let it land. Or, you might pick the feature that a site possesses by virtue of its representation on a map being the closest to your dart after you throw it. The same thing occurs when you choose based on whim—I mean you choose based on a feature that was determined by whim. You allow the criteria for choice to be determined by something other than your interests.

If you want to avoid arbitrariness, what you have to do is revise your interests rather than whimsically adding irrelevant features as decision criteria. Now, we're supposing that you are headed to the Yucatan planning to see a ruin. We're also supposing that you need to fill in this plan with an intention to visit a particular one of the several sites which you could hit. It might have been that there were some desiderata like that the sites cannot be too far from a certain central point, in order to minimize the traveling time. Things like that. But we're supposing that we're past all desiderata like that.

Supposing there are no such desiderata, there are two routes to go. One could be arbitrary. Choose the one you like, as we say; and I suppose we mean that one may choose based on whatever whimsical criterion comes to mind due to passing fancy.

The second route is to we adjust our interests. The assumption here is that we can settle the matter by articulating our desires. This is imaginable in the Mexico expedition case, though perhaps less obviously imaginable than articulating one's desires in the midst of deliberation about career or marriage. And here is what we can imagine: that you do not approach your decision with articulate and unrevisable desires in hand. Your desires are always open to revision. And that means that the desiderata are open to revision.

Unprincipled decisions are distinct from mere choices not because they have an element of arbitrariness, but because in making them we are not concerned merely with *selecting*, but with establishing the grounds for selection. A person trying to make up his mind is, I want to say, seeking to organize his desires in the hopes of resolving some confusion. It is a distortion of this process to try to fit it into any format designed to elucidate a process of selecting.

The notion that all decisions are mere choices presupposes a certain picture of deliberation, according to which it is a matter of bringing a range of motivations to bear on the problem of selecting one out of a range of options. The various motivations can be seen as pluses or minuses for the different options, and the task is to sort these pluses and minuses out. Now, many will grant that much of our deliberation does not begin with already weighted motives, which are then weighed against each other to determine which option they, on the whole, recommend. But I want to say something stronger than that: much deliberation does not even begin with motives with *un*settled weights. We might think of deliberation as aimed precisely at articulating various mental elements *into* motives. That is to say, the various mental elements are not competing and conspiring motives, for they are not yet motives. Deciding to desire is not self-deceptive, because our desires are, so to speak, interpretations of our intentions; and these aren't fixed but are continually subject to revision.

## Summary

Each of choice and unprincipled decision may be thought of as preceded by a sort of uncertainty. Whereas the picture of deliberation as the attempt to choose looks at the different considerations as conspiring and competing reasons which give the options between which you have to choose different scores, the picture of deliberation as an attempt to decide looks at them as moldable elements, given meaning when we shape them into coherent postures. Now, if the idea is that we are searching for an end to confusion, this is not a problem to be resolved by explicitly following principles. You search for an articulation of desires which is satisfying in the sense that it sits well with you and does not feel wrong. In this sense, then, deliberation is in part a search for desirable ends.

*Results*

The ideas that I present here come as part of an attempt to understand the sometimes puzzling relation between value and desire, and the roles of each in deliberation and decision. Thinking of deliberation as involving deliberation about *possibility* rather than, or at least in addition to, *obligation* leads to an interesting analysis of the relation between value and desire. By identifying value judgments with decisions, and therefore seeing them as the conclusions of deliberation about what it is possible to desire, I vindicate two platitudes about value judgment which are sometimes thought to conflict. These are: (1) that value judgments are internally related to changes in the state of one's will and (2) value judgments are not merely the expressions of pro-attitudes.[2]

## Value and desire

It is common to distinguish one's *motivational* system from one's *valuational* system, and to think of persons as the battleground of a struggle between these two systems, working perhaps, in the long run, towards equilibrium between them. Let us call this the orthodox view.

The orthodox view represents value and desire as if they are two types of consideration that come into deliberation; competition occurs not only among desires, but also between desire and value, or value judgment. The conflict between value and desire then takes on this form: you want y, and performing A will get you y, but you disapprove of performing A. Or perhaps you disapprove of wanting y, or you disapprove of the relative ranking that the desire for y has.

The program of the orthodox view is therefore to recognize two types of consideration, and then find a place for this distinction in a theory of mind by representing the conflict in terms of *competing attitudes*. So the big question concerns what sort of attitude is evaluation based on? Is evaluation *belief-like*? Or is the valuing attitude more *desire-like*? Either way, the attitudes are evaluative insofar as they concern behavior or motivation and either forbid, allow or praise it. An evaluative belief will do this all within the content of the belief. That is, the content of the belief will be that such and such behavior is praiseworthy, permissible or whatever. Thus a belief-like evaluative attitude will be a cognitive attitude towards the proposition that such and such behavior is praiseworthy, or blameworthy. A desire-like evaluative attitude, on the other hand, will be a non-cognitive attitude of praise or blame towards behavior of a certain sort. (A more complicated view that we have already mentioned is that evaluative desires are second-order desires, directed not at types of behavior, but at the 'shape' of one's first order desires.)

Either way, what you get is an explanation of evaluation and motivation in terms of attitudes in conflict. And, depending on which attitude you choose as the basis for evaluation, you get different sorts of problems. A view of evaluation as belief-like leads to trouble in explaining the relation between evaluative belief on the one hand, and intention and action on the other. A view of evaluation as desire-like has trouble explaining why evaluation is any different from any other desire. And a view of evaluation as second order desiring has both problems: like the belief case, it has the problem of finding an internal conceptual connection between valuation and intention; and like the desire case, it has trouble explaining why these second order desires are in any way special.

The drift of the view I'm trying to present is different: Instead of recognizing two types of consideration for deliberation to take into account, it recognizes two modes or components of deliberation. Instead of pitting evaluative against non-evaluative considerations, I'll say that *deliberation itself*, not the considerations taken into account in deliberation, is both evaluative and non-evaluative, corresponding to the unprincipled and principled aspects of deliberation.

Thus, my hypothesis provides a way of resisting the idea that we may clearly distinguish between *desire-based* and *value-based* reasons. I do not think that value *competes* with desire. Rather, I think that the notions of desire and value each capture an important and distinct aspect of deliberation. The attempt to move forward in the pursuit of one's interests is often as much an attempt to articulate one's desires as it is a pursuit of them. And my point has been that the concern to articulate one's desires just is the concern with value. To pursue value, then, is to attempt to articulate what one wants. And we do this by exploring what it is possible for us to desire.

### Judgment internalism

Another virtue of putting things this way is that it provides a nice interpretation of judgment internalism. Judgment internalism is a view about the nature of evaluative judgment. It holds that there is an internal conceptual connection between the sort of event that qualifies as an evaluative judgment and the state of one's motivational system.

Value judgments are often compared to the conclusions of reasoning about what we ought to do or intend (Smith, 1992). That is, we think of value judgments as belief-like. Because there is a constant conjunction between beliefs about what we ought to intend, and what we subsequently intend, it figures that we should not be surprised to find a similar connection between evaluative beliefs and intention. It is true that this should not be surprising. But even if there is such a constant conjunction, this does not establish an internal conceptual connection.

You can't get more than a contingent connection between reasoned conclusions—i.e., those which take reasons into account—and motivation. One thing that we mean when we say of someone that he has a reason to do such and such, is that if we are correct about his aims, then, if he deliberates correctly then he will choose to do the thing. That means that we have been able to reason for him, and that means that we think that his aims plus principles of decision lead to conclusion. And if we can see this, he should be able to see it, too. But we can only get him to see it by representing his interests and beliefs within a principled framework and showing how if we are right about his beliefs and desires, then he really ought to draw this conclusion. He may grant everything we say. And he may form the appropriate intention. But this last bit, the forming of the intention, is something quite separate from the ought belief. An intention needs support from desires; it cannot survive with support only from beliefs about what one ought to intend. And since that is so, there is no internal conceptual connection.

Suppose we can display to an agent that some action serves his ends. Does this necessarily motivate him? If he is rational, we might say. This confuses his acceptance of the argument, with his acquisition of a particular desire. The acquisition of desire, following an ought belief is just a contingent matter. But you might argue that here, too, there is a rationality constraint—namely, if he does not acquire the appropriate desire, he is irrational. But to believe that it is rational to believe or desire something is not yet to believe or desire it, even if in the usual case, such recognition is followed by a revision of belief or desire. If we can believe we should A, without desiring to A, then there's simply no conceptual connection.

When I say that an account of judgment is internalist, I mean that it displays an internal conceptual connection between judgments and certain relevant changes in the shape of one's desires. Since desires are the states which motivate and guide action, this connects deliberative judgments internally to motivation. I hope it is clear that the account of unprincipled decision I have given—in terms of trying on articulations of desire in a search for adequate and desirable desires—meets the internalist criterion. Unprincipled decisions, as I have described them, are, in fact, revisions of one's practical attitudes.

This is different from the way the internalist claim is taken on a more standard interpretation. There it is taken to be a thesis about a relation between a belief of a certain kind—a normative belief, like a belief about what acts or rankings of desires are right—and some inclination to think or act in the way appropriate to the normative content of the belief. Thus, with respect to the practical, one would form beliefs about what is right to do or what rankings of desires are good, and an internalist is expected to say that accompanying such beliefs—in the normal case—is a tendency to do that which one believes is right to do, or to adopt the desires one believes to be good. On such a view, judgment is a kind of belief. On the account I offer, a judgment is not a belief. It is a decision.

### Practical reflection

Some writers have made what can be called a 'constitutive claim' about certain sorts of description. This is the claim that there is a kind of description which somehow and in some sense constitutes its object. I mentioned that Velleman (1989) and Taylor (1985) make such a claim. The sort of description they have in mind is the interpretation of one's own psychological states, and the idea is something like this: in describing our psychological states, we provide a route for their expression, and we thereby end up constituting ourselves as we describe ourselves. One's understanding of oneself helps to shape oneself.

Now, this is a different sort of thing from the *evaluative* sort of self-reflection of the 'desired desires' story wherein one describes oneself and compares that description to an ideal which one holds up for oneself. The evaluative self-reflection does not imply anything like the constitutive claim.

Likewise with the theory of reasoning and deliberation: here also we may contrast evaluative and constitutive self-reflection. We may seek to describe our reasoning practices in order to criticize them from the point of view of our normative ideal; or we may seek authentic self-descriptive theories of reasoning and deliberation which mold our deliberative practice by displaying how we can exploit our capacity to reason.

My constitutive claim is not that the theory of reasoning and deliberation shapes us normatively, although of course it does if we produce normative frameworks for deliberation that we take seriously; rather, the claim is that our deliberative practices take on a different character depending on how we approach the study of reasoning and deliberation. If we think of the goal as one of articulating a normative framework for deliberation, then we give greater expression to our potential for principled decision. If we think of the goal as one of articulating maxims for exploring possibilities, then we give greater expression to our potential for unprincipled decision.

Now, if it is true that deliberation has the potential that I have been trying to bring out, then perhaps it is also true that we have a sort of freedom of mind, which we can either exercise or squander, depending on our awareness of it. That is, when we become aware of the dual nature of decision, we can make the most of it by toying with the balance between the two components. Not only have we more room to maneuver in the case of practical dilemma, but we can also refuse to be arbitrary. We don't have to be stymied by dilemma, because we can revise our interests. And we don't have to be arbitrary, because we can articulate our interests to the point where arbitrariness is unnecessary.

If you think of your deliberation as principled, and think of it only in that way—that is, where it is not principled, it is arbitrary—you will be hopeful of coming up with the proper idealization of deliberation. But this may, if I am right, result in cultivating a distinct and slightly impoverished phenomenology of deliberation.

*Birbeck College, London*

## NOTES

[1] Interestingly, David Wiggins (1987) disagrees with Kolnai about Aristotle, but seems nonetheless to have similar thoughts about deliberation. Everyone knows that doctors and statesmen are not simply faced with technical problems, and it is hard to believe that this is what Aristotle means. See also Schon (1983), especially pp. 46-7, for an interesting discussion of the 'reflective practitioner'.

[2] For this use of the expression 'the will', see Frankfurt (1988): 'To identify an agent's will is either to identify the desire (or desires) by which he is motivated in some action he performs or to identify the desire (or desires) by which he will or would be motivated when or if he acts...it is the notion of an *effective* desire—one which moves (or will or would move) a person all the way to action' (p. 14).

## REFERENCES

Frankfurt, H. (1988), 'Freedom of the Will and the Concept of a Person', in *The Importance of What We Care About* (Cambridge).
Harman, G. (1975), ' Practical Reasoning' , *Review of Metaphysics.*

Harman, G. (1986), *Change in View* (MIT).
Kolnai, A. (1977), 'Deliberation is of Ends', in *Ethics, Value and Reality: Selected Papers of Aurel Kolnai* (The Athlone Press).
Mill, J.S. (1861), *Utilitarianism.* Page numbers taken from 1957 Macmillan edition.
Moore G.E. (1903), *Principia Ethica* (Cambridge).
Rawls, J. (1971) *A Theory of Justice* (Harvard).
Schon, D.A. (1983), *The Reflective Practitioner* (Basic Books).
Smith, M. (1992), 'Valuing: Desiring or Believing?' in D. Charles and K. Lennon, eds., *Reduction, Explanation, and Realism* (Oxford).
Taylor, C. (1985), *Human Agency and Language: Philosophical Papers, vol. 1* (Cambridge).
Velleman, D. (1989), *Practical Reflection* (Princeton).
Watson, G. (1982), 'Free Agency', in G. Watson, ed., *Free Will* (Oxford).
Wiggins, D. (1987), 'Deliberation and Practical Reason', in his *Needs, Values, Truth* (Blackwell).
Williams, B. (1973), 'Deciding to Believe', in *Problems of the Self* (Cambridge).

JEANETTE KENNETT AND MICHAEL SMITH

# PHILOSOPHY AND COMMONSENSE: THE CASE OF WEAKNESS OF WILL

Here is a little story. As he has done a hundred times before, John heads off to the local shop to buy some chocolate bars. He knows that eating so much chocolate isn't good for him. Being over forty and doing no exercise a passion for chocolate simply adds to an already significant weight problem. But thoughts like this do not move him. Each day, fully cognizant of the effects of eating chocolate upon his health, John heads off to the local shop, arrives, buys several chocolate bars, unwraps one, and then proceeds to eat it, unwraps another, and then proceeds to it, and so on and so on and so on.

Now here is a bit of commonsense. In certain crucial respects the story we have just told is underdescribed. For as Gary Watson points out, a story like this can be filled out in at least three different ways, ways that in turn reflect our commonsense understandings of recklessness, weakness of will, and compulsion (Watson 1977). Moreover, whether we fill out the story in one or another of these ways is of great practical significance. For the allocation of moral responsibility is in large part determined by whether we think of John, in the story, as being either reckless, or weak, or compelled.

Consider recklessness first. To fill out the story in this way it is sufficient to imagine John being fully in control of what he does, eating chocolate because he judges it better to do so than to refrain and so regain his health, but making his judgement as a result of some sort of culpable mistake or error of reasoning. John is reckless if, knowing full well that it will harm him if he eats chocolate, he none the less freely acts upon his judgement that it is best to eat chocolate, where this judgement is not necessarily wrong, but is at least wrong by his own lights; a judgement he himself would not have made if only he had taken more care. Commonsense tells us that, when the case is filled out in this way, John is fully responsible for what happens to him. For responsibility attaches to the known consequences of what we freely choose to do and John freely chooses to eat chocolate bars while knowing the consequences.

Next consider compulsion. This time it is sufficient to imagine John judging the benefits of eating chocolate to be less than the benefits of refraining, but then eating the chocolate anyway, despite his judgement, because his desire for chocolate is simply irresistable. For John is compelled when he is literally out of control, and an irresistable desire is precisely a desire that an agent cannot control. Commonsense tells us that, in this case, John is not re-

*Michaelis Michael and John O'Leary-Hawthorne (eds.), Philosophy in Mind, pp.141-157.*

sponsible for what he does. For, again, responsibility attaches to the known consequences of what we freely choose to do and, in this case, John does not freely choose to eat chocolate bars.

Finally, there is weakness of will. When we fill out the story in this way we imagine a case interim between the two just described. It is like the case in which we imagine John being compelled in that we suppose he eats chocolate despite judging it more desirable to refrain. But it is also like the case in which he is reckless in that we imagine him having the capacity to control what he does. When John is weak, his problem is not that he is out of control, but rather that, despite having the capacity to control himself, he simply gives in to the temptation of the chocolate bars. He eats another chocolate bar despite his judgement that it would be better to refrain, and despite his ability to resist his desire. In this case, commonsense thus once again tells us that John is responsible. For, again, responsibility attaches to the known consequences of what we freely choose to do, and, since his actions are under his control, John evidently freely chooses to eat chocolate bars.

Now here is some philosophy. Though commonsense delivers up these distinctions between recklessness, weakness and compulsion, the distinctions should be accepted only if they can be vindicated philosophically. For commonsense, unlike philosophy, has no final authority of its own. The question is therefore whether these distinctions mark *real* differences, differences that can be articulated in any systematic way.

The problematic category is, of course, weakness of will. Being interim between recklessness and compulsion it occupies a potentially unstable middle ground. The instability emerges when we attempt to spell out the connections between better judgement, desire, and free and intentional action. For, as Donald Davidson points out (1980), the idea that someone may act freely and intentionally contrary to her better judgement is difficult to square with two apparently plausible principles connecting, on the one hand, desire with action:

P1. If an agent wants to do x more than she wants to do y and believes herself free to do either x or y, then she will intentionally do x if she does either x or y intentionally

and, on the other, better judgement with desire:

P2. If an agent judges that it would be better to do x than to do y, then she wants to do x more than she wants to do y

For, given P2, if an agent judges it best to do x, then she most wants to do x. And then, given P1, since she most wants to do x, x is what she will freely and intentionally do if she does anything freely and intentionally. So P1 and P2 together entail that an agent will never act freely and intentionally contrary to her *better* judgement. How, after all, can an agent judge it best to do x and yet *better* to do y?[1]

If this is right, then a question naturally arises about the status of the commonsense distinctions between recklessness, weakness and compulsion. For if the middle-ground between recklessness and compulsion disappears then we will have to decide whether to assimilate cases of what were thought to be weakness to recklessness, or rather to compulsion. But which way should we go?

If we assimilate weakness to recklessness, then though we hold on to the idea that the weak-willed are responsible, we do so at the price of thinking that they do not really judge to be best the actions that they say they deem best. We must conclude that, in addition to acting badly, they think it best to act in this way, contrary to their insincere reports about what they value in the way of action. And if we assimilate weakness to compulsion, then though we hold on to the idea that the weak really do judge it better to act in one way, despite acting in another, we do so at the price of thinking that they are not really responsible for what they do. Either way, a dissolution of the commonsense distinctions between recklessness, weakness and compulsion promises to bring with it a significant revision of our ordinary moral practices.

For the record, we doubt very much whether philosophy could ever force us to give up on commonsense distinctions like those between recklessness, weakness and compulsion, and we therefore doubt whether it could ever force us to revise our ordinary moral practices in the ways just described. However, and unfortunately, we do not know how to argue for these conclusions in full generality. Rather it seems to us that the arguments for such conclusions have to be given piecemeal. In this particular case, the arguments in favour of commonsense are, we think, relatively easy to provide. For the reasons so far given for thinking that there is no such thing as weakness of will are crucially flawed. Our task in this paper is thus to present our own solution to Davidson's problem, a solution that, in our view, consitutes one small but important part of an ongoing defence of commonsense in moral psychology (see also Pettit and Smith 1990, 1993, forthcoming; Smith 1992; Kennett 1991, 1993; Kennett and Smith forthcoming).

The remainder of the paper divides into four main sections. In the first we introduce a distinction between two sorts of reasons and argue that this distinction shows P2 to be false for more or less commonsense reasons. In the second section we argue that whatever plausibility attaches to P2 attaches to it because a closely related principle, P2*, is indeed true. However, as we show, P2*, together with P1, leaves room for all sorts of divergences between better judgement and desire. In the third section we show how P2*, together with P1, suggests a story about the nature and operation of self-control that allows us to distinguish an agent's having a capacity for self-control which she fails to exercise from her having no capacity for self-control at all. And in the fourth and final section we show how our account underwrites the commonsense distinctions between recklessness, weakness of will and compulsion.

## *1. Normative reasons vs motivating reasons*

In 'Reasons for Action and Desire' Michael Woods observes that 'the concept of a reason for action stands at the point of intersection, so to speak, between the theory of the explanation of actions and the theory of their justification' (1972, p.189). This suggests a more or less commonsense distinction between kinds of reasons depending on whether we use the concept in a way more at home in the theory of the explanation of action or more at home in the theory of their justification. When we say of someone that she has a reason to $\phi$, we might mean that she in a state that is potentially explanatory of her $\phi$-ing, whether or not she would be justified in $\phi$-ing. Or we might mean that she is justified in $\phi$-ing whether or not she is in a state potentially explanatory of her $\phi$-ing.

As we said, we think that this is a more or less commonsense distinction. Here is an example of Gary Watson's to prove the point (1982). Imagine a mother with a screaming baby. At her tether's end, she finds herself desiring to drown her screaming baby in the bathwater. To remove unnecessary complications, suppose that the desire is non-instrumental in character. The mother's anger and frustration produce the desire directly, much as anger can directly produce a desire to smash a glass or punch a hole in a wall (Hursthouse 1991).

What are we to say about such a woman? Suppose she succeeds in acting on her desire. It seems that we should then certainly say that she drowns her baby intentionally. But it also seems that we may do so without supposing that she acts with rational justification. In other words, we may suppose that she has a reason of the explanatory kind but entirely lacks a reason of the justificatory kind. Or consider the claim that what she would be rationally justified in doing, in her distraught state, is taking her baby out of the bathwater, drying it, dressing it, and then putting it out of harm's way while she goes into another room to calm down. It seems to us that this claim may well be true, and that this may accordingly be what she has a reason of the justificatory kind to do. But even if it is, it seems that we need not suppose that she has a reason of the explanatory kind to act in this way. For we may suppose not just that her anger and frustration produces aberrant desires, like the desire to drown her baby, but also that it destroys any desire whatsoever to care for her baby.

This ambiguity in our concept of a reason is, as we have said, more or less commonsensical. Let's therefore introduce a distinction to keep track of the ambiguities. Let's call reasons of the explanatory kind 'motivating' reasons, and reasons of the justificatory kind 'normative' reasons (Smith 1987). The question naturally arises whether we can give a more precise and systematic account of the distinction between motivating and normative reasons.

In virtue of their explanatory role, motivating reasons are best thought of as being psychological states, states with the potential for explaining, teleologically, and perhaps causally, our doing what we do. In our view, the best account of such states is therefore Humean in character. A motivating reason to $\phi$ is constituted by a desire for an outcome of a certain kind and a belief to the effect that $\phi$-ing will produce an outcome of that kind (Smith 1987, 1988).

Normative reasons, by contrast, have the role of justifying actions. In our view they are therefore best thought of not as psychological states at all - for psychological states do not justify anything (Pettit and Smith 1990) - but rather as propositions to the effect that this or that course of action is to some extent worth doing; propositions whose truth would justify, to a corresponding extent, acting in this way or that. In our view, the best account of such propositions is given by a dispositional theory of value (Johnston 1989, Lewis 1989, Smith 1989). Normative reasons are propositions concerning the desirability of acting in certain ways, where facts about desirability are in turn simply facts about our idealised desires. To say that we have a normative reason to $\phi$ in circumstances C, then, is simply to say that, under conditions of increasing information and rational reflection, we would come stably to desire that we $\phi$ in C. Or, for short, it is simply to say that, if we were fully rational, we would desire that we $\phi$ in C.[2]

This more systematic way of making out the distinction between motivating reasons and normative reasons allows us to make good sense of our examples. On the one hand, according to this account the angry and frustrated mother has a motivating reason to drown her baby just in case she has a desire to drown her baby combined with a suitable means-end belief. And of course this she does have. But it does not follow that she has a normative reason to drown her baby, because this would require that, under conditions of full rationality, she would desire that she drowns her baby in circumstances like those she currently faces. And on the plausible assumption that under conditions of full rationality she would no longer be angry and frustrated, there is no reason to suppose this to be so. For, recall, we are supposing that her desire to drown her baby is produced by her anger and frustration.

On the other hand, according to this account, the mother has a normative reason to take her baby out the bathwater, dry it, dress it, and then put it out of harm's way while she goes into another room to calm down, just in case, under conditions of full rationality, she would desire that this is what she does in her distraught state. And, again, we may well suppose this to be so. For, recall, though we are assuming that her anger and frustration destroys her desire to care for her baby, were her anger and frustration to go, as it would under conditions of full rationality, her desire to care for her baby would plausibly return. Our analysis thus suggests that she may indeed have normative reason to act in this way towards her baby, though she needn't have any desire at all to act in this way in her distraught state.

With this distinction between motivating and normative reasons in place we are finally in a position to give our argument against P2. P2 purports to be a principle connecting better judgement with desire. To the extent that P2 maps on to distinctions delivered up by commonsense, it should therefore be equivalent to a principle connecting an agent's beliefs about her normative reasons with her motivating reasons. In other words, P2 should be equivalent to the principle:

> If an agent believes that she has more normative reason to do x than to do y, then her motivating reason to do x is stronger than her motivating reason to do y.

But this principle is evidently implausible, as we can see by considering once again our example.

The mother with the screaming baby may well believe that she would desire to take her baby out the bathwater, dry it, dress it, put it out of harm's way, and then go into another room to calm down, if she were forming a desire about what to do in her distraught state under conditions of full rationality: that is, while not in a distraught state. Accordingly, this may be what she believes she has normative reason to do. But, not being fully rational, being rather in a distraught state, she may have no motivating reason, no actual desire, to act in this way at all. P2 is thus false.

(Just for the record, note that the converse principle is false as well. Though the mother does in fact desire to drown her screaming baby in the bathwater in her distraught state, it does not follow that she believes that she would desire herself to do this in her distraught state if she were in conditions of full rationality. For she may well, and indeed typically would, know that her desire is one which is entirely the result of the anger and frustration of the moment.)

P2 is, then, quite evidently false, false for more or less commonsense reasons. For once we distinguish normative from motivating reasons, we see that an agent's beliefs about her normative reasons may come apart from her motivating reasons in a quite systematic way. And in that case it follows that better judgement may come apart from desire in a quite systematic way as well.[3]

*2. Normative reasons and the effectiveness of deliberation*

Though P2 is false, we think that it is an attempt to articulate an important insight about the connection between our beliefs about our normative reasons and our motivating reasons. The insight emerges when we ask about the ways in which normative and motivating reasons relate to deliberation.

It is plausible to assume that action is the product of reasons that we discover through deliberation. But, if this is right, then we must ask how this is possible. For when we deliberate, we don't just try to find out what we actually want, we try to find out whether what we actually want is worth wanting. In other words, we attempt to find out whether what we actually want is desirable; something that we would come stably to want under conditions of ever increasing information and rational reflection. And so, in our view, when we deliberate we attempt to discover normative reasons.

But if this is right, if we are concerned to discover normative reasons rather than motivating reasons when we deliberate, then it follows immediately that deliberation is only *contingently* practical in its issue. For, as the distinction between motivating and normative reasons together with P1 makes plain, what we do is a matter not of the normative reasons we believe ourselves to have, but is instead a matter of the motivating reasons we actually have. A piece of deliberation favouring $\phi$-ing issues in an attempt to $\phi$ just in case, contingently, the deliberator's actual desires *match* the desires she believes she would have if she were fully rational.

Is this contingency objectionable? It might well be thought that it is. For, it might be said, if deliberation is only contingently effective, then the connection between deliberation and action is altogether *fortuitous*. We have no right to think, as we ordinarily do, that action is the product of the reasons that we discover through deliberation. Rather, our beliefs about our normative reasons are mere epiphenomena, causally irrelevant in the production of action.

In our view it is this line of thought that leads theorists to embrace P2. For one striking feature of P2 is that it guarantees the effectiveness of deliberation. If better judgement entails desire, then there is no gap between discovering that an action is worth performing, and desiring to act accordingly. However, and unfortunately, as we have seen, better judgement does not entail desire. It is thus wrong to suppose that we can plausibly capture the insight that the connection between deliberation and action is not altogether fortuitous by embracing P2. And so we must ask whether we can capture that insight in some other way.

We believe that the insight can be captured by a weaker principle than P2. Specifically, it is captured by the principle we get by adding a '*ceteris paribus*' clause to P2. According to this modified principle, the claim that deliberation is effective is equivalent to the idea that *ceteris paribus* action is the product of the reasons that we discover through deliberation. But is this weakened principle itself any more plausible than P2? What does the '*ceteris paribus*' clause amount to?

In our view, it amounts to the assumption that the deliberating agent whose deliberations are effective is a *rational* deliberator. And nor should this be suprising. For the agent who believes that it is desirable to $\phi$ and yet does not desire to $\phi$ *is* irrational (Smith 1992, Pettit and Smith 1993, forthcoming). In order to see this, suppose an agent believes that she would desire to $\phi$ in circumstances C if she were fully rational, but that she does not desire to $\phi$ in C. What can we say about this agent? The clearest thing to say is surely that she is *irrational by her own lights*. For she fails to have a desire that she believes she would have if she were more rational. And, so long as irrationality by one's own lights is a species of irrationality - an extension of the requirement of coherence, perhaps - it follows immediately that this agent is irrational *tout court*.

If this is right, then it would seem to follow that rationality demands of us that we desire to act in ways that we believe we have normative reason to act. Indeed, it would seem that rationality demands of us something even stronger. It demands of us not just that we desire to act in ways that we believe ourselves to have normative reason to act, but also that the relative strength of our desires *matches* the relative importance of these normative reasons. For suppose we believe ourselves to have a normative reason to $\phi$ in circumstances C: that is, we believe that amongst the desires we would have if we were fully rational is a desire that we $\phi$ in C. Importantly, this may well be only one of the desires we imagine ourselves having, for it is perfectly possible for us to believe that we would have other different, and perhaps even confliciting, desires concerning what to do in C if we were fully rational. Now suppose that we do. Then we presumably believe that our desire to $\phi$ would have a certain *strength* vis a vis these other desires we imagine ourselves having. This, the relative strength of our imagined desire to $\phi$ in C, fixes something important about the normative reason we believe ourselves to have for $\phi$-ing in C. Specifically, it fixes the *relative importance* of that normative reason. Or so we claim.

Thus, if our imagined desire to $\phi$ in C is the strongest of our imagined desires concerning what to do in C, then it follows that this is the normative reason to which we attach most importance. If we imagine it to be a relatively weak desire - suppose we believe that we would also have a much stronger desire to $\psi$ in C if we were fully rational - then, though we take ourselves to have some normative reason to $\phi$ in C, we judge ourselves to have more normative reason to $\psi$ in C. And so we could go on.

If this is right, if the connection between the normative reasons we believe ourselves to have after we have deliberated and our motivating reasons is a *rational*, albeit *contingent*, connection, then that suggests that we should accept the following principle connecting the normative reasons we believe ourselves to have with desire:

> If an agent believes that she has more normative reason to do x than to do y, then she *should* want to do x more than she wants to do y

or, equivalently,

> P2*. If an agent judges that it would be better to do x than to do y, then she *should* want to do x more than she wants to do y

For according to P2*, rationality demands of us that we have desires that match in both *content and strength* the desires we believe we would have if we were fully rational.

In our view P2* suffices to capture the insight that the connection between deliberation and action is not altogether fortuitous. But, if this is right, then it follows that there are all sorts of ways in which better judgement and desire may fail to coincide, even when we do desire in the way we judge best.

Suppose, for example, that we believe that we have a normative reason to $\phi$ and that we do desire to $\phi$, just as we should. Given P2*, our desire to $\phi$ may still be rationally criticizable. For, on the one hand, our desire to $\phi$ may be *stronger* than it should be. The strength of our desire to $\phi$ may be such that we are disposed to $\phi$ instead of $\psi$ even when we take ourselves to have more normative reason to $\psi$ than $\phi$; disposed to $\phi$ instead of $\kappa$ even when we take ourselves to have more normative reason to $\kappa$ than $\phi$; and so on. (The limiting case of this, of course, would be desiring to $\phi$ when we believe ourselves to have no normative reason to $\phi$.) Or, alternatively, our desire to $\phi$ may be *weaker* than it should be. For the strength of our desire to $\phi$ may be such that we are disposed to $\psi$ instead of $\phi$ even when we take ourselves to have more normative reason to $\phi$ than $\psi$; we are disposed to $\kappa$ instead of $\phi$ even when we take ourselves to have more normative reason to $\phi$ that $\kappa$; and so on. (The limiting case of this, of course, would be our having no desire at all to $\phi$, despite the fact that we judge ourselves to have some normative reason to $\phi$.)

To sum up: the reasons we discover through deliberation reliably lead to action just in case we are rational in the sense captured by P2*. And, accordingly, whatever plausibility attaches to P2 derives from the plausibility of P2*. Rationality demands of us not just that we actually desire to act in the ways we believe we have normative reason to act, but also that the strength of our actual desires be isomorphic with the importance of our believed reasons. And what this means is that a space may open up between the normative reasons an irrational agent believes herself to have, and the desires she actually has.

## 3. *Self-control*

It should now be clear what sorts of circumstances call for the exercise of self-control, in our view. When an agent deliberates, and decides which course of action she has most normative reason to pursue, she all too often finds that she is motivated to do something else instead. In such situations, she will in fact act otherwise if she does not take appropriate steps. This is when self-control is called for.

Consider again the example of John described at the outset. When we imagine John being weak or compelled we suppose that, though he judges it most desirable to refrain from eating chocolates, his desires are out of line with his judgement. Perhaps we imagine his desire to experience the taste of chocolate being much stronger than it should be, or his desire to be healthy being much weaker than it should be, so that he ends up desiring to eat chocolate more than he desires to regain his health, despite judging it far less desirable to experience the taste of chocolate than to regain his health. Without the exercise of self-control John will knowingly fail to do what he takes himself to have most reason to do. When we imagine John being weak, we imagine such an exercise of self-control to be possible. When we imagine him being compelled, we imagine him incapable of such an exercise of self-control.

But what does it mean to say that John can or cannot exercise self-control? Talk of exercising self-control sounds like talk of action, and, as we have seen, action is motivated by an antecedent desire. But in that case, doesn't a regress threaten? If the exercise of self-control is itself to be explained by the presence of desires that may themselves be either too weak or too strong, then won't an agent's exercise of self-control itself need to be under her control?

According to some, questions like these suggest a paradox in the very idea of self-control (Mele 1988). But in our view they suggest something rather different. When we talk of an agent exercising self-control we may seem to be supposing that every exercise of self-control is itself an action. But, if we do, that supposition is false. Some exercises of self-control must themselves be, not actions, but rather manifestations of our cognitive dispositions. In order to see how they can be manifestations of our cognitive dispositions, we need to consider in some detail how John might exercise self-control in the case described.

Let's suppose that John believes that it is most desirable to refrain from eating chocolate because he believes that it is more desirable to be healthy than to get pleasure, and because he believes that, though it will also lead to less in the way of pleasure, refraining from eating chocolate will cause him to be healthy. And let's suppose further that he doesn't have any actual desire to be healthy at all, despite his judgement. He believes that it is desirable to be healthy because he believes that, under conditions of full rationality, he would desire to be healthy. But, not being in conditions of full rationality in fact, he does not have any actual desire to be healthy.

In order to make the case as simple and uncontroversial as possible, let's assume further that conditions of full rationality may simply be understood as conditions of full information. In the past, whenever John has thought long and hard about what health is and involves, he has found that he ends up desiring very strongly to be healthy. But he has also found that this desire is difficult to keep. When thoughts about what health is and involves are not before his mind, he finds that his desire to be healthy simply lapses. This is why John believes that he would desire very strongly to be healthy, if he were fully rational, though he does not in fact desire to be healthy.

If this is John's situation, what might happen when he exercises self-control? One obvious answer suggests itself. Before he is about to enter the shop and buy some chocolate John might simply think about what health is and involves. For if John has these thoughts, and if his having these thoughts causes him to desire to be healthy, and if the desire it causes in him is strong enough, then he will find himself with a desire to be healthy strong enough to resist the temptation of the chocolate bars (Pettit and Smith forthcoming, Kennett and Smith forthcoming).

Importantly, note that this answer does not suggest any sort of regress. For John's having certain thoughts about the nature of health need not itself be thought of as an action, and so we need not suppose that John has any antecedent desires which explain why they are had. His having certain thoughts may rather be just what they seem to be: thoroughly cognitive matters to be explained in terms of his cognitive dispositions, not actions that require explanation in terms of antecedent desires that might themselves be either weaker or stronger than they should be, and which themselves must therefore be capable of being brought under his control.

And note that talk of cognitive dispositions is important here, for having thus distanced ourselves from the idea that an exercise of self-control is itself an action, we don't want to commit ourselves to the suggestion that an agent's having certain thoughts is all there is to the exercise of self-control. After all, to return to John's case, if his having the thoughts about health that cause him to desire health is simply a matter of luck, then this could hardly constitute his exercise of self-control. For his having these thoughts would then reflect no credit upon him, whereas his exercise of self-control does.

The answer lies in the idea that John's exercise of self-control is the manifestation of a cognitive disposition. In other words, John's having thoughts that cause him to desire health is no mere accident, and in turn reflects credit upon him, if his having those thoughts is itself a manifestation of a disposition he has to have such thoughts when he is otherwise disposed to act on his desires and contrary to his better judgement. For, as we have already seen, it is rational for an agent to do what she believes she has most normative reason to do. And, since this is so, her being disposed to have such thoughts is simply a way of shoring up her tendency to act rationally. What reflects credit upon an agent, when she exercises self-control, is thus the rationality of her having the thoughts she has in her difficult circumstances. John does well when he has thoughts that cause him to desire health because that it is a rational thing for him to think in circumstances where having such thoughts will cause him to have the desire that will in turn cause him to act in the way he believes he has most normative reason to act.

There is a problem looming here, however. For, as we have seen, the capacity for self-control is supposed to be one that an agent *may or may not* exercise. This, you will recall, is the crux of the distinction between weakness and compulsion: the weak agent is supposed to have a capacity for self-control that she fails to exercise whereas the compulsive agent is supposed not to have the capacity for self-control at all. But if the capacity for self-control is simply a dispostion to have the right thoughts at the right times, then what is the difference between failing to exercise self-control and having no capacity for self-control at all? An agent who fails to exercise self-control fails to have the right thoughts at the right times. And an agent who has no capacity for self-control at all also fails to have the right thoughts at the right times. The agent who fails to exercise self-control and the agent who fails to have the capacity for self-control would therefore seem to be one and the same. We have lost the distinction between weakness and compulsion. Or so it might seem.

The problem here is more apparent than real. We have said that an agent who has the capacity for self-control has a disposition to have those thoughts that will cause her to act in accordance with her beliefs about what she has most normative reason to do, when she is otherwise disposed to act on her desires and contrary to her beliefs. But for this to be so it is not necessary for her to have the right thoughts on each and every occasion that she is disposed otherwise to lose control. It is enough that, in circumstances where she will act contrary to her better judgement, she could have had such thoughts.

Thus, consider two agents who fail to exercise self-control, one of whom we we would ordinarily describe as weak, the other as compelled. We just do ordinarily distinguish a sense of 'could' in which we can truly say of one that she *could* have had thoughts that would cause her to act in accordance with her beliefs about what she has most normative reason to do, whereas the other *could not* have had such thoughts. The truth of the claim that an agent could have had such thoughts, in this ordinary sense of 'could', is thus to be determined in a largely *pragmatic* fashion. We might, for example, appeal to her past behaviour. If an agent has, in the past, had similar thoughts in similar situations then, for ordinary purposes, this might be enough to show that she could have had such thoughts in the situation she in fact faces here and now. The pragmatic character of the 'could' is evident once we notice that it is our interests in an agent and her predicament that fix the standards of similarity.

For example, if in the past John has had thoughts about the nature of health when, say, he went into yet other shops to buy chocolate, and his having these thoughts then caused him to desire health with sufficient strength to resist the temptation of the chocolates, then we might take this to be sufficient to show that he could have had such thoughts in the situation that he in fact faces here and now. Or perhaps it is sufficient that he had similar thoughts when he went into yet other shops in the past to buy cake, something else he has a passion for, and on those occasions his thoughts led him to desire health strongly enough to overcome the temptation of the cakes. Which similarity counts depends on which similarity is salient for us, given our interest in John and the way we think about his current predicament. A full account of these similarites would constitute a significant part of our understanding of rational agency. Ideally, an analysis of self-control would need to spell out these details. But the general idea should be plain enough.

Armed with this sense of 'could', we see that we can indeed distinguish between an agent who possesses the capacity for self-control, but fails to exercise it, and an agent who has no capacity for self-control at all. In short, an agent has a capacity for self-control that she fails to exercise when there is a nearby possible world in which she has the thoughts that cause her to have the desire that causes her to act in accordance with her better judgement, whereas an agent has no capacity for self-control at all when there is no nearby possible world in which she has the thoughts that cause her to have the desire that causes her to act in accordance with her better judgement. And whether or not this is so is in turn a function of our ordinary, pragmatically determined, standards of similarity, standards that in turn fix which possible worlds are to be deemed 'nearby'.

## 4. *How to distinguish recklessness, weakness and compulsion*

Let's return to the beginning. Commonsense tells us that we can distinguish between recklessness, weakness and compulsion. Our own account of the distinction between normative and motivating reasons, and our story about the nature and operation of self-control, allows us to make these distinctions in a more or less commonsense way.

An agent acts recklessly when, in forming her beliefs about what she has most normative reason to do - that is, in deciding what she would most want to do in the circumstances she faces if she were fully rational - she takes insufficient care, making a judgement she would not have made if only she had taken her time and thought about matters more carefully. But, having made her judgement, her desires match her beliefs perfectly. When an agent is reckless, her capacity for self-control is thus not at issue. She acts freely and intentionally in accordance with her sloppily formed beliefs about what she has most normative reason to do.

By contrast, when agents are weak or compelled, their capacity for self-control is precisely what is at issue. An agents acts weakly when she acts on her desires, and contrary to her beliefs about what she has most normative reason to do, but it is still the case that she could have had thoughts which would have led her to act in accordance with her beliefs, despite her desires. She has the capacity for self-control, but she fails to exercise it. And an agent acts compulsively when she acts on her desires, and contrary to her beliefs about what she has most normative reason to do, and it is not the case that she could have had thoughts which would have led her to act in accordance with her beliefs despite her desires. She lacks the capacity for self-control altogether. Thus, just as commonsense tells us, the weak agent gives in to her desires, whereas the compulsive agent is overcome by them.

Indeed, our own account of the distinctions between recklessness, weakness and compulsion suggests that there is a rich diversity of ways in which agents may be reckless, or weak, or compelled. Consider the case of compulsion, just by way of illustration. In addition to being compelled by their desires, as for instance we can imagine John being compelled by his desire to eat chocolate, our account suggests that agents may also be compelled by their own decision-making processes. Imagine an agent who is brain-washed and is therefore incapable of assessing evidence in forming her judgement about what it is desirable to do; someone who would have ended up deciding that a certain course of action is desirable no matter what evidence came before her. Our own account suggests that this agent may be just as compelled, and so just as lacking in responsibility for what she does when she acts on her judgement, as John is, even though, unlike John, she may have and exercise the capacity for self-control when she acts: that is, even though she may have the capacity to have the right thoughts at the right times so as to ensure that she acts in accorance with her compulsive judgement should she find that she has desires that are out of line with her judgement.

In our view this is a desirable consequence of our view. For the idea that recklessness, weakness of will and compulsion are richly diverse in their nature, richly diverse in just the ways that our own account predicts, is, after all, just more good commonsense.[4]

*Monash University*

## NOTES

[1] Davidson's own solution to this problem, which turns on his distinction between a conditional and an unconditional version of better judgement, has been criticized by others (Watson 1977, Peacocke 1985, Hurley 1985). We think that these criticisms decisively refute Davidson's own solution. Since they are familiar enough, we will not repeat them here.

[2] For an exploration of some of the complications of this formulation see Pettit and Smith forthcoming.

[3] In essence our own view is like that developed by Gary Watson (1977, 1982). Watson insists that we distinguish between an agent's *valuational* system and her *motivational* system, and this is much like our distinction between an agent's beliefs about her normative reasons and her motivating reasons. However Watson thinks that in embracing this distinction we are forced to accept a Platonic, rather than a Humean, conception of beliefs and desires, a conception according to which beliefs about what is desirable *are* desires. But for the reasons given in the text, we think that such a conception is implausible. Since an agent may have beliefs about what is desirable without having corresponding desires at all, beliefs about what is desirable cannot themselves be desires. See also Smith 1987, 1988, 1992, and, especially, Pettit and Smith 1993.

[4] Earlier versions of this paper were read at the University of Auckland, the University of Melbourne, and a conference on *Philosophy in Mind* held at the University of New South Wales. We are grateful to all of those who participated in useful discussions on these occasions, and to Gerald Dworkin and Philip Pettit who subsequently gave us very helpful comments .

## REFERENCES

Davidson, D. (1980), 'How is Weakness of the Will Possible?' in his *Essays on Actions and Events* (Oxford University Press).

Hurley, S.L. (1985), 'Conflict, Akrasia and Cognitivism' in *Proceedings of the Aristotelian Society*.

Hursthouse, R. (1991),'Arational Actions', *Journal of Philosophy*.

Johnston, M. (1989), 'Dispositional Theories of Value', *Proceedings of the Aristotelian Society*, Supplementary Volume.

Kennett, J. (1991), 'Decision Theory and Weakness of Will', *Pacific Philosophical Quarterly*.

Kennett, J. (1993), 'Mixed Motives', *Australasian Journal of Philosophy*.

Kennett, J. and Smith, M. (forthcoming), 'Self-Control in Action'.

Lewis, D. (1989), 'Dispositional Theories of Value', *Proceedings of the Aristotelian Society*, Supplementary Volume.

Mele, A. (1988), 'Irrationality: a precis', *Philosophical Psychology*.

Peacocke, C. (1985), 'Intention and Action' in Bruce Vermazen and Merrill B.Hintikka, eds., *Essays on Davidson: Actions and Events* (Oxford University Press).

Pettit, P. and Smith, M. (1990), 'Backgrounding Desire', *Philosophical Review*.
Pettit, P. and Smith, M. (1993), 'Practical Unreason', *Mind*.
Pettit, P. and Smith, M.. (forthcoming), 'Brandt on Self-Control' in Brad Hooker, ed., *Rules, Utility and Rationality* (Westview Press).
Smith, M. (1987), 'The Humean Theory of Motivation', *Mind*.
Smith, M. (1988), 'On Humeans, Anti-Humeans and Motivation: A Reply to Pettit', *Mind*.
Smith, M. (1989), 'Dispositional Theories of Value', *Proceedings of the Aristotelian Society*, Supplementary Volume.
Smith, M. (1992), 'Valuing: Desiring or Believing?' in D. Charles and K. Lennon, eds., *Reduction, Explanation and Realism* (Oxford University Press).
Watson, G. (1977), 'Skepticism about Weakness of Will', *Philosophical Review*.
Watson, G. (1982), 'Free Agency', in Gary Watson, ed., *Free Will* (Oxford University Press).
Woods, M. (1972), 'Reasons for Action and Desire', *Proceedings of the Aristotelian Society*, Supplementary Volume.

ROBERT BRANDOM

# REASONING AND REPRESENTING

## I

### 1

One useful way of dividing up the broadly cognitive capacities that constitute our mindedness is to distinguish between our sentience and our sapience. Sentience is what we share with nonverbal animals such as cats—the capacity to be *aware* in the sense of being *awake*. Sentience, which so far as our understanding yet reaches is an exclusively biological phenomenon, is in turn to be distinguished from the mere reliable differential responsiveness we sentients share with artifacts such as thermostats and land mines. Sapience, on the other hand, concerns *understanding* or intelligence, rather than irritability or arousal. One is treating something as sapient insofar as one explains its behavior by attributing to it intentional states such as belief and desire as constituting *reasons* for that behavior. Sapients act as though reasons matter to them. They are rational agents in the sense that their behavior can be made intelligible, at least sometimes, by attributing to them the capacity to make practical inferences concerning how to get what they want, and theoretical inferences concerning what follows from what.

Besides thinking of sapience in terms of reasons and inference, it is natural to think of it in terms of truth. Sapients are believers, and believing is taking-true. Sapients are agents, and acting is making-true. To be sapient is to have states such as belief, desire, and intention, which are contentful in the sense that the question can appropriately be raised under what circumstances what is believed, desired, or intended would be *true*. Understanding such a content is grasping the conditions that are necessary and sufficient for its truth.

These two ways of conceiving sapience, in terms of inference and in terms of truth, have as their common explanatory target contents distinguished as intelligible by their *propositional* form. What we can offer as a reason, what we can take or make true, has a propositional content, a content of the sort that we express by the use of declarative sentences and ascribe by the use of 'that' clauses. Propositional contents stand in inferential relations, and they have truth conditions.

*Michaelis Michael and John O'Leary-Hawthorne (eds.), Philosophy in Mind, pp.159-178.*
*Kluwer Academic Publishers, 1994,* 

Propositional contentfulness is only part of the story about sapience, however. When we try to understand the thought or discourse of others, the task can be divided initially into two parts: understanding what they are thinking or talking about and understanding what they are thinking or saying about it. My primary aim here is to present a view about the relation between what is *said* or *thought* and what it is said or thought *about*. The former is the propositional dimension of thought and talk, and the latter is its *representational* dimension. The question I will address is why any state or utterance that has propositional content also should be understood as having representational content. (For this so much as to be a question, it must be possible to characterize propositional content in nonrepresentational terms.)

The answer I will defend is that the representational dimension of propositional contents should be understood in terms of their *social* articulation—how a propositionally contentful belief or claim can have a different significance from the perspective of the individual believer or claimer, on the one hand, than it does from the perspective of one who attributes that belief or claim to the individual, on the other. The context within which concern with what is thought and talked *about* arises is the assessment of how the judgements of one individual can serve as reasons for another. The representational content of claims and the beliefs they express reflects the social dimension of the game of giving and asking for reasons.

2

It may be remarked at the outset that it will not do just to think of the representational dimension of semantic contentfulness according to a designational paradigm—that is, on the model of the relation between a name and what it is a name of. For that relation is a *semantic* relation only in virtue of what one can go on to *do* with what is picked out by the name—what one can then *say* about it. Merely picking out an object or a possible state of affairs is not enough. What about it? One must say something about the object, claim that the state of affairs obtains or is a fact.

One of Kant's epoch-making insights, confirmed and secured for us also by Frege and Wittgenstein, is his recognition of the *primacy of the propositional*. The pre-Kantian tradition took it for granted that the proper order of semantic explanation begins with a doctrine of concepts or terms, divided into singular and general, whose meaningfulness can be grasped independently of and prior to the meaningfulness of judgements. Appealing to this basic level of interpretation, a doctrine of judgements then explains the combination of concepts into judgements, and how the correctness of the resulting judgements depends on what is combined and how. Appealing to this derived interpretation of judgements, a doctrine of consequences finally explains the combination of judgements into inferences, and how the correctness of inferences depends on what is combined and how.

Kant rejects this. One of his cardinal innovations is the claim that the fundamental unit of awareness or cognition, the minimum graspable, is the *judgement*. Thus concepts can only be understood as abstractions, in terms of the role they play in judging. A concept just is a predicate of a possible judgement (Critique of Pure Reason, A69/B94) which is why

> *The only use which the understanding can make of concepts is to form judgements by them.*(Critique of Pure Reason, A68/B93).

For Kant, any discussion of content must start with the contents of judgements, since anything else only has content insofar as it contributes to the contents of judgements. This is why his transcendental logic can investigate the presuppositions of contentfulness in terms of the categories, that is, the 'functions of unity in judgement' (Critique of Pure Reason, A69/B94). This explanatory strategy is taken over by Frege, for whom the semantic notion of conceptual content ultimately has the theoretical task of explaining pragmatic *force*—the paradigmatic variety of which is *assertional* force, which attaches only to declarative sentences. As the later Wittgenstein puts the point, only the utterance of a sentence makes a move in the language game. Applying a concept is to be understood in terms of making a claim or expressing a belief. The concept 'concept' is not intelligible apart from the possibility of such application in *judging*.

The lesson is that the relation between designation and what is designated can only be understood as an aspect of judging or claiming *that* something (expressed by a declarative sentence, not by a singular term or predicate by itself) is so—i.e. is *true*. That is judging, believing, or claiming *that* a proposition or claim is true (expresses or states a fact), *that* something is *true of* an object or collection of objects, *that* a predicate is *true of* something else. Thus one must be concerned with what is said or expressed, as well as what it is said *of* or true *of*—the thought as well as what the thought is *about*.

### 3

Accordingly we start our story with an approach to propositional contents—what can be *said*, or *believed*, or *thought*, in general what can be *taken* (to be) *true*. The guiding idea is that the essential feature distinguishing what is propositionally contentful is that it can serve both as a premise and as the conclusion of *inferences*. Taking (to be) true is treating as a fit premise for inferences. This is exploiting Frege's semantic principle—that good inferences never lead from true premises to conclusions that are not true—not in order to define good inferences in terms of their preservation of truth, but rather to define truth as what is preserved by good inferences.

On the side of propositionally contentful *intentional states*, paradigmatically *belief*, the essential inferential articulation of the propositional is manifested in the form of intentional interpretation or explanation. Making behavior intelligible according to this model is taking the individual to act for *reasons*. This is what lies behind Dennett's slogan: 'Rationality is the mother of intention'. The role of belief in imputed pieces of practical reasoning, leading from beliefs and desires to the formation of intentions, is essential to intentional explanation—and so is reasoning in which both premise and conclusion have the form of believables.

On the side of propositionally contentful *speech acts*, paradigmatically assertion, the essential inferential articulation of the propositional is manifested in the fact that the core of specifically *linguistic* practice is the game of giving and asking for *reasons*. Claiming or asserting is what one must do in order to give a reason, and it is a speech act that reasons can be demanded for. Claims both serve as and stand in need of reasons or justifications. They have the contents they have in part in virtue of the role they play in a network of inferences.

Indeed, the *conceptual* should be distinguished precisely by its inferential articulation. This is a point on which traditional empiricism needed instruction by traditional rationalism. What is the difference between a parrot or a thermostat that represents a light as being red or a room as being cold by exercising its reliable differential responsive disposition to utter the noise 'That's red' or to turn on the furnace, on the one hand, and a knower who does so by applying the concepts *red* and *cold*, on the other? What is the knower able to *do* that the parrot and the thermostat cannot? After all, they may respond differentially to *just* the same range of stimuli. The knower is able to *use* the differentially elicited response in *inference*. The knower has the practical know-how to situate that response in a network of inferential relations—to tell what follows from something being red or cold, what would be evidence for it, what would be incompatible with it, and so on. For the knower, taking something to be red or cold is making a move in the game of giving and asking for reasons—a move that can justify other moves, be justified by still other moves, and that closes off or precludes still further moves. The parrot and the thermostat lack the concepts in spite of their mastery of the corresponding noninferential differential responsive dispositions, precisely because they lack the practical mastery of the inferential articulation in which grasp of conceptual content consists.

The idea, then, is to start with a story about the sayable, thinkable, believable (and so propositional) contents expressed by the use of declarative sentences and 'that' clauses derived from them—a story couched in terms of their roles in *inference*.[1] Conceptual content is in the first instance *inferentially* articulated. To approach the representational dimension of semantic content from this direction, it is necessary to ask about the relation between *inference* and *reference*. This is to ask about the relation between what is said or thought and what it is said or thought *about*. How can the representational dimension of conceptual content be brought into the inferential picture of propositional contents? The thesis to be elaborated here is that the representational dimension of discourse reflects the fact that conceptual content is not only *inferentially* articulated, but *socially* articulated. The game of giving and asking for reasons is an essentially *social* practice.

## 4

The rationale for such a claim emerges most clearly from consideration of certain very general features of discursive practice. Here it is useful to start with another of Kant's fundamental insights, into the *normative* character of the significance of what is conceptually contentful. His idea is that judgements and actions are above all things that we are *responsible* for. Kant understands concepts as having the form of *rules*, which is to say that they specify how something *ought* (according to the rule) to be done. The understanding, the conceptual faculty, is the faculty of grasping rules, of appreciating the distinction between correct and incorrect application they determine. Judgings and doings are acts that have contents that one can take or make true and for which the demand for reasons is in order. What is distinctive about them is the way they are governed by rules. Being in an intentional state or performing an intentional action has a normative significance. It counts as undertaking (acquiring) an obligation or commitment; the content of the commitment is determined by the rules that are the concepts in terms of which the act or state is articulated. Thus Kant picks us out as distinctively normative or rule-governed creatures.

Descartes inaugurated a new philosophical era by conceiving of what he took to be the ontological distinction between the mental and the physical in epistemological terms, in terms of accessibility to cognition, in terms, ultimately, of certainty. Kant launched a new philosophical epoch by shifting the center of concern from *certainty* to *necessity*. Where Descartes' descriptive conception of intentionality, centering on certainty, picks out as essential our grip on the concepts employed in cognition and action, Kant's normative conception of intentionality, centering on necessity, treats their grip on us as the heart of the matter. The attempt to understand the source, nature, and significance of the norms implicit in our concepts—both those that govern the theoretical employment of concepts in inquiry and knowledge and those that govern their practical employment in deliberation and action—stands at the very center of Kant's philosophical enterprise. The most urgent question for Kant is how to understand the *rulishness* of concepts, how to understand their *authority, bindingness*, or *validity*. It is this normative character that he calls 'Notwendigkeit', necessity.

The lesson to be learned from this Kantian normative conceptual pragmatics is that judging and acting are distinguished from other doings by the kind of *commitment* they involve. Judging or claiming is staking a claim—undertaking a commitment. The conceptual articulation of these commitments, their status as distinctively *discursive* commitments, consists in the way they are liable to demands for *justification*, and the way they serve both to justify some further commitments and to preclude the justification of some other commitments. Their propositional contentfulness consists precisely in this inferential articulation of commitments and entitlements to those commitments.

Specifically *linguistic* practices are those in which some performances are accorded the significance of assertions or claimings—the undertaking of inferentially articulated (and so propositionally contentful) commitments.[2] Mastering such linguistic practices is a matter of learning how to keep score on the inferentially articulated commitments and entitlements of various interlocutors, oneself included. Understanding a speech act—grasping its discursive significance—is being able to attribute the right commitments in response. This is knowing how it changes the score of what the performer and the audience are committed and entitled to.

One way of thinking about the claims by which discursive commitments are expressed is in terms of the interaction of inferentially articulated *authority* and *responsibility*. In making an assertion one lends to the asserted content one's *authority*, licensing others to undertake a corresponding commitment. Thus one essential aspect of this model of discursive practice is *com-*

*munication*: the interpersonal, intracontent inheritance of entitlement to commitments. In making an assertion one also undertakes a *responsibility*, to justify the claim if appropriately challenged, and thereby to redeem one's entitlement to the commitment acknowledged by the claiming. Thus another essential aspect of this model of discursive practice is *justification*: the intrapersonal, intercontent inheritance of entitlement to commitments.

## II

### 1

One can pick out what is *propositionally* contentful to begin with as whatever can serve both as a premise and as a conclusion in *inference*—what can be offered as and itself stand in need of *reasons*. Understanding or grasping such a propositional content is a kind of know-how—practical mastery of the game of giving and asking for reasons, being able to tell what is a reason for what, distinguish good reasons from bad. To play such a game is to keep *score* of what various interlocutors are committed and entitled to. Understanding the content of a speech act or a belief is being able to accord the performance of that speech act or the acquisition of that belief the proper practical significance—knowing how it would change the score in various contexts. Semantic, that is to begin with, inferential, relations are to be understood in terms of this sort of pragmatic scorekeeping. Taking it that the claim expressed by one sentence entails the claim expressed by another is treating anyone who is committed to the first as thereby committed to the second. We typically think about inference solely in terms of the relation between premise and conclusion, that is, as a monological relation among propositional contents. Discursive practice, the giving and asking for reasons, however, involves both inter*content* and inter*personal* relations. The claim is that the representational aspect of the propositional contents that play the inferential roles of premise and conclusion should be understood in terms of the social or dialogical dimension of communicating reasons, of assessing the significance of reasons offered by others.

If whatever plays a suitable role in inference is propositionally contentful, and whatever is propositionally contentful therefore also has representational content, then nothing can deserve to count as specifically *inferential* practice unless it at least implicitly involves a representational dimension. Nonetheless, one can give sufficient conditions for a social practice to qualify as according *inferentially articulated* significances to performances, that is, to be a practice of making claims that can serve as reasons for others, and for which reasons can be demanded, without using any specifically representational vocabulary. That is what the model of discursive practice as keeping score on commitments and entitlements does. The story I want to tell is then how the implicit representational dimension of the inferential contents of claims arises out of the difference in social perspective between *producers* and *consumers* of reasons. The aim is an account in nonrepresentational terms of what is expressed by the use of explicitly representational vocabulary.

The connection between *representation*, on the one hand, and *communication* or the *social* dimension of inferential practice, on the other, is sufficiently unobvious that I want to start with a quick point that may help show why one could so much as think that representation could be understood in these terms. The claim is that assessment of what people are talking and thinking *about*, rather than what they are saying about it, is a feature of the essentially *social* context of *communication*. Talk about representation is talk about what it is to secure communication by being able to use each other's judgements as reasons, as premises in our own inferences, even just hypothetically, to assess their significance in the context of our own collateral commitments.

One way to get a preliminary taste for how one could think that representational semantic talk could be understood as expressing differences in social perspective among interlocutors, consider how assessments of *truth* work. Perhaps the central context in which such assessments classically arise is attributions of *knowledge*. According to the traditional *JTB* account, knowledge is justified true belief. Transposed into a specification of a normative status something could be taken to have by interlocutors who are keeping score of each others commitments and entitlements, this account requires that in order for it to be *knowledge* that a scorekeeper takes another to have, that scorekeeper must adopt three sorts of practical attitude: First, the scorekeeper must *attribute* an inferentially articulated, hence propositionally contentful *commitment*. This corresponds to the *belief* condition on knowledge. Second, the scorekeeper must *attribute* a sort of inferential *entitlement* to that commitment. This corresponds to the *justification* condition on knowledge.

What is it that then corresponds to the *truth* condition on knowledge? For the scorekeeper to take the attributed claim to be true is just for the scorekeeper to endorse that claim. That is, the third condition is that the scorekeeper himself *undertake* the same commitment attributed to the candidate knower.

Undertaking a commitment is adopting a certain *normative stance* with respect to a claim; it is not attributing a property to it. The classical metaphysics of truth properties misconstrues what one is doing in endorsing the claim as *describing* in a special way. It confuses *attributing* and *undertaking* or *acknowledging* commitments, the two fundamental social flavors of deontic practical attitudes that institute normative statuses. It does so by assimilating the third condition on treating someone as having knowledge to the first two. Properly understanding truth talk in fact requires understanding just this difference of social perspective: between *attributing* a normative status to another, and *undertaking* or adopting it oneself.[3] It is the practice of assessing the truth of claims that underlies the idea that propositional contents can be understood in terms of truth conditions. What I want to do is to show how this idea of *truth* claims as expressing differences in social perspective can be extended to representation more generally.

## 2

The prime explicitly representational locution of natural languages is *de re ascriptions of propositional attitudes*. It is their use in these locutions that make the words 'of' and 'about' express the intentional directedness of thought and talk—as distinct from their use in such phrases as 'the pen of my aunt' and 'weighing about five pounds'. Thus in order to identify vocabulary in alien languages that means what 'of' and 'about' used in this sense do, one must find expressions of de re ascriptions of propositional attitudes. It is these ascriptions that we use to *say* what we are talking and thinking *about*. My strategy here is to address the question of how to understand what is expressed by representational vocabulary by asking how expressions must be *used* in order to qualify as *de re* ascriptions of propositional attitudes.

The tradition distinguishes two readings of or senses that can be associated with propositional attitude ascriptions. Ascriptions *de dicto* attribute belief in a *dictum* or saying, while ascriptions *de re* attribute belief about some *res* or thing. The distinction arises with sentential operators other than 'believes'; consider to begin with the claim:

*The President of the United States will be black by the year 2000.*

Read *de dicto*, this means that the dictum or sentence

*The President of the United States is black.*

will be true by the year 2000. Read *de re*, it means that the *res* or thing, the present President of the United States, namely Bill Clinton, will be black by the year 2000. Our concern here is with how this distinction applies to ascriptions of propositional attitude—though it is a criterion of adequacy on the account offered here that it can be extended to deal with these other contexts as well.

In ordinary parlance the distinction between *de dicto* and *de re* readings is the source of systematic ambiguity. Sometimes, as in the case above, one of the readings involves a sufficiently implausible claim that it is easy to disambiguate. It is best, however, to regiment our usage slightly in order to mark the distinction explicitly. This can be done with little strain to our ears by using 'that' and 'of' in a systematic way. Consider:

> *Henry Adams believed the popularizer of the lightning rod did not popularize the lightning rod.*

It is quite unlikely that what is intended is the *de dicto*

> *Henry Adams believed **that** the popularizer of the lightning rod did not popularize the lightning rod.*

Adams would presumably not have endorsed the *dictum* that follows the 'that'. It is entirely possible, however, that the *de re* claim

> *Henry Adams believed **of** the popularizer of the lightning rod **that** he did not popularize the lightning rod.*

is true. For since the popularizer of the lightning rod is the inventor of bifocals (namely Benjamin Franklin), this latter claim could be true if Henry Adams had the belief that would be ascribed *de dicto* as

> *Henry Adams believed **that** the inventor of bifocals did not popularize the lightning rod.*

Quine emphasizes that the key grammatical difference between these two sorts of ascriptions concerns the propriety of *substitution* for singular terms occurring in them. Expressions occurring in the *de re* portion of an ascription—within the scope of the 'of' operator in the regimented versions—have in his terminology *referentially transparent* uses: coreferential terms can be intersubstituted *salva veritate*, that is, without changing the truth value of the whole ascription. By contrast, such substitution in the *de dicto* portion of an ascription—within the scope of the 'that' operator in the regimented versions—may well change the truth value of the whole ascription. Syntactically, *de re* ascriptions may be thought of as formed from *de dicto* ones by *exporting* a singular term from within the 'that' clause, prefacing it with 'of', and putting a pronoun in the original position. Thus the *de dicto* form

*S believes that* $\Phi$ *(t),*

becomes the *de re*

*S believes of t that* $\Phi$ *(it).*

The significance of Quine's fundamental observation that the key difference between these two sorts of ascription lies in the circumstances under which the substitution of coreferential expressions is permitted was obscured by considerations that are from my point of view extraneous:

1. Quine's idiosyncratic view that singular terms are dispensable in favor of the quantificational expressions he takes to be the genuine locus of referential commitment leads him to look only at quantified ascriptions, embroils his discussion in issues of existential commitment, and diverts him into worries about when 'exportation' is legitimate.
2. This emphasis led in turn—Kaplan bears considerable responsibility here—to ignoring the analysis of ordinary de re ascriptions in favor of what I call *epistemically strong de re* ascriptions, which are used to attribute a privileged epistemic relation to the object talked or thought about. This detour had fruitful consequences for our appreciation of special features of the behavior of demonstratives (and as a result, of proper name tokenings anaphorically dependent on them), particularly in modal contexts. But from the point of view of understanding aboutness in general—my topic here—it was a detour and a distraction nonetheless.

The important point is, as the regimentation reminds us, that it is *de re* propositional attitude ascribing locutions that we use in everyday life to express what we are talking and thinking *of* or *about*. One way of trying to understand the representational dimension of propositional content is accordingly to ask what is expressed by this fundamental sort of representational locution. What are we *doing* when we make claims about what someone is talking or thinking *about*? How must vocabulary be used in order for it to deserve to count as expressing such *de re* ascriptions? Answering that question in a way that does not itself employ representational vocabulary in specifying that use is then a way of coming to understand representational relations in nonrepresentational terms.

3

The rest of this essay is about the expressive role of *de re* ascriptions. I'm going to present it in the technical vocabulary I prefer, which is in some ways idiosyncratic; but the basic point about the way the use of this paradigmatic representational locution expresses differences in social perspective does not depend on the details of that idiom.[4]

Recall that I think we should understand discursive practice in terms of the adoption of practical attitudes by which interlocutors keep score on each other's commitments (and entitlements to those commitments, but we can ignore them here). Claiming (and so, ultimately, judging) is *undertaking* or *acknowledging* a commitment that is propositionally contentful in virtue of its *inferential* articulation. The large task is to show what it is about that inferential articulation in virtue of which claimable contents are therefore also *representational* contents. This is to move from propositional contents introduced as potential premises and conclusions of inferences, via the social dimension of inferential articulation that consists of giving and asking for reasons of each other in communication, to propositions as talking of or about objects, and saying of them how they are. [I'll give short shrift here to the *objectivity* part of the claim, but think about how assessments of *truth* were presented above as distinct from assessments of *belief* and *justification*.]

*Undertaking* a commitment is doing something that makes it appropriate for others to *attribute* it. This can happen in two different ways. First, one may *acknowledge* the commitment, paradigmatically by being disposed to *avow* it by an overt assertion. Or one may acknowledge it by employing it as a premise in one's theoretical or practical reasoning. This last includes being disposed to *act* on it *practically*—taking account of it as a premise in the practical reasoning that stands behind one's intentional actions. Second, one may undertake the commitment *consequentially*, that is, as a conclusion one is committed to as an inferential consequence entailed by what one *does* acknowledge. These correspond to two senses of 'believe' that are often not distinguished: the sense in which one only believes what one takes oneself to believe, and the sense in which one believes willy nilly whatever one's beliefs commit one to. [The fact that people often move back and forth between belief in the empirical sense, which does not involve inferential closure, and belief in the logical or ideal sense that does, is one of the reasons that when being careful I prefer to talk in terms of commitments rather than beliefs—I don't officially believe in beliefs.] The second sense is the one in which if I believe Kant revered Hamann, and I believe Hamann was the Magus of the North, then whether the question has ever arisen for me or not, whether I know it or not, I in fact believe Kant revered the Magus of the North.

*Attributing* beliefs or commitments is a practical attitude that is *implicit* in the scorekeeping practices within which alone anything can have the significance of a claim or a judgement. *Ascribing* beliefs or commitments is making that *implicit* practical attitude *explicit* in the form of a claim. In a language without explicit attitude ascribing locutions such as the 'believes that' or 'claims that' operator, attributing commitments is something one can only *do*. Propositional attitude ascribing locutions make it possible explicitly to *say* that that is what one is doing: to express that practical deontic scorekeeping attitude as a propositional content—that is, as the content of a claim. In this form it can appear as a premise or conclusion of an inference; it becomes something which can be offered as a reason, and for which reasons can be demanded. The paradigm of the genus of *explicitating* vocabulary, of which propositional attitude ascribing locutions are a species, is the conditional. The use of conditionals makes explicit as the content of a claim, and so something one can *say*, the endorsement of an *inference*—an attitude one could otherwise only manifest by what one *does*. Ascriptional vocabulary such as 'believes' or 'claims' makes *attribution* of doxastic commitments explicit in the form of claimable contents.

## 4

In asserting an ascriptional claim of the form

*S believes (or is committed to the claim) that* $\Phi(t)$,

one is accordingly doing two things, adopting two different sorts of deontic attitude: one is *attributing* one doxastic commitment, to $\Phi(t)$, and one is *undertaking* another, namely a commitment to the ascription. The explicitating role of ascriptional locutions means that the content of the commitment one *undertakes* is to be understood in terms of what one is doing in *attributing* the first commitment.

The ascription above specifies the content of the commitment attributed by using an unmodified 'that' clause, which according to our regimentation corresponds to an ascription *de dicto*. A full telling of my story requires that quite a bit be said about how these ascriptions work, but I'm not going to do that here. Roughly, the ascriber who specifies the content of the attributed commitment in the *de dicto* way is committed to the target being prepared to *acknowledge* the attributed commitment in essentially the terms specified—that is, to endorse the *dictum*.[5]

I want to take an appropriate account of *de dicto* ascriptions of propositional attitudes for granted, and show what is different about *de re* ascriptions, those that are regimented in the form:

> *S claims of t that $\Phi$ (it).*

I think that the beginning of wisdom in this area is the realization that (once what I have called 'epistemically strong de re ascriptions' have been put to one side) the distinction between *de dicto* and *de re* should not be understood to distinguish two kinds of *belief* or belief-contents, but two kinds of *ascription*—in particular two different *styles* in which the *content* of the commitment ascribed can be *specified*.[6] (Dennett is perhaps the most prominent commentator who has taken this line. See Dennett, 1982.)

In specifying the content of the claim that is attributed by an ascription, a question can arise as to who the ascriber takes to be responsible for this being a way of *saying* (that is, making explicit) what is believed, the content of the commitment. Consider the sly prosecutor, who characterizes his opponent's claim by saying:

> *The defense attorney believes a pathological liar is a trustworthy witness.*

We can imagine that the defense attorney hotly contests this characterization:

> *Not so; what I believe is that the man who just testified is a trustworthy witness.*

To which the prosecutor might reply:

> *Exactly, and I have presented evidence that ought to convince anyone that the man who just testified is a pathological liar.*

If the prosecutor were being fastidious in characterizing the other's claim, he would make it clear who is responsible for what: the defense attorney claims that a certain man is a trustworthy witness, and the prosecutor claims that that man is a pathological liar. The disagreement is about whether this guy is a liar, not about whether liars make trustworthy witnesses. Using the regimentation suggested above, the way to make this explicit is with a *de re* specification of the content of the belief ascribed. What the prosecutor *ought* to say (matters of courtroom strategy aside) is:

> *The defense attorney claims of a pathological liar that he is a trustworthy witness.*

This way of putting things makes explicit the division of responsibility involved in the ascription. That someone is a trustworthy witness is part of the commitment that is *attributed* by the ascriber, that that individual is in fact a pathological liar is part of the commitment that is *undertaken* by the ascriber.

Ascription always involves attributing one doxastic commitment and, since ascriptions are themselves claims or judgements, undertaking another. My suggestion is that the expressive function of *de re* ascriptions of propositional attitude is to make explicit which aspects of what is said express commitments that are being *attributed* and which express commitments that are *undertaken*. The part of the content specification that appears within the *de dicto* 'that' clause is limited to what, according to the ascriber, the one to whom the commitment is ascribed would (or in a strong sense should) *acknowledge* as an expression of what that individual is committed to. The part of the content specification that appears within the scope of the *de re* 'of' includes what, according to the *ascriber* of the commitment (but not necessarily according to the one to whom it is ascribed) is acknowledged as an expression of what the target of the ascription is committed to. (This is what the target should, according to the ascriber, acknowledge only in a much weaker sense of 'should'.) Thus the marking of portions of the content-specification of a propositional attitude ascription into *de dicto* and *de re* portions makes explicit the essential deontic scorekeeping distinction of *social* perspective between commitments attributed and those undertaken.

## 5

The difference expressed by segregating the content specification of a propositional attitude ascription into distinct *de re* and *de dicto* regions, marked in our regimentation by 'of' and 'that', can be thought of in terms of *inferen-*

*tial* and *substitutional* commitments. According to the model I started with, propositional, that is, assertible, contents are inferentially articulated. Grasping such a content is being able to distinguish in practice what should follow from endorsing it, and what such endorsement should follow from. But the consequences of endorsing a given claim depends on what other commitments are available to be employed as auxiliary hypotheses in the inference. The ascriber of a doxastic commitment has got two different perspectives available from which to draw those auxiliary hypotheses in specifying the content of the commitment being ascribed: that of the one to whom it is *ascribed* and that of the one *ascribing* it. Where the specification of the content depends only on auxiliary premises that (according to the ascriber) the target of the ascription *acknowledges* being committed to, though the ascriber may not, it is put in *de dicto* position, within the 'that' clause. Where the specification of the content depends on auxiliary premises that the *ascriber* endorses, but the target of the ascription may not, it is put in *de re* position.

More particularly, the use of expressions as singular terms is governed by *substitution*-inferential commitments.[7] The rule for determining the scorekeeping significance and so the expressive function of *de re* ascriptions that I am proposing is then the following. Suppose that according to *A*'s scorekeeping on commitments, *B* acknowledges commitment to the claim $\Phi(t)$. Then *A* can make this attribution of commitment explicit in the form of a claim by saying

*B claims* ***that*** $\Phi(t)$.

If in addition *A* acknowledges commitment to the identity $t=t'$, then whether or not *A* takes it that *B* would acknowledge that commitment, *A* can also characterize the content of the commitment ascribed to *B* by saying

*B claims of t' that* $\Phi(it)$.

Again, the question just is whose substitutional commitments one is permitted to appeal to in specifying the consequences someone is committed to by acknowledging a particular doxastic commitment. Where in characterizing the commitment the ascriber has exfoliated those consequences employing only commitments the ascriptional target would acknowledge, the content specification is *de dicto*. Where the ascriber has employed substitutional commitments he himself, but perhaps not the target, endorses, the content specification is *de re*.

Understood in this way, what is expressed by *de re* specifications of the contents of the beliefs of others are crucial to *communication.* Being able to understand what others are saying, in the sense that makes their remarks available for use as premises in one's own inferences, depends precisely on being able to specify those contents in *de re*, and not merely *de dicto* terms. If the only way I can specify the content of the shaman's belief is by a de dicto ascription

> *He believes malaria can be prevented by drinking the liquor distilled from the bark of that kind of tree,*

I may not be in a position to assess the truth of his claim. It is otherwise if I can specify that content in the *de re* ascription

> *He believes of quinine that malaria can be prevented by drinking it,*

for 'quinine' is a term with rich inferential connections to others I know how to employ. If he says that the seventh god has just risen, I may not know what to make of his remark. Clearly he will take it to have consequences that I could not endorse, so nothing in my mouth could *mean* just what his remark does. But if I am told that the seventh god is the sun, then I can specify the content of his report in a more useful form:

> *He claims of the sun that it has just risen,*

which I can extract *information* from, that is, can usc to generate premises that I can reason with. Again, suppose a student claims that

> *The largest number that is not the sum of the squares of distinct primes is the sum of at most 27 primes.*

He may have no idea what that number is, or may falsely believe it to be extremely large, but if I know that

> *17163 is the largest number that is not the sum of the squares of distinct primes,*

then I can characterize the content of his claim in *de re* form as:

> *The student claims of 17163 that it is the sum of at most 27 primes,*

and can go on to draw inferences from that claim, to assess its plausibility in the light of the rest of my beliefs. (It is true, but only because *all* integers are the sum of at most 27 primes.) Identifying what is being talked about permits me to extract information across a doxastic gap.

We saw originally in the treatment of truth assessments the crucial difference between *attributing* a commitment and *undertaking* or acknowledging one. We now see what is involved in moving from the claim that

> *It is true that Benjamin Franklin invented bifocals,*

which is the undertaking of a commitment to the effect that Benjamin Franklin invented bifocals, via the undertaking of a commitment to the claim that Benjamin Franklin is the popularizer of the lightning rod, to the claim that

> *It is true of the popularizer of the lightning rod that he invented bifocals.*

(It is through this 'true of' locution that the earlier remarks about the essentially social structure of truth assessments connects with the account just offered of the social structure that underlies propositional attitude ascriptions *de re*.) Extracting information from the remarks of others requires grasping what is expressed when one offers *de re* characterizations of the contents of their beliefs—that is to be able to tell what their beliefs would be true *of* if they were true. It is to grasp the *representational* content of their claims. The point I have been making is that doing this is just mastering the *social* dimension of their inferential articulation.

### Conclusion

I have claimed that the primary representational locution in ordinary language, the one we use to talk about the representational dimension of our thought and talk, to specify what we are thinking and talking *about*, is *de re* ascriptions of propositional attitude. It is the role they play in such ascriptions that gives their meanings to the 'of' or 'about' we use to express intentional directedness. I have also claimed that the expressive role of these locutions is to make explicit the distinction of social perspective involved in keeping our books straight on who is committed to what. The social dimension of inference involved in the communication to others of claims that must be available as reasons both to the speaker and to the audience, in spite of differences in collateral commitments, is what underlies the representational dimension of discourse.

Beliefs and claims that are *propositionally* contentful are necessarily *representationally* contentful because their inferential articulation essentially involves a *social* dimension. That social dimension is unavoidable because the inferential significance of a claim, the appropriate antecedents and consequences of a doxastic commitment, depend on the background of collateral commitments available for service as auxiliary hypotheses. Thus any specification of a propositional content must be made from the perspective of some such set of commitments. One wants to say that the *correct* inferential role is determined by the collateral claims that are *true*. Just so; that is what *each* interlocutor wants to say—each has an at least slightly different perspective from which to evaluate inferential proprieties. Representational locutions make explicit the sorting of commitments into those attributed and those undertaken—without which communication would be impossible, given those differences of perspective. The *representational* dimension of propositional contents reflects the *social* structure of their *inferential* articulation in the game of giving and asking for reasons.

*University of Pittsburgh*

## NOTES

[1] This idea is motivated and explored at greater length in Brandom (1988).

[2] By this criterion, the 'Slab' Sprachspiel that Wittgenstein describes early in the *Investigations*, for instance, does not qualify as a genuinely *linguistic* practice. For further discussion of why this is a good way to talk, see Brandom (1983).

[3] There are a myriad of technical details that need to be cleared up in order to make an analysis of truth talk along these lines work. I've addressed those difficulties elsewhere—that is where the prosentential or anaphoric account of truth comes in. See Brandom (1988A). For present purposes, those details can be put to one side.

[4] The approach pursued here (including both a treatment of *de dicto* ascriptions, and of epistemically strong *de re* ascriptions) is presented at length in my *Making It Explicit: Reasoning, Representing, and Discursive Commitment*, forthcoming, especially in Chapter Eight 'The Social Route From Reasoning to Representing'.

[5] Obviously, such an account requires emendation to handle the cases where the one to whom a propositional attitude is ascribed would use indexicals, or a different language, to express that attitude. See *Making It Explicit*, Chapter Eight.

[6] One way to see that this is right is that the ascription-forming operators can be *iterated*: S' can claim of t that S claims of it that Φ(it). Thus there would in any case not be *two* different kinds of

belief (*de dicto* and *de re*), but an infinite number.

[7] This line of thought is worked out in detail in 'What Are Singular Terms, and Why Are There Any?', which is Chapter Six of *Making It Explicit.*

## REFERENCES

Brandom, R. (1983), 'Asserting', *Nous*, XVII #4 (November), pp. 637-650.

Brandom, R. (1988), 'Inference, Expression, and Induction: Sellarsian Themes', *Philosophical Studies* 54, pp. 257-285.

Brandom, R. (1988A), 'Pragmatism, Phenomenalism, and Truth Talk', *Midwest Studies in Philosophy vol. XII: Realism* (University of Minnesota Press), pp. 75-93.

Brandom, R. (forthcoming), *Making It Explicit: Reasoning, Representing and Discursive Commitment* (Harvard University Press).

Dennett. D. (1982), 'Beyond Belief' in A. Woodfield (ed.), *Thought and Object* (Oxford University Press).

Kant, I. *Critique of Pure Reason.*

JOHN R. SEARLE

# THE PROBLEM OF CONSCIOUSNESS[1]

The most important scientific discovery of the present era will come when someone — or some group — discovers the answer to the following question: How exactly do neurobiological processes in the brain cause consciousness? This is the most important question facing us in the biological sciences, yet it is frequently evaded, and frequently misunderstood when not evaded. In order to clear the way for an understanding of this problem. I am going to begin to answer four questions: 1. What is consciousness? 2. What is the relation of consciousness to the brain? 3. What are some of the features that an empirical theory of consciousness should try to explain? 4. What are some common mistakes to avoid?

## *I. What is consciousness?*

Like most words, 'consciousness' does not admit of a definition in terms of genus and differentia or necessary and sufficient conditions. Nonetheless, it is important to say exactly what we are talking about because the phenomenon of consciousness that we are interested in needs to be distinguished from certain other phenomena such as attention, knowledge, and self-consciousness. By 'consciousness' I simply mean those subjective states of sentience or awareness that begin when one awakes in the morning from a dreamless sleep and continue throughout the day until one goes to sleep at night or falls into a coma, or dies, or otherwise becomes, as one would say, 'unconscious'.

Above all, consciousness is a biological phenomenon. We should think of consciousness as part of our ordinary biological history, along with digestion, growth, mitosis and meiosis. However, though consciousness is a biological phenomenon, it has some important features that other biological phenomena do not have. The most important of these is what I have called its 'subjectivity'. There is a sense in which each person's consciousness is private to that person, a sense in which he is related to his pains, tickles, itches, thoughts and feelings in a way that is quite unlike the way that others are related to those pains, tickles, itches, thoughts and feelings. This phenomenon can be described in various ways. It is sometimes described as that feature of consciousness by way of which there is something that it's like or something that it feels like to be in a certain conscious state. If somebody asks me what it

*Michaelis Michael and John O'Leary-Hawthorne (eds.), Philosophy in Mind, pp.179-190.*
*Kluwer Academic Publishers, 1994,* 

feels like to give a lecture in front of a large audience I can answer that question. But if somebody asks what it feels like to be a shingle or a stone, there is no answer to that question because shingles and stones are not conscious. The point is also put by saying that conscious states have a certain qualitative character; the states in question are sometimes described as 'qualia'.

In spite of its etymology, consciousness should not be confused with knowledge, it should not be confused with attention, and it should not be confused with self-consciousness. I will consider each of these confusions in turn.

Many states of consciousness have little or nothing to do with knowledge. Conscious states of undirected anxiety or nervousness, for example, have no essential connection with knowledge. Consciousness should not be confused with attention. Within one's field of consciousness there are certain elements that are at the focus of one's attention and certain others that are at the periphery of consciousness. It is important to emphasize this distinction because 'to be conscious of' is sometimes used to mean 'to pay attention to'. But the sense of consciousness that we are discussing here allows for the possibility that there are many things on the periphery of one's consciousness — for example, a slight headache I now feel or the feeling of the shirt collar against my neck — which are not at the centre of one's attention. I will have more to say about the distinction between the center and the periphery of consciousness in Section III.

Finally, consciousness should not be confused with self-consciousness. There are indeed certain types of animals, such as humans, that are capable of extremely complicated forms of self-referential consciousness which would normally be described as self-consciousness. For example, I think conscious feelings of shame require that the agent be conscious of himself or herself. But seeing an object or hearing a sound, for example, does not require self-consciousness. And it is not generally the case that all conscious states are also self-conscious.

### *II. What are the relations between consciousness and the brain?*

This question is the famous 'mind-body problem'. Though it has a long and sordid history in both philosophy and science, I think, in broad outline at least, it has a rather simple solution. Here it is: Conscious states are caused by lower level neurobiological processes in the brain and are themselves higher level features of the brain. The key notions here are those of *cause* and *feature*. As far as we know anything about how the world works, variable rates of neuron firings in different neuronal architectures cause all the enormous variety of our conscious life. All the stimuli we receive from the external world are converted by the nervous system into one medium, namely, variable rates of neuron firings at synapses. And equally remarkably, these variable rates of neuron firings cause all of the colour and variety of our conscious life. The smell of the flower, the sound of the symphony, the thoughts of theorems in Euclidian geometry — all are caused by lower level biological processes in the brain; and as far as we know, the crucial functional elements are neurons and synapses.

Of course, like any causal hypothesis this one is tentative. It might turn out that we have overestimated the importance of the neuron and the synapse. Perhaps the functional unit is a column or a whole array of neurons, but the crucial point I am trying to make now is that we are looking for causal relationships. The first step in the solution of the mind-body problem is: brain processes *cause* conscious processes.

This leaves us with the question, what is the ontology, what is the form of existence, of these conscious processes? More pointedly, does the claim that there is a causal relation between brain and consciousness commit us to a dualism of 'physical' things and 'mental' things? The answer is a definite no. Brain processes cause consciousness but the consciousness they cause is not some extra substance or entity. It is just a higher level feature of the whole system. The two crucial relationships between consciousness and the brain, then, can be summarized as follows: lower level neuronal processes in the brain cause consciousness and consciousness is simply a higher level feature of the system that is made up of the lower level neuronal elements.

There are many examples in nature where a higher level feature of a system is caused by lower level elements of that system, even though the feature is a feature of the system made up of those elements. Think of the liquidity of water or the transparency of glass or the solidity of a table, for example. Of course, like all analogies these analogies are imperfect and inadequate in various ways. But the important thing that I am trying to get across is this: there is no metaphysical obstacle, no logical obstacle, to claiming that the relationship between brain and consciousness is one of causation and at the same time claiming that consciousness is just a feature of the brain. Lower level elements of a system can cause higher level features of that system, even though those features are features of a system made up of the lower level elements. Notice, for example, that just as one cannot reach into a glass of water and pick out a molecule and say 'This one is wet', so, one cannot point to a single synapse or neuron in the brain and say 'This one is thinking about my grandmother'. As far as we know anything about it, thoughts about grandmothers occur at a much higher level than that of the single neuron or synapse, just as liquidity occurs at a much higher level than that of single molecules.

Of all the theses that I am advancing in this article, this one arouses the most opposition. I am puzzled as to why there should be so much opposition, so I want to clarify a bit further what the issues are: First, I want to argue that we simply know as a matter of fact that brain processes cause conscious states. We don't know the details about how it works and it may well be a long time before we understand the details involved. Furthermore, it seems to me an understanding of how exactly brain processes cause conscious states may require a revolution in neurobiology. Given our present explanatory apparatus, it is not at all obvious how, within that apparatus, we can account for the causal character of the relation between neuron firings and conscious states. But, at present, from the fact that we do not know *how* it occurs, it does not follow that we do not know *that* it occurs. Many people who object to my solution (or dissolution) of the mind-body problem, object on the grounds that we have no idea how neurobiological processes could cause conscious phenomena. But that does not seem to me a conceptual or logical problem. That is an empirical/theoretical issue for the biological sciences. The problem is to figure out exactly how the system works to produce consciousness, and since we know that in fact it does produce consciousness, we have good reason to suppose that are specific neurobiological mechanisms by way of which it works.

There are certain philosophical moods we sometimes get into when it seems absolutely astounding that consciousness could be produced by electrobiochemical processes, and it seems almost impossible that we would ever be able to explain it in neurobiological terms. Whenever we get in such moods, however, it is important to remind ourselves that similar mysteries have occurred before in science. A century ago it seemed extremely mysterious, puzzling, and to some people metaphysically impossible that life should be accounted for in terms of mechanical, biological, chemical processes. But now we know that we can give such an account, and the problem of how life arises from biochemistry has been solved to the point that we find it difficult to recover, difficult to understand why it seemed such an impossibility at one time. Earlier still, electromagnetism seemed mysterious. On a Newtonian conception of the universe there seemed to be no place for the phenomenon of electromagnetism. But with the development of the theory of electromagnetism, the metaphysical worry dissolved. I believe that we are having a similar problem about consciousness now. But once we recognize the fact that conscious states are caused by neurobiological processes, we automatically convert the issue into one for theoretical scientific investigation. We have removed it from the realm of philosophical or metaphysical impossibility.

## *III. Some Features of Consciousness*

The next step in our discussion is to list some (not all) of the essential features of consciousness which an empirical theory of the brain should be able to explain.

### Subjectivity

As I mentioned earlier, this is the most important feature. A theory of consciousness needs to explain how a set of neurobiological processes can cause a system to be in a subjective state of sentience or awareness. This phenomenon is unlike anything else in biology, and in a sense it is one of the most amazing features of nature. We resist accepting subjectivity as a ground floor, irreducible phenomenon of nature because, since the seventeenth century, we have come to believe that science must be objective. But this involves a pun on the notion of objectivity. We are confusing the *epistemic* objectivity of scientific investigation with the *ontological* objectivity of the typical subject matter in science in disciplines such as physics and chemistry. Since science aims at objectivity in the epistemic sense that we seek truths that are not dependent on the particular point of view of this or that investigator, it has been tempting to conclude that the reality investigated by science must be objective in the sense of existing independently of the experiences in the human individual. But this last feature, ontological objectivity, is not an essential trait of science. If science is supposed to give an account of how the world works and if subjective states of consciousness are part of the world, then we should seek an (epistemically) objective account of an (ontologically) subjective reality, the reality of subjective states of consciousness. What I am arguing here is that we can have an epistemically objective science of a domain that is ontologically subjective.

### Unity

It is important to recognize that in non-pathological forms of consciousness we never just have, for example, a pain in the elbow, a feeling of warmth, or an experience of seeing something red, but we have them all occurring simultaneously as part of one unified conscious experience. Kant called this feature 'the transcendental unity of apperception'. Recently, in neurobiology it has been called 'the binding problem'. There are at least two aspects to this unity that require special mention. First, at any given instant all of our experiences are unified into a single conscious field. Second, the organization of our consciousness extends over more than simple instants. So, for example, if I begin speaking a sentence, I have to maintain in some sense at least an iconic memory of the beginning of the sentence so that I know what I am saying by the time I get to the end of the sentence.

### Intentionality

'Intentionality' is the name that philosophers and psychologists give to that feature of many of our mental states by which they are directed at, or about states of affairs in the world. If I have a belief or a desire or a fear, there must always be some content to my belief, desire or fear. It must be about something even if the something it is about does not exist or is a hallucination. Even in cases when I am radically mistaken, there must be some mental content which purports to make reference to the world. Not all conscious states have intentionality in this sense. For example, there are states of anxiety or depression where one is not anxious or depressed about anything in particular but just is in a bad mood. That is not an intentional state. But if one is depressed about a forthcoming event, that is an intentional state because it is directed at something beyond itself.

There is a conceptual connection between consciousness and intentionality in the following respect. Though many, indeed most, of our intentional states at any given point are unconscious, nonetheless, in order for an unconscious intentional state to be genuinely an intentional state it must be accessible in principle to consciousness. It must be the sort of thing that could be conscious even if it, in fact, is blocked by repression, brain lesion, or sheer forgetfulness.

### The distinction between the center and the periphery of consciousness

At any given moment of non-pathological consciousness I have what might be called a field of consciousness. Within that field I normally pay attention to some things and not to others. So, for example, right now I am paying attention to the problem of describing consciousness but very little attention to the feeling of the shirt on my back or the tightness of my shoes. It is sometimes said that I am unconscious of these. But that is a mistake. The proof that they are a part of my conscious field is that I can at any moment shift my attention to them. But in order for me to shift my attention to them, there must be something there which I was previously not paying attention to which I am now paying attention to.

### The gestalt structure of conscious experience.

Within the field of consciousness our experiences are characteristically structured in a way that goes beyond the structure of the actual stimulus. This was one of the most profound discoveries of the Gestalt psychologists. It is most obvious in the case of vision, but the phenomenon is quite general and extends beyond vision. For example, the sketchy lines drawn in Fig. 1 do not physically resemble a human face.

Figure 1.

If we actually saw someone on the street that looked like that, we would be inclined to call an ambulance. The disposition of the brain to structure degenerate stimuli into certain structured forms is so powerful that we will naturally tend to see this as a human face. Furthermore, not only do we have our conscious experiences in certain structures, but we tend also to have them as figures against backgrounds. Again, this is most obvious in the case of vision. Thus, when I look at the figure I see it against the background of the page. I see the page against the background of the table. I see the table against the background of the floor, and I see the floor against the background of the room, until we eventually reach the horizon of my visual consciousness.

### The aspect of familiarity

It is a characteristic feature of non-pathological states of consciousness that they come to us with what I will call the 'aspect of familiarity'. In order for me to see the objects in front of me as, for example, houses, chairs, people, tables, I have to have a prior possession of the categories of houses, chairs, people, tables. But that means that I will assimilate my experiences into a set of categories which are more or less familiar to me. When I am in an extremely strange environment, in a jungle village, for example, and the houses, people and foliage look very exotic to me, I still perceive that as a house, that as a person, that as clothing, that as a tree or a bush. The aspect of familiarity is thus a scalar phenomenon. There can be greater or lesser degrees of familiarity. But it is important to see that non-pathological forms of consciousness come to us under the aspect of familiarity. Again, one way to consider this is to look at the pathological cases. In Capgras's syndrome, the patients are unable to acknowledge familiar people in their environment as the people they actually are. They think the spouse is not really their spouse but is an imposter, etc. This is a case of a breakdown in one aspect of familiarity. In non-pathological cases it is extremely difficult to break with the aspect of familiarity. Surrealist painters try to do it. But even in the surrealist painting, the three-headed woman is still a woman, and the drooping watch is still a watch.

### Mood

Part of every normal conscious experience is the mood that pervades the experience. It need not be a mood that has a particular name to it, like depression or elation; but there is always what one might call a flavour or tone to any normal set of conscious states. So, for example, at present I am not especially depressed and I am not especially ecstatic, nor indeed, am I what one would call simply 'blah'. Nonetheless, there is a certain mood to my present experiences. Mood is probably more easily explainable in biochemical terms than several of the features I have mentioned. We may be able to control, for example, pathological forms of depression by mood-altering drugs.

**Boundary conditions**

All of my non-pathological states of consciousness come to me with a certain sense of what one might call their 'situatedness'. Though I am not thinking about it, and though it is not part of the field of my consciousness, I nonetheless know what year it is, what place I am in, what time of day it is, the season of the year it is, and usually even what month it is. All of these are the boundary conditions or the situatedness of nonpathological conscious states. Again, one can become aware of the pervasiveness of this phenomenon when it is absent. So, for example, as one gets older there is a certain feeling of vertigo that comes over one when one loses a sense of what time of year it is or what month it is. The point I am making now is that conscious states are situated and they are experienced as situated even though the details of the situation need not be part of the content of the conscious states.

## *IV. Some Common Mistakes about Consciousness*

I would like to think that everything I have said so far is just a form of common sense. However, I have to report, from the battlefronts as it were, that the approach I am advocating to the study of consciousness is by no means universally accepted in cognitive science nor even neurobiology. Indeed, until quite recently many workers in cognitive science and neurobiology regarded the study of consciousness as somehow out of bounds for their disciplines. They thought that it was beyond the reach of science to explain why warm things feel warm to us or why red things look red to us. I think, on the contrary, that it is precisely the task of neurobiology to explain these and other questions about consciousness. Why would anyone think otherwise? Well, there are complex historical reasons, going back at least to the seventeenth century, why people thought that consciousness was not part of the material world. A kind of residual dualism prevented people from treating consciousness as a biological phenomenon like any other. However, I am not now going to attempt to trace this history. Instead I am going to point out some common mistakes that occur when people refuse to address consciousness on its own terms.

The characteristic mistake in the study of consciousness is to ignore its essential subjectivity and to try to treat it as if it were an objective third person phenomenon. Instead of recognizing that consciousness is essentially a subjective, qualitative phenomenon, many people mistakenly suppose that its essence is that of a control mechanism or a certain kind of set of dispositions to behavior or a computer program. The two most common mistakes about consciousness are to suppose that it can be analysed behavioristically or computationally. The Turing test disposes us to make precisely these two mistakes, the mistake of behaviorism and the mistake of computationalism. It leads us to suppose that for a system to be conscious, it is both necessary and sufficient that it has the right computer program or set of programs with the right inputs and outputs. I think you have only to state this position clearly to enable you to see that it must be mistaken. A traditional objection to behaviorism was that behaviorism could not be right because a system could behave as if it were conscious without actually being conscious. There is no logical connection, no necessary connection between inner, subjective, qualitative mental states and external, publicly observable behavior. Of course, in actual fact, conscious states characteristically cause behavior. But the behavior that they cause has to be distinguished from the states themselves. The same mistake is repeated by computational accounts of consciousness. Just as behavior by itself is not sufficient for consciousness, so computational models of consciousness are not sufficient by themselves for consciousness. The computational model of consciousness stands to consciousness in the same way the computational model of anything stands to the domain being modelled. Nobody supposes that the computational model of rainstorms in London will leave us all wet. But they make the mistake of supposing that the computational model of consciousness is somehow conscious. It is the same mistake in both cases.

There is a simple demonstration that the computational model of consciousness is not sufficient for consciousness. I have given it many times before so I will not dwell on it here. Its point is simply this: *Computation is defined syntactically*. It is defined in terms of the manipulation of symbols. But the syntax by itself can never be sufficient for the sort of contents that characteristically go with conscious thoughts. Just having zeros and ones by themselves is insufficient to guarantee mental content, conscious or unconscious. This argument is sometimes called 'the Chinese room argument' because I originally illustrated the point with the example of the person who goes through the computational steps for answering questions in Chinese but does not thereby acquire any understanding of Chinese (Searle, 1980). The point of the parable is clear but it is usually neglected. *Syntax by itself is not sufficient for semantic content.* In all of the attacks on the Chinese room argument, I have never seen anyone come out baldly and say they think that syntax is sufficient for semantic content.

However, I now have to say that I was conceding too much in my earlier statements of this argument. I was conceding that the computational theory of the mind was at least false. But it now seems to me that it does not reach the level of falsity because it does not have a clear sense. Here is why.

The natural sciences describe features of reality that are intrinsic to the world as it exists independently of any observers. Thus, gravitational attraction, photosynthesis, and electromagnetism are all subjects of the natural sciences because they describe intrinsic features of reality. But such features such as being a bathtub, being a nice day for a picnic, being a five dollar bill or being a chair, are not subjects of the natural sciences because they are not intrinsic features of reality. All the phenomena I named — bathtubs, etc. — are physical objects and as physical objects have features that are intrinsic to reality. But the feature of being a bathtub or a five dollar bill exists only relative to observers and users.

Absolutely essential, then, to understanding the nature of the natural sciences is the distinction between those features of reality that are intrinsic and those that are observer-relative. Gravitational attraction is intrinsic. Being a five dollar bill is observer-relative. Now, the really deep objection to computational theories of the mind can be stated quite clearly. Computation does not name an intrinsic feature of reality but is observer-relative and this is because computation is defined in terms of symbol manipulation, but the notion of a 'symbol' is not a notion of physics or chemistry. Something is a symbol only if it is used, treated or regarded as a symbol. The Chinese room argument showed that semantics is not intrinsic to syntax. But what this argument shows is that syntax is not intrinsic to physics. There are no purely physical properties that zeros and ones or symbols in general have that determine that they are symbols. Something is a symbol only relative to some observer, user or agent who assigns a symbolic interpretation to it. So the question, 'Is consciousness a computer program?', lacks a clear sense. If it asks, 'Can you assign a computational interpretation to those brain processes which are characteristic of consciousness?' the answer is: you can assign a computational interpretation to anything. But if the question asks, 'Is consciousness intrinsically computational?' the answer is: nothing is intrinsically computational. Computation exists only relative to some agent or observer who imposes a computational interpretation on some phenomenon. This is an obvious point. I should have seen it ten years ago but I did not. .

*University of California,*
*Berkeley*

## NOTES

[1] This paper was originally published in *Experimental and Theoretical Studies of Consciousness*, CIBA Foundation Symposium 174 (Wiley, Chichester, 1993), p. 61-80. It is reprinted here with the kind permission of the CIBA Foundation. The theses advanced in this paper are presented in more detail and with more supporting argument in John Searle (1992).

## REFERENCES

Searle, J.R. (1980), 'Minds, Brains, and Programs', *Behavioral and Brain Sciences* 3, 417-57.
Searle, J.R. (1992), *The Rediscovery of the Mind* (Cambridge, Mass: MIT Press, 1992).

GRAHAM PRIEST

# GÖDEL'S THEOREM AND THE MIND ... AGAIN

## *0 Introduction*[1]

Attempts to establish, a priori, the nature and structure of the mind go back to the very origins of philosophy, and have continued apace ever since. The things on the basis of which, people have tried to establish this are many and varied: the phenomenology of consciousness, the nature of action, end even the nature of the state (to take but a few examples). Logic, as traditionally (though incorrectly) conceived, is precisely a compendium of the laws of thought. Hence it is unsurprising that some philosophers have tried to use logic as a basis in the inquiry. The high point of this enterprise was undoubtedly Kant and his successors. Kant thought that the structure of the mind could be read off from the categories of logic - which was, for him, the somewhat bowdlerised form of Aristotelian Logic of his day; and under Hegel, not only the number of categories blossomed, but so did the nature of the mind in question; until it came to encompass everything - literally.

The generally accepted failure of the enterprise of German Idealism has made modern philosophers a little wary of trying to deduce the nature of the mind from logic. But though their aims may have become more modest, the enterprise has continued. The boldest modern move in this direction is undoubtedly the attempt to refute mechanism - by which I mean, here, the view that the mind is essentially a computer - by an appeal to Gödel's first Incompleteness Theorem. Though various philosophers have been tempted by this thought, the locus classicus of the argument is the (in)famous paper 'Minds Machines and Gödel' by J.R.Lucas (1961). In this paper, Lucas argued that a mind will always be able to prove mathematical results that a (computational) machine cannot, and so cannot be such a machine.

Over the last 30 years the paper has occasioned a considerable literature[2] mainly critical; and one might have thought the issue closed. But at the Turing conference at the University of Sussex in 1990 in a paper entitled 'Minds, Machines and Gödel: a retrospect' Lucas still stoutly defended his argument against the literature. Moreover, Roger Penrose has recycled the argument as the dialectical centre-piece of his book *The Emperor's New Mind* (1990), which has stirred up the hornet's nest of critics again.[3] Since the argument has refused to lie down and die quietly, I think it worth thinking through it again

*Michaelis Michael and John O'Leary-Hawthorne (eds.), Philosophy in Mind, pp.191-201.*

from first principles. That is the purpose of this paper.

I will start with what I take to be a fair statement of Lucas' argument; I will then evaluate it. In the process, this will require a number of important clarifications, in particular, over what it is to give a proof. Both the original paper and many subsequent discussions are flawed by confusions concerning this notion. I do not intend this to be a scholarly paper, and so will not discuss the large literature. In fact, many of the standard objections to the argument can be seen to fail, or to be beside the point, once the argument is spelled out carefully. On the other hand, I make no claim to originality for a number of the points I shall make. What originality this paper has, it seems to me, is in spelling out the argument more clearly than is commonly done, with the attendant benefits of this.

## *1 Lucas' Argument*

So, what is Lucas' argument? It can be put very simply as follows. Take any mind, M, and computer, C:

> There is a mathematical truth of which C cannot give a proof but of which M can. Hence M is not C.

The argument is an instance of the indiscernibility of identicals (or rather, its converse, the difference of discernibles), and whilst one might have certain doubts about the unrestricted validity of this form of argument when the discerning property is intensional, these are of no concern here: the property of having the ability to do so and so is quite extensional. Hence there is no problem about the validity of the argument, and any problem must reside in the premise.

Is this true? Well, obviously not. The minds of dogs, newborn babes and mathematical illiterates cannot give any kind of mathematical proof. Of course, Lucas never meant his argument to apply to any mind. He has in mind, here, a mind of reasonable mathematical sophistication. So let us assume henceforth that M is of this kind, whilst noting that, if the argument works, mathematicians are not computers, but the rest of you still may be.

## *2 Output and Ability*

Given that M is a mind of this kind, why should we suppose the premise to be true? It is supposed to follow from an application of Gödel's Theorem. We will look at the details in a moment, but first there is the preliminary question of how, given M and C, we are to decide what they can and cannot do. (Note that we cannot simply replace 'can' with 'does' in the argument, since there is no reason to suppose that M will actually prove the statement.) Fortunately we do not need to address the issue as far as M is concerned, but for C there is no escaping the question. The strategy adopted by Lucas is never spelled out very clearly, but in effect, I take it, it comes to this. We suppose C set in motion and demonstrate that the proof of the formula in question will not be given (or, if the machine is non-deterministic, will never be given whichever of the non-determined paths is followed); we conclude from this that C cannot give a proof of the formula.

There are two points that should be made about this strategy straight away. The first is that what is actually given as output by a computer depends, of course, on its input (in the case of a Turing Machine, the initial state of the tape). Hence, in the context of the argument 'computer' should be taken to mean 'device + input'; and what the argument proves, if it works, is therefore that there is no machine with input that is identical to M.

The second point concerns the fact that the argument moves from what the machine doesn't do, to what it can't do. If it were a person that were in question here the inference would be hotly contested by compatibilists (soft determinists). Philosophers of this stripe point out that the fact that someone does not do something does not entail that they cannot (i.e., have not got the ability to), and argue that this is so even if it is, in fact, determined that they do not to it. Since what is at issue here is whether computers may be minds, can we not make out the same case for C?

There are two possible strategies for reply here. The first is to confront the arguments for compatibilism head on. The second is to argue that even if compatibilism is true in general, there is something specific about the computer case which rules it out: the machine is, after all, doing everything it can. I do not think it appropriate to take up the first strategy here; and I do not know how to pursue the second successfully. So I intend to leave the issue there, assume that in this case cannot follows from does not, and simply note the weak point of the argument.

### *3 Gödel's Theorem*

We must now address the central question of what the formula is, of which C, supposedly, never gives a proof but of which M can. This is to be delivered by an application of Gödel's famous Incompleteness Theorem, although how, never gets spelled out very carefully. Let us start with Gödel's Theorem itself. Statements of this come in various shapes and sizes. The relevant one in the present context is as follows.[4]

Let T be an axiomatic theory that can represent all recursive functions. Then there is a formula, $\phi$, such that i) if T is consistent $\phi$ is not in T and ii) if the axioms and rules of T are intuitively correct, we can establish $\phi$ to be true by an intuitively correct argument.

The statement of the theorem contains various technical notions. To understand the discussion it is not necessary to have a complete grip on them, but it is necessary to have a reasonable understanding, so I will spend a little time explaining them.

A theory is a set of sentences of some (formal) language. In the case in question we must suppose the language to be a language with numerals, function symbols or predicates for addition, multiplication, etc. To be a theory the set must be closed under deducibility. That is, any logical consequence of its members must also be a member. It is pertinent to inquire what notion of deducibility is in question here. In fact, it is sufficient to assume that it validates little more than modus ponens, substitutivity of identicals and some simple quantifier inferences.

To say that the theory can represent all recursive functions is to say that certain arithmetic facts are in the theory. It suffices, as Gödel showed, to suppose that the theory contains statements of the basic properties of addition and multiplication.

To say that the theory is axiomatic is to say that there is a decidable set of axioms such that the members of the theory are exactly the logical consequences of the axioms. If this is true then the members of the set can be effectively generated (by applying the rules systematically to the axioms). In the jargon of recursion theory, they are recursively enumerable (re). Conversely, as Craig's Theorem shows, any theory that is recursively enumerable has a decidable set of axioms, and is therefore axiomatic. Hence, being axiomatic and being recursively enumerable are the same thing.

### 4 *. . . and Computers*

Gödel's theorem tells us that under certain conditions a true formula is not provable in a certain theory. But this does not yet give us what is required. We need a formula for which C cannot give a proof. How do we get this? Lucas hoped to obtain this in virtue of the close connection between computations and axiom systems.

Let us take some computational device, say a Turing Machine (though in virtue of Church's Thesis, essentially the same points will hold of any computational device). We can think of a computational state as a pair comprising the non-blank part of the tape (with the square being scanned marked in some way) and the machine state. Starting from some initial state, the computational state is modified by the application of certain effective rules to generate a potentially infinite set of states. By their nature the set of states generated is re. Or, to be slightly more precise, if we were to code the states arithmetically, the set of codes would be re.[5]

It should be noted that the existence of connectionist machines casts some doubt on this conclusion. Provided such a machine is set up as a discrete state system, as they normally are (and indeed, must be, if they are to be implemented on standard machines) then the conclusion holds. In theory, at least, however, they could be set up as analog machines, with outputs being continuous functions of inputs over real-valued time. In this case, there is no reason to suppose that the sequence of output states is re. Indeed, it is not even clear what this would mean any more. Such a possibility therefore poses a very radical challenge to Lucas' argument. However, an analog notion of computation might threaten Church's Thesis itself, and would therefore occasion a radically novel situation that I do not wish to go into here.

Leaving this issue aside, we can conclude that the set of computational states is re. But we cannot apply anything like Gödel's Theorem yet. To do this we need to know about what proofs the machine can give, that is, to look at its output. So let us suppose that the machine is hooked up to some output device. For want of a better word, let us call this a mouth. We can suppose that every time the machine state takes (a) certain determined value(s) some part of the tape which can be effectively determined is output through the mouth. The sequence of outputs can be effectively culled from an re set, and so is re too.

Now, of these outputs some will be proofs. Let the set of theorems of which these are the proofs be T. We are at last in a position to see whether we can apply Gödel's Theorem to this T. First, let us consider whether the general conditions are satisfied. That is, can T represent all recursive functions; is it deductively closed; and is it re?

*5 Representability and Closure*

Is there, for a start, any reason to suppose that T can represent all recursive functions. For all we have said so far, of course not. Recall, however, that we are in the process of attempting to show that C is not M, where M is a mind of reasonable mathematical sophistication. Now a mind of such sophistication can clearly establish the basic properties of addition and multiplication. Hence, if C cannot do the same it is not M. We may therefore suppose, without loss of generality, that T can represent all recursive functions.

Next, we turn to deductive closure. Is T deductively closed? Again, for all we have said so far, no. We can argue, as before, however, that we can restrict ourselves to the case where T is deductively closed: the set of theorems that M can establish is deductively closed, and so if T is not deductively closed C is not M. It might be doubted that the set of theorems a mind can establish is deductively closed. The set of theorems any human mind will establish is, of course, finite, and so not deductively closed. But the set of theorems a human mind can establish would seem deductively closed, at least in principle: for example, if a mind can establish a and a|b then it can establish b. The principle is that the mind is given sufficient time and secondary memory space - which is a principle we also have to apply to the computer, of course.

*6 Recursive Enumerability*

Finally let us turn to the crucial question of whether T is re. The output of C is re. To get at T we need to disentangle the set of things given as proofs from the rest of the output, and then extract T from this. For T to be re both of these procedures must be effective. If they are not, there is no reason why T should itself be re. An re set - such as the natural numbers - can have non-re subsets - e.g., the set of codes of true arithmetical statements.

Let us start with the question of whether we can effectively disentangle proofs from the rest of the output. To answer this we have to address the question of what, exactly, it is to give a proof. In fact, a number of things might be meant by this. Let us consider them in turn, and see whether any of them will do what is required.

First, the obvious sense in which M can give a proof is an intensional one. M produces a statement with the intention that it be understood in a certain way, namely as establishing a certain mathematical statement as true. Now, it is not clear that a computer can have intensional states in the same way. But someone who denied this would have a much more fundamental objection against the identity of M and C. So suppose that it can. Is there an effective procedure for telling when the output of the computer is being given as a proof in this sense? Clearly not; the computer might not be giving a proof at all: it might be joking, lying or just doodling. And there is certainly no effective procedure for ruling out these cases. If this is what is meant by giving a proof the argument therefore folds here.

Alternatively, we might interpret the notion of giving a proof simply as the outputting of a proof (whatever intention - if any - is behind it). After all, M does this too. But in this case the output (which might just be a sequence of 1s and 0s) has no intrinsic meaning at all; neither, therefore, does the question of how one can recognise a proof in the output. Such a question makes sense only relative to some scheme for decoding the output. We can get around this problem as follows. M will, presumably, speak some language and so can give a proof of the formula in question in that language, or even in some canonical part of it, say a certain formal language, L. If C cannot even output strings which are, syntactically, sentences of L then it cannot do what M does. Hence C is not M. Thus, we may restrict our attention to the case where C outputs sentences of L, and take the decoding scheme to be the normal semantics of L. Can we now effectively recognise a proof in L when one is output? This depends on still further disambiguations.

A proof is a deductive argument; but obviously not all deductively valid arguments are proofs (0=1; hence 1=0). Suppose we take a proof to be a sound deductive argument, i.e., a valid argument with true premises. In this case there is no way that recognition of a proof can be effective. For even assuming that the question of whether or not an inference is valid is decidable, the question of whether or not a premise is true is certainly not. Hence, again, if we use this notion of proof the argument folds.

A final sense of the notion of proof (and the only one that, as far as I can see, is capable of advancing the argument further) is that according to which a proof is a sequence of statements that appears to us (or, rather to an L speaker, say M) to be sound. This, plausibly, can be recognised effectively. Henceforth I shall interpret proof (and cognate notions such as consistency) in this way. We may now take it that the set of proofs is re. Assuming, that L is such that we can effectively determine the theorem from the proof, T is re too. The assumption is not toothless. It rules out referring to the conclusions by names such as 'Euclid's Prime Theorem'. But provided we take it, as we may, that all statements are spelled out (e.g., $\forall x \varepsilon N \exists y \varepsilon N(y>x \wedge prime(y))$), and not just named, the assumption is satisfied.

*7 Consistency*

Having seen that the general conditions of Gödel's Theorem are satisfied on one (and only one) understanding of proof, we must now see whether the particular conditions for parts i) and ii) of it are satisfied (on the same understanding). Let us take things in that order. To establish that $\phi$ is not provable, we need to establish that T is consistent. Are there any reasons to suppose so? Clearly not. Nor will it help to argue that we can restrict ourselves to those computers where T is consistent, since an inconsistent machine cannot be M. Real minds are frequently inconsistent in the sense of providing proofs of inconsistent things. Most people have produced a proof that 0=1 when doing elementary algebra; and nearly everybody has applied the algorithm for adding two numbers (which is a proof of sorts) and got the wrong answer.

Are there any avenues of repair here? One argument Lucas uses in his original paper is that if T is deductively closed and inconsistent then it contains a proof of everything; and therefore C cannot be M because people, even if inconsistent, do not offer proofs of everything. This reply, however, will not work. There is no reason to suppose that the logic in question is explosive. It may well be paraconsistent; in which case triviality does not follow from inconsistency. (As I observed, to prove Gödel's Theorem, we need to make very few assumptions about what the logic in question is.)

Another move that Lucas makes in his original paper is to argue that although people may be inconsistent, they are not essentially so, in the sense that when they discover that they have proved inconsistent things they will take back at least one of the proofs. However, this does not really help, since there is no reason to suppose that C may not be essentially consistent in the same sense. It might be suggested that we grant this, but take for T the set of all those theorems that are proved and never taken back, which is consistent. Whether or not this is so, the argument now collapses. Even granting that taking back is something that can be effectively recognised when it occurs, there is no effective way of telling when a theorem is going to be taken back. There is therefore no longer any reason to suppose that T is re.

The only way, it seems to me, that offers any hope of getting T to be consistent is to suppose that M (and so any C which is supposed to be M) is not only a mathematical mind but an ideal mathematical mind, that never makes mistakes of any kind: either of memory, inference, judgment or output. But this is sufficient to destroy the argument. After all, the only candidate for a mind of this kind is God('s?). So at best, we have a (theo)logical proof that God is not a computer.

But I am sceptical of even this repair. I doubt that even the ideal mathematical mind is (mathematically) consistent. Simple Peano arithmetic may be consistent (we hope); but once one passes beyond these bounds contradictions are wont to arise. Take Berry's Paradox, for example. There are only a finite number of numerical descriptions of some preassigned length, say 100 words. Hence there must be numbers that are not so described. In particular, the least such is not so described. But we have just so described it. This and many similar logical paradoxes threaten consistency even for an ideal mind.

It might be replied that there is something wrong with this proof. I do not think this is the case, though I shall not enter into the discussion here. Let me just say that if there is, we have not yet found it (at least to most people's satisfaction). Alternatively, it might be argued that the phenomenon of logical paradoxes is irrelevant in the present context since we may take L to be non-self-referential. Nothing, however, could be further from the truth. The proof of Gödel's Theorem makes notorious use of self-reference. Indeed, the very theorem that is claimed to be unprovable is, intuitively, a logical paradox. Roughly, the sentence in question, $\phi$, says of itself that it is not provable. Now suppose that it is false. Then it is provable, and so true. Hence it is true, and so unprovable. But we have just proved this; hence it is provable. Thus, it would seem, we cannot apply Gödel's Theorem to infer that $\phi$ is unprovable, since the theory in question is inconsistent; and the inconsistency is precisely $\phi \& \neg\phi$.[6] In any case, there is therefore no hope of trying to argue that the logical paradoxes are an irrelevant phenomenon that may be safely cordoned off.

### *8 The Mathematician's Proof*

Next, and finally, we can turn to the question of whether $\phi$ is indeed provable by M, in other words, whether part ii) of the theorem can be applied. In order for this to be so we require that the axioms and rules of T be intuitively correct. Are they? First, what are they?[7] With a bit of rational reconstruction, we can always suppose that the sole rule of inference is modus ponens, which is certainly intuitively correct. What of the axioms? Given any proof in L that C comes up with we can effectively pick out its ultimate premises, just as we picked out its ultimate conclusion. By our assumption of what a proof is, each ultimate premise is intuitively correct. This does not quite give us what we want, however. Since the collection of proofs is re, the collection of ultimate premises is re, but need not be decidable, and so need not be a decidable set of axioms for T. However, Craig's Theorem (Craig, 1953) shows us how to construct a decidable and logically equivalent set. If $\alpha$ is the nth member of the enumeration, we simply take $\alpha\& \ldots \&\alpha$ (with n conjuncts) as an axiom. Clearly, if a is intuitively correct, so is this. So T does have a set of axioms all of which are intuitively correct, and we can apply the second part of Gödel's Theorem to establish that $\phi$ is provable by M.[8]

*9 Conclusion*

We have now considered the whole of Lucas' argument. By carefully spelling it out and trimming it we have seen how a number of its problems can be avoided. But, as we have also seen, in the last analysis it fails: however one spells out the notion of proof concerned, the argument breaks down somewhere. Gödel's Theorem, it would seem, has no more to teach us about the nature of the mind than do the categories of Aristotelian Logic.

*University of Queensland*

NOTES

[1] Apart from the introduction and the conclusion, this paper is the same as 'Creativity and Gödel's Theorem', which appeared in *Creativity*, T.Dartnell (ed), Kluwer Academic Publishers (forthcoming). I am grateful to the editor and to Kluwer for their permission to reprint the material.
[2] For a reasonably comprehensive but by no means complete bibliography see D.Boyer (1983).
[3] See Brain and Behavioural Sciences 13 (1990), pp. 643-705.
[4] I take it essentially from ch. 3 of Priest (1987); a proof of the theorem in this form can also be found there.
[5] If the device is a non-deterministic Turing Machine then there is no reason why the set of actual states generated is re. But in this case we consider a suitable deterministic Turing Machine that generates all the possible states of the non-deterministic machine in some systematic order.
[6] For a further discussion of all these issues, see Priest (1987).
[7] It is sometimes suggested that Lucas' argument fails since M may not be able to determine what the axioms of T are, and so may not be able to 'formulate its own Gödel sentence', $\phi$. This is an ignoratio. It is sufficient for the argument that $\phi$ exist; it is not necessary that M can determine that $\phi$ is the Gödel sentence in question.
[8] It is sometimes argued that M may not be able to establish $\phi$ on the ground that M may not be able to establish that T is sound. (Usually, the point is put in terms of consistency since people fail to distinguish between these notions.) However, if T is defined in the way required for the rest of the argument to work (as the set of theorems whose proofs are intuitively correct), the soundness of T can be established in an intuitively correct way quite trivially, as follows: all the axioms are true, all the rules of inference are valid; so (by recursion) all the theorems are true.

## BIBLIOGRAPHY

Boyer, D. (1983), 'Lucas Gödel and Astaire', *Philosophical Quarterly* 33, pp. 147-59.
Craig, W. (1953), 'On Axiomatizability within a System', *Journal of Symbolic Logic* 18, pp. 30-2.
Lucas, J.R. (1961), 'Minds, Machines and Gödel', *Philosophy* 36, pp. 112-27.
Penrose, R. (1989), *The Emperor's New Mind: concerning computers minds and the laws of physics* (Oxford University Press, Oxford).
Priest, G. (1987), *In Contradiction* (Nijhoff).

GILBERT HARMAN

# EPISTEMOLOGY AND THE DIET REVOLUTION

Epistemology, the theory of knowledge as pursued by philosophers, has changed a great deal in the last fifty years. This is how things look today (or at least how they looked the other day).

## *Logic*

The most important point in contemporary epistemology concerns the relation (or should I say the lack of a relation) between logic and epistemology. The point is that principles of logic are not principles of reasoning.

Until recently, there has been a persistent tendency in epistemology to confuse logical issues about implication and consistency with epistemological issues about reasoning and change in view. But it is easy to see that these are distinct issues.

For one thing, principles of logic like *modus ponens* do not have a psychological or epistemological subject matter. They do not say anything about what someone should believe or infer. Nor is there any obvious connection between principles of logic and principles about what people should believe. The fact that a proposition is implied by your beliefs does not guarantee that you should infer it, since, for one thing, it may conflict with other beliefs of yours, and in any event, you should not clutter your thought by making inferences for the sake of making inferences (see Harman 1986, ch. 1 and 2, and also Goldman 1986).

*Modus Ponens* is the principle that a conditional proposition, *if P then Q*, and the proposition comprising its antecedent, *P*, jointly imply the consequent, *Q*. The principle is not (as it is sometimes called) a *rule of inference*. It does not say that from *if P then Q* and *P* you can infer *Q*. It says nothing about inference at all. It says nothing about what you or anyone else may or can do. It is not about that. The principle may be understood, perhaps, to say that the implication in question is a 'valid implication' But it should not be understood to say anything about 'valid inference.' The whole notion of 'valid inference' is a category mistake, confusing implication and inference. Anyone who tells you that *modus ponens* is a rule of inference is not your friend.

It may be useful to observe that none of the following principles has the strict generality that is characteristic of principles of logic.

- Your beliefs should be deductively closed.

*Michaelis Michael and John O'Leary-Hawthorne (eds.), Philosophy in Mind, pp.203-214.*

- Your beliefs should be consistent.
- Your degrees of belief should be probabilistically coherent.

One way to appreciate the lack of strict generality in such principles is to observe that you have limited resources. You cannot afford to devote all your resources to trying to believe all the consequences of your beliefs or to making your beliefs consistent or probabilistically coherent. You have other things to do with your time.

This is not to deny the importance for epistemology of *commitment*, where commitment is in part to be explained in terms of something like implication. If you are committed to something that you also have reason to disbelieve, that can be a reason to modify your view. But this reason may not carry much weight in relation to others. You may wish to have lunch first, before figuring out how best to modify your view. After lunch you may want to take a nap. After your nap you may have other things to do.

It is not implication in its full ancestral glory but obvious or immediate implication that is relevant to epistemology. You do not have any reason at all to believe something whose proof is beyond the powers of anyone now alive, even though it is implied by things you presently accept. At best you have a reason to accept the first steps of this unknown proof. And you may not even have a reason to do that. You do not always have a reason to infer what follows immediately from what you now accept. It is not the case that you should clutter your mind with trivia, for example. You need a reason to be interested in one or another immediate consequence of your view. (We will come back to this.)

Ignoring the difference between logic and the theory of reasoning has had and continues to have serious bad effects in the theory of knowledge and the philosophy of science. For one thing, there is the ill fated enterprise of "inductive logic." The thought behind inductive logic goes something like this:

- Deductive logic is the theory of deductive reasoning.
- Much reasoning is not deductive reasoning.
- Call that reasoning inductive reasoning.
- We need a theory of inductive reasoning.
- So, we need an inductive logic

This goes wrong, because as I have already mentioned, a logic is not the same thing as a theory of reasoning. In particular, there are no deductive principles of reasoning; there are only deductive principles of implication.

So, one bad result of this persistent error is the search for an inductive logic. Related bad results include the search for relevance logic and the search for the practical syllogism.

A second kind of bad result, also related to the first, is the exaggeration of probability theory as a cure for epistemological ills. Probability theory offers an extended logic, but says nothing about reasoning, in the same way that logic says nothing about reasoning. Probability theory has neither a normative nor a psychological subject matter.

Probability theory can be developed as a theory of coherence for degrees of belief, but such coherence is a kind of consistency, and a theory of consistency is not a theory of reasoning.

A third bad consequence of confusing logic and the theory of reasoning is the lingering idea that logic provides a model of what we should expect from a theory of reasoning. This idea leads to the thought that a theory of reasoning must have the formal and precise character that formal logic has. In the grip of this idea, some philosophers of science have treated inference to the best explanation as a nonstarter in the search for an acceptable account of reasoning. They object that the relevant principle of inference has not been formally stated, has not been expressed in a way that can be mechanically applied, and so is not a "serious proposal" in the way that new logics and Bayesian decision theory are "serious".

To counter this sort of objection to the study of inference to the best explanation, it is important to see that, since logic and decision theory are not proposals about inference, they are not "serious" proposals about inference. There is no reason to think that serious proposals about inference are going to look anything like a formal calculus.

### *Naturalized Epistemology*

Yet another bad consequence of the formal logical ideal is the occasional objection that one or another principle of a proposed theory of reasoning, for example, a principle of simplicity, lacks *a priori* justification in contrast to basic principles of logic and probability theory.

Actually, this points to a second important point in contemporary theory of knowledge. (You will remember that the first point is that principles of logic are not principles of reasoning.) The second point is that we do not have intuitive insight into the basic principles of reasoning. This point becomes clear enough once principles of logic are no longer confused with principles of reasoning. It may be that the basic principles of logic are obviously true, but that is completely irrelevant to whether principles of reasoning are obviously true. And, indeed, our only access to principles of reasoning is indirect, by considering what considerations actually do affect reasoning.

This is connected to the old issue of *Verstehen*—the issue that divides positivists and behaviorists like Comte (1830-42), Watson (1913), Hempel (1949), and Skinner (1974) from *Verstehen* theorists like Dilthey (1883), Quine (1960), and Nagel (1974).[1] Positivists say that we understand reasoning only if we can see it as an instance of a familiar general principle. Advocates of *Verstehen* reply that we can appreciate someone else's reasoning only to the extent that we can understand it from within, as reasoning we can imagine ourselves doing. Positivism is clearly wrong about reasoning, since we sometimes do understand someone's reasoning even though we do not at present have available an external characterization of the principles of reasoning in the way we have a characterization of principles of logic. Although we cannot simply check to see whether another person's reasoning is in accord with certain formally stated principles, we can sometimes understand it from within.

This means that the philosophical theory of knowledge has to be continuous with psychological studies of reasoning, perception, and memory. The theory of knowledge also connects with artificial intelligence, especially studies of machine learning, nonmonotonic reasoning, and probability networks. We all need all the help we can get!

Recognition of the continuity between these subjects is sometimes called 'naturalized epistemology.' Fifty years ago, some analytic philosophers took philosophy to be entirely distinct from psychology.

> Psychology is an empirical subject. Philosophy is an *a priori* subject, involved with the analysis of meanings. The two enterprises are therefore quite different. And that is that!

Even fifty years ago, this was unconvincing. (For example, Quine's famous essay, 'Truth by Convention', had been published in 1936!) There was at that time already considerable doubt about the alleged distinction between *a priori* and empirical theoretical principles. In any event, it has long been clear that philosophy is not confined to the analysis of meanings and, where philosophy does concern itself with the analysis of meanings, philosophy is continuous with (and usually refuted by) developments in linguistics and psycholinguistcs.

There is no special philosophical method used by philosophers and not by scientists. The point is not merely that scientists are sometimes philosophers. Rather, the point is that philosophy involves the same kinds of analysis and speculation that occur in theoretical science.

Where philosophy does differ from certain sciences—for example, psychology—it differs in not possessing an experimental paradigm. There are no controlled experiments in philosophy, at least as philosophy is normally practiced.

Of course, there are no controlled experiments in certain sciences either—for example, economics, or linguistics. Controlled experiments are not part of the defining essence of science. Indeed, it is possible that the heavy experimental paradigm in psychology has actually hindered progress in that subject.

One thing that has become extremely clear is that academic department boundaries should not be treated as serious disciplinary limits. The impressive progress currently underway in the theory of knowledge is not limited to any single traditional department.

So much then for the objection that philosophy is *a priori* and science is *empirical*. Consider now a related objection, by way of the normative-descriptive distinction.

> Psychology can be concerned only with describing how we actually do reason, whereas philosophy is concerned to specify how we ought to reason. Psychology is descriptive, philosophy is normative. So, there is a significant difference in the way philosophy and psychology approach issues about reasoning.

I say in reply that the distinction between how people reason and how they ought to reason is similar to the distinction between how people digest food and how they ought to do so. The normative-descriptive distinction in epistemology is like the normative-descriptive distinction in digestion. As far as normative advice goes, we are limited to dietary advice, not advice about what principles of reasoning or digestion to follow. (I might remark here that the original title of this talk was "Epistemology and the Diet Revolution.")

Much is currently being learned about the information processing in low-level visual perception — how edges are detected and analyzed into shapes, how the images from the two eyes are combined into a single percept, how shapes are perceived as objects, etc. (see Hildreth and Ullman, 1989). We can try to understand this processing but we cannot in any straightforward way modify it. We can give people advice about what to look at, even what aspects of what they look at to attend to, but we cannot give people advice about how to detect edges or analyze shapes.

The same point holds for the information processing involved in inference. We can give advice to people about what to keep in mind, but we cannot tell them what rules to follow in their thinking.

*Pragmatism*

Naturalized epistemology tends toward pragmatism and an emphasis on such practical aspects of reasoning as that you should not clutter your thinking with random conclusions that merely follow from your present beliefs. Your reasoning is directed toward answering questions in which you interested. That is one reason why there are no purely deductive principles of inference. You may have no reason to accept a trivial consequence of things you accept, because you may have no reason to be interested in whether that trivial consequence is true.

In reasoning toward the best explanation of a hypothesis, you are influenced by considerations of simplicity that themselves appear to have a practical aspect: the simplest hypothesis is the hypothesis that is simplest to use in certain respects (Harman, 'Simplicity as a Pragmatic Criterion for Deciding What Hypotheses to Take Seriously', forthcoming.).

Pragmatic considerations also support the epistemic conservatism that I will talk about in the next section.

Some of the deepest issues in contemporary epistemology involve the relations between theoretical and practical reasoning. Clearly, there is a distinction between epistemic reasons to believe something and more practical reasons (*e.g.*, research grants only available to people with certain beliefs). How do the practical aspects of theoretical reasoning allow for such a distinction?

Some philosophers have suggested that purely epistemic reasons are practical reasons that derive from purely epistemic goals, goals like believing all and only what is true. One difficulty in developing this idea is that people do not normally have general purely epistemic goals of this sort. Curiosity is almost always more limited. You wonder what the answer to a particular question is. And your concern often, if not always, has a practical and not purely epistemic source: knowing the answer will help advance some practical project.

A second difficulty with the suggestion (that purely epistemic reasons are practical reasons that derive from purely epistemic goals) is that the most straightforward way to develop the suggestion leads to the absurd consequence that you should believe anything whose probability is greater than 0.5! (Hempel, 1965).[2]

Actually, we can use the notion of conditional subjective probability to explain a good part of the intuitive distinction between epistemic and nonepistemic reasons. You take *R* to be an epistemic reason to believe *P* to the extent that your conditional probability of *P* given *R* is greater than your conditional probability of *P* given *not R*. You take *R* to be a nonepistemic reason to believe *P* to the extent that you take *R* to be a reason to believe *P* but not an epistemic reason.

Why do we suppose that the restriction of research grants to people with certain beliefs provides only a nonepistemic reason for those beliefs? Because we suppose that such a restriction does not affect the likelihood that those beliefs are true. Suppose *R* is that research grants are available only to those who believe *P*. Then our conditional probability of *P* given the existence of such a restriction is no greater than our conditional probability of *P* given no such restriction. Although we might take the grant restrictions to be a reason to believe *P*, it's a nonepistemic reason, not an epistemic reason.

This criterion allows that some pragmatic considerations might serve as epistemic reasons. Considerations of simplicity, for example, might be incorporated into your conditional probabilities.

Many interesting issues arise in thinking about nonepistemic reasons. For example, a person who is committed to supposing that there are nonepistemic reasons (by the suggested criterion) has incoherent commitments in the sense of being subject to a temporal Dutch book.[3] Since a commitment to nonepistemic reasons can be perfectly rational, it follows that one can be perfectly rational while having that sort of incoherent commitments![4]

Belief is a pragmatic matter. Beliefs are functional states and, in particular, they are states that in part function to help an agent reach his or her goals. Bas van Fraassen has stressed pragmatic aspects of explanation and of acceptance (1980, ch. 5). Van Fraassen distinguishes belief from acceptance, which he takes to have pragmatic aspects not possessed by belief. But this is to overlook the fact that all belief has pragmatic aspects because of the sort of functional state that belief is. Belief cannot be distinguished from acceptance because belief is one kind of acceptance. There are other kinds as well, distinguished in part by how they relate to action. Normally, to accept something as a belief is to be prepared to use it as a starting point in further theoretical and practical reasoning. This is not so for accepting something as a working hypothesis however, or for accepting something for the sake of argument.

*Conservatism*

Finally, it is necessary to say something about (what I take to be) the most important aspect of contemporary epistemology, the recognition of the role of epistemic conservatism. Conservatism has various aspects, the most fundamental of which can be expressed in the following principle: 'What requires justification is changing beliefs, not continuing to believe them.' If you believe something, you are justified in continuing to believe it in the absence of special reasons challenging that belief. The mere fact that you cannot justify your belief is not a reason to stop believing it.

The basic conservative principle has a long history. It figures in C. S. Peirce's rejection of Cartesian doubt (1955, pp. 228-9). Goodman relies on it in his discussion of how to justify principles of reasoning in *Fact, Fiction, and Forecast*. (1973, pp. 62-2). Rawls invokes conservatism in his *Theory of Justice* when he appeals to the method of reflective equilibrium (1971, pp. 46-53).

The Cartesian approach to epistemology that was dominant through the first half of the twentieth century, assumed that current beliefs always require justification whether or not you have special reasons to challenge them. If no justification for a given belief of yours is forthcoming, you should abandon that belief, according to Cartesianism.

The Cartesian approach tended toward foundationalism, requiring that all beliefs be justified ultimately in terms of basic beliefs that are self-evident or intuitively justified on the basis of sensory experience. In the first half of the twentieth century, self-evidence was often taken to have a linguistic aspect with "analytic" basic principles that were taken to be justified solely because of their very meaning.

The Cartesian approach is now history. Starting in the late thirties, and certainly by the fifties, the analytic-synthetic distinction was on its way out. Foundationalism followed.

Contrary to what is sometimes said by people who should know better, there are no perennial problems of philosophy. Philosophy does not stand still. The philosophical issues of today are not the issues of ten years ago or fifty years ago and certainly not the issues of two hundred or two thousand years ago.

In particular, general scepticism has ceased to be an interesting philosophical issue. General scepticism appeared to be worth taking seriously only within the Cartesian approach. Now that the Cartesian approach is no longer viable, scepticism has lost its philosophical interest.

But more needs to be said about conservatism, which is occasionally misunderstood. Here is a standard objection sometimes heard:

> Suppose that there are two competing hypotheses, *G* and *H*, that are equally supported by the evidence, equally simple, and otherwise the same in other respects relevant to choosing between them. Then there can be no reason to accept one of these hypotheses rather than the other. The thing to do is to suspend judgment between them.
>
> But now suppose that Elizabeth has thought of only one of these hypotheses, *G*. Of the hypotheses she takes into consideration, that one hypothesis, *G*, is clearly best, so she accepts *G* and rejects the alternatives. Then, after she has accepted that first hypothesis, *G*, she becomes aware of the other hypothesis, *H*. Doesn't conservatism imply that Elizabeth should at this point prefer *G* to *H* simply because she already believes *G* rather than *H*? And isn't that wrong?

Answer: Yes, that is wrong, but no, conservatism does not have that implication. Conservatism tells Elizabeth not to reject a belief simply because she has no justification for it. Conservatism says a belief needs no justification in the absence of a challenge to that belief. But one very compelling challenge to a belief is that the reasons that led to the belief were insufficient. When Elizabeth becomes aware of the second hypothesis *H*, she becomes aware that her reasons for believing *G* did not favor *G* over *H*. So, she should at this point give up her belief in *G* and suspend judgment between *G* and *H*.

However, Elizabeth is not required to remember in detail her reasons for believing *G*. Indeed, people cannot keep track of all their detailed reasons and often lose track altogether of their reasons for certain beliefs. If Elizabeth has lost track of her reasons for *G*, when she later learns about the hypothesis *H*, she may not be in a position to realize that her reasons for accepting *G* were inadequate to distinguish *G* from *H*, so there may be no reason for her at that later stage to abandon *G*.

Here there is a difference between individual epistemology and scientific epistemology. In science, there is a greater keeping track of reasons than there is in ordinary life. Reasons are written down and records are kept in science.

A related aspect of conservatism arises from the fact that we can't be inquiring into the truth of everything at the same time. Active inquiry uses time, energy, and other resources. Decisions have to be made about how these resources are to be used. Sometimes we decide that some issues are not worth pursuing, because we believe that we will not be able to resolve those issues. Sometimes we decide that we have done enough investigation into a given question and should simply fully accept the answer we have currently arrived at. Fully to accept something in this way involves ending inquiry into whether it is true. So full acceptance of *P* involves two things: (1) ending inquiry into whether *P* is so, and (2) a readiness to use *P* as a premise in further inquiry.

Inquiry can be reopened after it has been closed, but this is a relatively serious matter. There has to be a strong case. A consideration against a belief that might have carried some weight while inquiry was ongoing may not be strong enough to lead to restarting an inquiry that has been ended.

Inquiry is often a group affair, so the decision to undertake an investigation or to end one can be a group decision, in the sense of a decision affecting what the members of the group do. Here the group might be an investigative agency, like the police force, or a lab or other sort of research group.

A whole science (physics, or chemistry, or psychology, for example) does not constitute a group of this sort, a group that has a procedure for initiating or ending inquiry into one or another issue. It is always possible for individual scientists or labs to proceed on their own, apart from what others are doing.

This is relevant to thinking about 'division of labor' in science. There is the sort of division of labor that occurs within a given research group, as different people take on different functions within a single inquiry. There is also the rather different sort of division of labor that occurs as different labs investigate different problems, or the same problems in different ways.

Notice that, in accepting something, a person is often not just accepting it for himself or herself, but for others as well. A person accepts something in a way that can be transmitted to others, not just as a report of the speaker's belief but as something that the hearer can rely on. Such acceptance for others invokes special responsibilities. The speaker gives his or her word that something is true.

In science, there is a distinction between data and conclusions. Data are presented in an authoritative way; other scientists are allowed simply to assume the truth of data; the scientists presenting data take full responsibility and can get into a great deal of trouble if they have misrepresented any data. Conclusions, on the other hand, are presented with accompanying reasons that can be evaluated by others, so the conclusions need not be accepted in the same authoritative way, 'for others,' with which data are accepted. Reports of data involve an implicit, 'You can take my word for it,' whereas conclusions do not; conclusions carry an implicit, 'You can see for yourself.'

This sort of distinction holds when scientists talk to scientists. When scientists report scientific conclusions to nonscientists, they cannot expect the nonscientists to understand the reasons, so they assume a greater responsibility for getting things right.[5]

Full acceptance ends inquiry and involves being prepared to use what has been accepted as a premise or starting point in further theoretical or practical inquiry. Full acceptance differs from tentative acceptance. Tentative acceptance keeps inquiry open. Tentaive acceptance is like supposition, or accepting something for the sake of the argument. What is tentatively accepted serves as a premise or starting point in further theoretical or practival reasoning (possibly with limits, to the extent to which this is allowed.) The point is to see where one gets by doing this, so with tentative acceptance it is important to keep track of reasons. It is hard to keep track of reasons and in practice tentative acceptance can tend to become full acceptance.

## *Conclusion*

In summary, then, the main principles of contemporary epistemology are the Logic-Epistemology distinction, Naturalized Epistemology, Pragmatism, and Conservatism.[6]

*Princeton University*

## NOTES

[1] I include Quine on the Verstehen side because of the fundamental role he assigns translation in his account of meaning. I discuss this further in Harman (1990).
[2] I discuss this and related issues further in 'Realism, Antirealism and Reasons for Belief' (forthcoming).
[3] The argument is attributed to David Lewis in Paul Teller (1973).
[4] I discuss this further in 'Realism, Antirealism and Reasons for Belief'.
[5] I discuss social aspects of acceptance in somewhat more detail in *Change in View*, pp. 50-52.
[6] The preparation of this paper was supported in part by a grant from the James S. McDonnell Foundation to Princeton University.

## REFERENCES

Comte, A. (1830-42), *Cours de philosophie positive.*
Dilthey, W. (1883), *Einleitung in die Geisteswissenschaften*, translated as *Introduction to the human sciences*, edited by Rudolf A. Makkreel and Frithjof Rodi (Princeton, N.J. : Princeton University Press, 1989)
Goldman, A. (1986), *Epistemology and Cognition* (Cambridge, Mass.: Harvard University Press).

Goodman, N. (1973), *Fact, Fiction and Forecast,* Third edition (Indianapolis, Bobbs-Merril).
Harman, G. (1986), *Change In View: Principles of Reasoning* (Cambridge, Mass., MIT Press).
Harman, G. (1990), 'Immanent and transcendent approaches to the theory of meaning', in R. Gibson and R. B. Barrett, eds., *Perspectives on Quine* (Oxford: Blackwell), pp. 144-157.
Harman, G. (forthcoming), 'Simplicity as a Pragmatic Criterion for Deciding What Hypotheses to Take Seriously', to appear in a volume edited by Douglas Stalker.
Harman, G. (forthcoming), 'Realism, Antirealism and Reasons for Belief', to appear in the conference proceedings for a conference on 'The Implications of Realism and Anti-realism for Epistemology' held in Santa Clara, February 28-29, 1992.
Hempel, C.G. (1949), 'The logical analysis of psychology', in H. Feigl and W. Sellars, eds., *Readings in Philosophical Analysis* (New York: Appleton-Century Crofts), pp. 373-384 (translated from a 1935 French version by Sellars).
Hempel, C.G. (1965), 'Inductive Inconsistencies', reprinted in *Aspects of Scientific Explana tion and Other Essays in the Philosophy of Science* (New York: Free Press).
Hildreth, E.C. and Ullman. S. (1989), 'The computational study of vision', in M. I. Posner, ed., *Foundations of Cognitive Science* (Cambridge, Mass.: The MIT Press), pp. 681-730.
Nagel, T. (1974), 'What is it like to be a bat?', *Philosophical Review* 83, pp. 435-50.
Peirce, C.S. (1955), 'Some consequences of four incapacities', in J. Buchler, ed., *Philosophical Writings of Peirce* (New York, Dover)..
Quine, W.V. (1936), 'Truth by Convention', in O. H. Lee, (ed.) *Philosophical Essays for A. N. Whitehead.* (New York: Longman's).
Quine. W.V. (1960), *Word and Object* (Cambridge, Mass.: MIT Press).
Rawls, J. (1971), *A Theory of Justice* (Cambridge, Mass.: Harvard Univ. Press)..
Skinner, B. F. S. (1974), *About Behaviorism* (New York).
Teller, P. (1973), 'Conditionalization and Observation', *Synthese* 26, pp. 218-58.
Van Fraassen, B. (1980), *The Scientific Image* (Oxford University Press).
Watson, J. B. (1913). 'Psychology as a behaviorist views it', *Psychological Review* 20, pp.158-77

JOHN O'LEARY-HAWTHORNE

# TRUTH-APTNESS AND BELIEF[1]

What can the philosophy of mind tell us about the furniture of the world? Those who wish to place philosophy of mind at the helm of the philosophical enterprise will likely want to answer 'Quite a lot'. Those who are sceptical about the power of philosophical reflection about mind and meaning to deliver on broad philosophical issues will want to answer 'Not very much'. In what follows, I discuss one initially promising way in which philosophical reflection on mind — and belief, in particular — might be felt to shed some light here. The strategy in question attempts to vitiate or secure the claim of various sorts of discourse to truth-aptness by exploiting those ways in which the concept of truth-aptness is linked to that of belief.

First, some preliminary remarks about the topic of truth-aptness are in order. When confronted with the disquotational schema 'P iff and only "P" is true', one might reasonably inquire as to the proper range of its instances. What English sentences can be plugged in? It is because this question demands an answer that we can be certain that the disquotational schema doesn't tell us all we need to know about truth (see Jackson, Smith and Oppy, 1994). If the 'iff' above is understood extentionally and classically, then instances of the schema will be true only if the substituends take a truth value. We thus need, at the very least, a theory about what sorts of utterances enjoy a truth value. Let us call such a theory, a theory of 'truth-aptness'.

In many cases, we can be confident concerning which substituends may be inserted in the schema. Questions can't. Various descriptions of shape can. Orders can't. But there are large chunks of discourse which have induced hesitation on the part of many philosophers on the issue of truth aptness: morality, aesthetics, mathematics, modality, probability, metaphysics and religion, to name some notable examples. Unlike the case of paradigm orders and questions, the claims belonging to these problem areas have at least a strong *prima facie* claim on truth aptness. Both the unproblematic and disputed discourses seem to serve as vehicles for thinking and reasoning; both can serve as answers to questions; both seem to combine happily with logical operators such as negation; both figure as subjects for a truth predicate in ordinary life; both can serve as premises and conclusions of arguments.[2]

The issue of truth aptness remains one area of philosophical perplexity where there seems little hope of answers apart from those that philosophers themselves can contrive. Certainly, it is hard to see how natural science, the

*Michaelis Michael and John O'Leary-Hawthorne (eds.), Philosophy in Mind, pp.215-241.*

favorite fall-back for English speaking philosophers, will be forthcoming on the bounds of truth-aptness. By default, philosophers have apparently been left to settle the problems that they have set for themselves in this area. (Sure enough, some philosophers take reducibility to natural science as the ultimate test of truth-aptness. But the claim 'All truth-apt claims are reducible to natural science' is certainly not the sort of thing about which the natural scientist pretends to have considered views and hence not a claim that springs *from* natural science, as it is ordinarily understood.)

How, then, are the limits of truth-aptness properly to be demarcated? Minimalists about these questions think that all the problematic cases — ethics, mathematics and so on — can be established as truth-apt in a fairly straightforward way. How to articulate minimalism satisfactorily is less clear, however. Some minimalists will insist on taking ordinary practice at face value, insisting that it will always be unreasonable to deny truth-aptness to any claim that is ordinarily treated as having a truth value. The trouble here is that there are cases where a sentence is treated by folk as truth-apt, but where, on pain of irresoluble paradox, it cannot be so treated. (Thus the fattest man in Alaska, not realizing that he enjoys that distinction, may claim 'Everything said today by the fattest man in Alaska is false', perhaps on reasonable grounds.) Alternatively, the minimalist will claim that some fairly trivial test for truth-aptness can be achieved by reflecting on the principles underlying our use of 'true' and 'false' and will then use those folk principles to argue that while, perhaps, variations on the Cretan liar are not truth-apt, all the *non-paradoxical* candidates — ethics, mathematics and so on — clearly count as truth-apt. The problem for the minimalist is to say what these principles are. Candidates are likely to be rejected on the grounds that they are not good tests or that they are not trivial tests (or both). Presumably, the test can't be merely syntactic, owing to the fact that one can't formulate a criterion of meaningfulness in syntactic terms and since meaningfulness is necessary for truth-aptness (see Jackson, Smith and Oppy, 1994). On the other hand, candidates that invoke such notions as belief and assertion may be reckoned plausible but non-trivial by so-called non-minimalists, who will not be willing to concede that such properties as being an assertion and being a belief are open to view.[3] Seeing no *trivial* criterion at hand, non-minimalists may feel forced to dig rather deeper in order to settle such disputes.[4]

One approach here is what we might call the metaphysical approach. On this approach, one simply addresses some subject matter head on and tries to convince oneself for or against its factuality by thinking hard about the putative properties and structures that seem to constitute it. An excellent case in point is Mackie's (1977) effort to show that goodness is just too queer to enjoy a place in the nature of things and hence that there are no facts about goodness for moral language to correspond to; while Mackie inferred the falsity of moral claims from these metaphysical reflections, one could well imagine a philosopher denying truth-aptness on such grounds. Such a philosopher would presumably be led to embrace some sort of emotivism about moral discourse. Another example of the metaphysical approach in action trades on Hume's idea that there cannot be necessary connections between distinct existences: for if moral properties did exist, it would seem that they would be distinct from and yet supervenient upon (and thus necessarily connected to) physical properties.[5]

One way that the metaphysical approach might proceed is to discover antinomies buried within some assumption of truth-aptness for some subject matter. The idea is here is to derive some contradiction from an assumption of truth-aptness and thence to conclude that the assumption is out of place. Kant' antinomies had a somewhat similar function in that they were used to relegate the objects of experience from the status of things in themselves Another, more common, way of proceeding is to bring a certain metaphysical picture — say, that of physicalism — to these issues, with the intent of seeing what sorts of discourse can be accommodated within that metaphysical picture on the assumption of their truth-aptness.

We may note at the outset one difficult problem that besets these debates, namely, getting clear about just what is at stake in questions of truth-aptness. In particular, we may reasonably wonder whether securing truth-aptness for a discourse will resolve all questions about the factuality and objectivity of that discourse in the affirmative. To the extent that we think it will, we will regard truth-aptness as a valuable prize. To the extent that it won't, the victory will seem less complete. (We may note in passing that Kant is a relevant example here: it is clear that for him, noumenal status is one thing, truth-aptness another, with the antinomies bearing only on the former. See also Wright, 1992.) Gideon Rosen's paper in this volume discusses precisely this issue. I shall not be pursuing it here.

The metaphysical approach has not been the only one to enjoy the attention of philosophers concerned with truth-aptness. Another point of entry has been seen to be provided by some fairly uncontroversial platitudes connecting truth-aptness with assertion on the one hand, belief on the other. The purported platitudes might include such claims as the following: that whatever we can assert is truth-evaluable, that things we can assert are things we can believe, that someone's asserting something typically gives one reason to believe something, that things we can believe are truth evaluable, and so on. What these platitudes seem to indicate is that if we are prepared to treat the sentences in some segment of discourse as assertoric or else are prepared to use them within the scope of belief ascriptions, then we ought to regard them as truth-apt. Meanwhile, if sentences can't be asserted or believed, then that is a good sign that they are not truth-apt. Not surprisingly, then, the metaphysical approach to truth-aptness is rivalled by what I shall call the semantic approach on the one hand, and the psychological approach on the other. Of course, none of these three approaches need be mutually exclusive. Yet in practice, a philosopher's work on the truth-aptness of some domain tends to accord primacy to one of them

The linguistic approach places philosophy of language at the centre of discussions of truth-aptness. Given the platitudes, the most obvious way for the philosopher of language to be of service here is to provide us with a theory of assertion, which will presumably take its place in some more general theory of speech acts. If we have a good theory of assertion that confirms what seems to be the case — that sentences in some problematic area are asserted all the time — then we will feel considerable pressure to conceding truth-aptness. If, however, our favorite theory of assertion delivers a surprising conclusion about a seemingly assertoric discourse, that will encourage us to deny truth-aptness to it. Of course, the contribution of philosophy of language to these issues will not be limited to the theory of speech acts. The theory of meaning will have a role to play here too: perhaps our favorite story about the nature of meaning and the limits of meaningfulness will help to resolve some issues of truth-aptness. Michael Dummett's efforts to convince us that evidence-transcendent sentences are meaningless constitutes an excellent example of how the philosophy of meaning can attempt to intervene in a decisive way here. (See, for example, Dummett 1976). Alternatively, the philosopher of language might focus on the truth predicate itself: by obtaining a deeper understanding of its contribution and point, we might reasonably hope to place issues of truth-aptness in a clearer light (see, for example, Price 1989).

The psychological approach hopes to bring the theory of mind to bear on issues of truth-aptness. As with the linguistic approach, there is more than one broad strategy that might be adopted here. Corresponding to the distinction between theories of reference, meaning and force that is commonly applied to linguistic acts, we can tentatively distinguish theories of reference, content, and attitude with regard to the mental realm. I say 'tentatively' because it is not clear to what extent these theories can be kept distinct. If one has a referential view of content, the first and second will collapse. If one thinks that a theory of content is, basically, a theory of belief and desire, the second and third will collapse. Yet, while tentative, the distinction is still a useful one. Corresponding to these three aspects of the theory of mind, there are three ways of making headway on issues of truth-aptness.

One might try to adjudicate upon these disputes by drawing on one's favourite account of how mental states succeed in referring to/denoting objective properties. In this context, one could imagine causal, counterfactual, explanatory and even biological considerations coming to the fore.

Alternatively one might hope to make progress by drawing on some account of how mental states succeed in having content, that is, express genuine thoughts. Often such accounts imply that certain inner states that appear to have content nevertheless do not, a result that clearly has potential for truth-aptness disputes. A good example is provided by Christopher Peacocke's recent *A Study of Concepts* (1992) Peacocke there wields the following principle as a means of identifying certain thoughts as illusory and hence, presumably, not truth-apt:

Discrimination Principle: For each content a thinker may judge, there is an adequately individuating account of what makes it the case that he is judging that content rather than any other. (p. 203).

The focus of this paper will be on a third strategy, which attempts to understand truth-aptness by thinking about the nature of psychological attitudes. The third strategy attempts to exploit the the connection, implicit in the platitudes, between truth-aptness and belief. (One may be led here directly or via the semantic approach: one's favorite theory of assertion may turn out, like Grice's (1957), to take intentions to produce belief as constitutive of assertion. If so, then any proper assesment of whether a segment of discourse is assertoric may well require an extended inquiry into the propositional attitudes.) How the psychological approach might attempt to proceed is clear enough: taking belief as the hallmark of truth-aptness, one will try either to secure the truth-aptness of a domain by showing that it provides fit objects of belief or render dubious its claim to truth-aptness by showing that its members are not the sort of thing that could be believed. Here, then, is one way that philosophical reflection on mind might be thought to have far reaching consequences.

We may note in passing one area where the psychological approach will be particularly well-placed to defend the truth-aptness of a domain — namely the domain of psychological talk itself. A denial of truth-aptness is not without precedent here: it was certainly evident, for example, in Daniel Dennett's early ruminations about the instrumental value of the intentional stance (1978). In response, we might well wonder 'What clearer case is there of fit objects of belief than claims about the mind?' The Dennettian of this stripe will be awkwardly placed to respond. He can't very well assign falsity to the claim that we have beliefs about beliefs. After all, the point of his view is that such claims are not truth apt. He will have to deny truth-aptness to that claim, and indeed to the sorts of platitudes about truth, assertion, and belief that we have mentioned. For, after all, his claim that psychological talk was not truth apt was supposed to be true. This would not make sense if the sorts of platitudes with which we began were also true.

The preceeding discussion brings out a distinction between what we might call a conservative as opposed to a radical approach to the issue of truth-aptness. The conservative sets great store in the platitudes about truth-aptness, insists upon respecting them as true. The radical is happy to jettison the purported platitudes if certain broad theoretical virtues are to be achieved by doing so. As we have seen, the non-factualist about psychological discourse would seem to be, ipso facto, a radical.[6]

This paper will operate largely from within a broadly conservative approach. I thus want to ask, 'If we assume that there are a set of platitudes connecting truth-aptness, assertion and the mind and insist on repecting them, is it reasonable to expect that the philosophy of belief can intervene in a decisive way to secure or negate the truth-aptness of various problem areas?' In section one, I reflect a little more carefully about what the platitudes actually are that connect truth-aptness with belief. Such reflection will immediately point to certain limitations connected with the approach to truth-aptness under consideration. In section two, I discuss the project of denying truth-aptness to a domain on the basis of reflections about belief. In section three, I reflect upon the plan of securing truth-aptness for a domain on the basis of such reflections.

## *1. What are the Platitudes Connecting Truth-aptness and Belief?*

### 1.1

My initial concerns stem from a worry that I may have been a little too glib about what exactly the platitudes are connecting belief, assertion and truth-aptness.

Let me begin with an observation concerning assertion and truth-aptness that will prove useful for thinking about how truth-aptness connects with belief. I have in mind the philosophical debate concerning such statements as

'The present king of France is bald'. Russell (1905) said it was truth apt, Strawson (1950) said it wasn't. Russell took it to be a disguised false existential quantification. Strawson argued that it contained a false presupposition and was precluded from truth-aptness on that score. Now I want to ask: are there any platitudes that resolve this debate? Do either of these philosophers clearly offend common sense? It seems to me that the answer is no. Let us inquire further. Does common sense tell us that if Strawson is right, then no one can assert that the present king of France is bald? Again, it seems that the answer is no. These questions are all to be settled, if at all, by general theoretical considerations concerning one's overall view of language. One needs to see the implications of each view for other segments of linguistic theory before forming a judgment upon the matter. Or so it seems to me. Thus, whatever platitudes there are, they do not settle whether an utterance of 'The present king of France is bald' is truth-apt. Common sense does not preclude our endorsing Strawson's theory of presupposition here. Nor does it settle whether we can assert that the present king of France is bald. Now by parity of reasoning, common sense does not preclude our saying both that someone can believe that the present king of France is bald and that such a belief is neither true nor false, where perhaps something analogous to Strawson's theory of presupposition might be brought into play.

We thus come to see that one can't very well offer it as platitudinous that anything that can be believed is truth-apt. Whatever platitudes there are, they do not seem able by themselves to rule out Strawson's approach. Now it might be felt that appeal to the Russell-Strawson debate constitutes something of a trick, to be fixed up by some modest adjustment to the platitudes. Such a reaction would be far too hasty, I contend. After all, the Strawson-Russell debate concerned what to say in cases where denotation failed for singular terms. Isn't that at least the sort of issue that confronts us in, say, the moral sphere? What we are precisely interested in here is how assertion and belief connect with a failure of some predicate to denote a genuine property.[7] So it can hardly be thought that the Russell-Strawson debate concerning denotationless singular terms is utterly irrelevant.

There is, admittedly, the following important difference between the debate about denotationless singular terms and the views about ethics, mathematics and so on that concern us here. Let us call a claim weakly truth-apt just in case, though it lacks a truth value, it could have one. Strawson will allow that 'The present king of France is bald' is weakly truth-apt, whereas most of those who deny the truth-aptness of whole domains, such as ethics, will not typically allow that the claims belonging to that domain are even weakly truth-apt. Granting this difference, one can still imagine a broadly Strawsonian approach plausibly taken by one who denied even weak truth-aptness to the moral sphere: 'Many people believe and assert moral claims. Such claims carry a certain sort of presupposition — that there are real features of the world answering to the basic moral concepts of goodness, duty, obligation, right, virtue and so on. Since these presuppositions fail, the relevant beliefs all turn out to be truthvalueless. Moreover, since these presuppositions necessarily fail, those beliefs are necessarily truthvalueless. Thus moral beliefs are radically unfit for either truth or falsity.' (There is a tendency to think that, unlike Mackie's error theory, a view of ethics that flags its claims as non-truth-apt avoids assigning epistemic blame to the users of ethical discourse. That line of thought over-simplifies things rather: even if the sentences of a discourse are not false, epistemic blame may accrue to the users of that discourse on account of a false presupposition that they are guilty of.)

Let us then distinguish:

P1 If one believes P, then P is truth-apt
P2 If one believes P, then one is committed to the truth-aptness of P.

I want to suggest that while P2 might plausibly be regarded as a platitude, P1 isn't. Of course, this ramifies for any attempt to secure the truth-aptness of a certain domain by showing that its members are things some people believe. Showing that people believe ethical claims will not, straightforwardly, show that ethical claims are truth-apt.

Still, P2 leaves open a certain sort of pragmatic argument in those cases where you yourself believe something and where it is evident that you will not relinquish that belief. Suppose, for example, you yourself believe ethical claims and will continue to do so. It would then be inconsistent for you to deny the truth-aptness of ethical claims, since you presuppose their truth-aptness by believing them. Of course if P1 were right, then one could not consistently deny truth-aptness even if one were to cease believing ethical claims oneself. P1 tells us that truth-aptness is guaranteed by anyone's ever believing the relevant claim. P2 gets us something weaker though not something so weak as to be uninteresting.

So far, I have been pondering in a general way the project of securing the truth-aptness of a domain by appealing the platitudes connecting belief and truth-aptness. Let us now turn to those arguments that attempt to secure the non-truth-aptness of a certain domain by using considerations about belief. Here again, we ought to to refine our account of the platitudes.

1.2

The putative platitude given in my original sketch of anti-truthaptness arguments that trade on the philosophy of belief — that any truth-apt claim will be a fit object of belief — is too strong. A claim might, for example, be too long and not believable, yet be perfectly truth apt for all that. Similarly, there are evidence-transcendent statements that we can scarcely imagine someone believing — say, that the number of apples past, present and future is odd. Finally there are claims that are so obviously false that we could not imagine anyone believing one of them: 'Bachelors are all married', for example. Of course, one might try to defend the claim that any truth-apt claim is a fit object for belief in the face of these examples, appealing to idealized individuals, oracular advice, madmen and so on. Yet even if such a position could be defended, one can hardly take seriously the idea that it is platitudinous.

Fortunately, these anti-truth-aptness arguments can get going without relying on anything so strong as the idea that all truth-apt propositions can be believed. In most of the domains under dispute, it does seem clear that if we found that all its claims, both short and long, fail as fit objects of belief, then that would seem to be good reason to judge that they are not truth apt. Of course, this line of thought won't be convincing in every domain. The argument that certain evidence-transcendent statements are not truth-apt because they are not believable will be unlikely to convince us. In that case, there is a good explanation of why we don't believe any member of the domain, an explanation that doesn't throw its truth-aptness into question. But in most disputed cases, we can agree that were the sentences of some discourse truth-apt, then it is quite obvious that many of its statements are believable. Perhaps something like the following can plausibly represent the operative platitude here:

P3 Unless obviously false or obviously evidence-transcendent, any truth-apt proposition that a human is capable of entertaining will be one that it is possible for a human to believe.

(Both occurrences of 'obviously' are important here. Suppose one took some religious beliefs to be evidence-transcendent. They still are not obviously evidence-transcendent. Some people regard testimony, experience, sacred texts, natural theology, cosmic order and so on as evidence for religious views whether or not they are right to do so.)

What does this leave us with? As far as anti-truth-aptness arguments go, we have left open the possibility of arguments that deploy, inter alia, P3 in order to secure a lack of truth-aptness. As far as pro-truth-aptness arguments go, we still have the possibility of pragmatic arguments that, deploying P2, run from our believing, say, ethical claims to their truth-aptness. In what follows, I engage in turn with each of these argumentative strategies.

## *2. Anti-truth-aptness Arguments*

### 2.1

Any attempt to show, using psychological considerations, that we don't believe sentences in one of the disputed areas will clearly rely heavily on the view that the mental is rather opaque. If we don't believe, say, ethical claims, that is certainly not immediately apparent to us. Worse still, the opposite will be thought immediately apparent. How is a philosopher to shake the Moorean confidence enjoyed by myself and others concerning such belief ascriptions as 'I believe that 2+2=4', 'I believe that things could have been such that my parents never met', 'I believe that it is better to be generous than selfish' and so on? Many of us will be tempted to think that the only puzzle about belief here is whether the philospher who says 'We have no moral beliefs' can really believe what he is saying! It is unlikely that the proponent of anti-truth-aptness arguments will have sufficient courage in his convictions to declare all ordinary ascriptions of the form 'I believe P' in the problematic area either false or senseless. After all, it would be a little perverse for a view that is supposed to be respecting platitudes about the mind to declare false a set of psychological ascriptions that included claims as platitudinous as any. More likely, the following standard sort of ploy will be adopted: 'When people say something of the form "I believe P" in the problematic area, they don't say something false. That is because they are only speaking loosely. Nevertheless, in the strict and philosophical sense of "believes", people don't have beliefs in the problematic domain.' More on this later.

It is important to bear in mind that we are focusing here upon denials of truth-aptness that are grounded mainly in reflections about the nature of belief. We thus should contrast psychological arguments for that conclusion from much more metaphysical arguments. Here is an example. One line of argument for the conclusion that we do not believe ethical claims might proceed using considerations of charity, which I shall put in its crudest form for the sake of clarity. Suppose Donald Davidson (1984) is right that it is constitutive of our concept of belief that we ascribe beliefs that are mostly true. Suppose we convinced ourselves that it would be outrageously optimistic to believe that there was such a property as goodness out there in the world. It might then be a violation of charity to ascribe moral beliefs: we might do well to see moral talk as expressing some other attitude. Indeed, we might be driven to say something stronger on the basis of charity considerations, namely, that it is impossible that there be a being with myriad moral beliefs. In this example, reflections upon belief are doing some modest work. In particular the principle of charity, applied to belief, is used to explain why metaphysical pessimism about a domain ought to lead one to prefer a denial of truth-aptness over an error theory about that domain. But the argument only gets going once metaphysical pessimism has been independently justified. My aim here is to see whether the psychological approach — and the philosophy of belief in particular — can make headway under its own steam.

In what follows, I shall be looking at one promising candidate for the philosophy of belief — that of ethics — seeing how an anti-truth-aptness argument from platitudes about belief might proceed and seeing why it is likely to founder. Even a comprehensive treatment of that argument is beyond the scope of this paper. Yet I hope to offer some general remarks about our concept of belief that, although sketchy, point to some instructive problems with that argument and others like it.

The argument that I have in mind issues from a strand common to much philosophising about belief. We commonly hear it claimed that there is a sort of triangle that is constituted by belief, desire and behavior. Beginning with the observation that one does what one believes will satisfy one's desires, many have contended that folk psychological explanation implicitly trades on some sort of decision theory whereby people's actions are explained on the basis of degrees of confidence assigned to various propositions together with expected utility/preference assignments to various propositions. On this story, states get to earn the title 'belief' by virtue of having this explanatory role in connection with action.

We are now invited to envisage the possibility that the states which are expressed by some area of problematic discourse are not states that play the sort of explanatory role associated with those high degrees of confidence that figure in decision theory and are thus not states that deserve the label 'belief' (unless that label is being used idiomatically). It is not all that hard to see how this argument might proceed in the case of ethics. When people make positive ethical remarks about a certain state of affairs, this points us to the fact that they value that state of affairs. Roughly, someone's valuing a state of affairs is constituted by, inter alia, their being disposed, ceteris paribus, to bring about that state of affairs when they believe the opportunity for doing so has presented itself. So described, valuing seems to function not as a belief but rather as a state that is partly or wholy constituted by a desire. Or so the proponent of this sort of argument will maintain. An excellent statement of this line of thought is provided by Simon Blackburn:

> The belief that a bottle contains poison does not by itself explain why someone avoids it; the belief coupled with the normal desire to avoid harm does. So if moral commitments express attitudes, they should function to supplement beliefs in the explanation of action. If they express beliefs, they should themselves need supplementing by mention of desires in a fully displayed explanation of action (fully diplayed because, of course, we often do not bother to mention obvious desires and beliefs, which people will presume each other to have). It can then be urged that moral commitments fall in the right way of the active, desires, side of the fence. If someone feels moral distaste or indignation at cruelty to animals, he only needs to believe that he is faced with a case of it to act or be pulled towards acting. It seems to be a conceptual truth that to regard something as good is to feel a pull towards promoting or choosing it, or towards wanting other people to feel the pull towards promoting or choosing it. Whereas if moral commitments express beliefs that certain truth-conditions are met, then they could apparently co-exist with any kind of attitude meeting the truth-conditions (1984, p. 188).

Blackburn goes on to recognise that the argument needs refinement, acknowledging that the link between moral commitment and desire is not straightforward. He then maps out a strategy for making good on the argument: one would show that while moral attitudes do not have the direct link to action that the initial line of thought would suggest, they nevertheless have an indirect effect that they would not have were they states of belief.

I want to argue that even if proponents of such arguments succeed in showing that discourse in some problematic area has a very different explanatory role in connection with action than, say, beliefs in natural science, their arguments will be radically indecisive. In the end, I shall argue, such arguments founder on an overly simplified conception of belief, one that is unfortunately symptomatic of much recent philosophy of belief.

I begin with an intuition pump. Is it really plausible to suppose that in order to represent any objective state of affairs whatsoever, a being needs to have states that connect with action in a certain sort of way? Does a mind's grasping truths require in general that it be disposed to act on those truths in certain ways? Put this way, the claim strikes me as very bold. Certainly, from the inside, the claim does not seem plausible; conscious apprehension of how things are does not seem, from the inside, to in general require action. I can imagine myself a permanently disembodied being, deceived into thinking that it is able to control a body, and yet is in truth radically unfit to do any such thing.

Nevertheless, it has to be admitted that from a third person point of view, the pill is much easier to swallow. Our evidence for attributing representational states and the explanatory point of our doing so seems to have an awful lot to do with how an agent acts in the world.

If some sort of harmony is to be achieved here, we shall need to think harder about the constitutive links that are alleged to hold both between belief and action. The root of the problem, it seems to me, lies in the fact that 'belief' is something of a catch-all that encompasses at least two sorts of related though distinct phenomena.

I want to begin by saying a bit about episodic judgments. A number of philosophers have felt the force of Norman Malcolm's (1972) suggestion that while dogs may think P for some value of P, they cannot have the thought that P. Psychological ascriptions in which 'the thought' figure tend to point to episodic mental states for which such figurative langauge as 'crossed my mind' 'dawned on me' 'occurred to me' and so on are appropriate (see Ginnane, 1960 for more on this). The expression 'judgment' is most at home in connection with mentalistic language of this sort. Judgments are, at least some of the time, conscious phenomena that are introspectable and with which each reader will be very familiar.

One point to note is emphasised by, inter alia, de Sousa (1971) and Dennett (1978): judgment that P is an all or nothing affair. Unlike the states that figure in decision-theoretic explanation, the notion of a degree of judgment is scarcely intelligible.

There is also a very intimate tie between assertion and judgment. This intimacy has been conceptualized in various ways. Some writers tend to conceive of judgment as the internalization of outward assertion, reckoning judgment impossible in cats and dogs. Malcolm seems to be thinking along these lines. Those more open to the possibility of judgment without public language will more likely conceptualize assertion as, paradigmatically, the outward expression of inner judgment.

Just as the practice of assertion is caught up in the game of what Sellars called 'the game of giving and asking of reasons' so judgments can serve as reasons for other judgments, and can be based on reasons. Indeed, the game of giving and asking of reasons is something we sometimes play alone: we inquire into our own reasons for a judgment and then try to answer those demands, sometimes satisfying ourselves, sometimes changing our minds. Not surprisingly, then, the connection of truth-aptness with judgment is just as strong as that with assertion: to judge that P is to be committed to P's being true.[8]

Let us return now to our intuition pump. It is not obvious that episodic judgments need in general connect with action in any distinctive sort of way. The reader might find the following thought experiment useful: imagine a being that enjoyed roughly the same sort of representation forming mechanisms as ourselves. That being would thus have sense organs rather like ourselves and whose ability to generate representations on the basis of perceptual imput was underwritten by intrinsic processes would be rather like ourselves. Suppose that being enjoyed nothing like our output systems, was not a being fit for acting and interacting in the world (unless one viewed its representation forming activity as itself a kind of action). Would that being be be incapable of making any judgments whatsover? I find this conclusion very implausible.

Is our concept of belief grounded in the twin concepts of judgment and commitment? One of the several good reasons that have been offered for thinking that this cannot be the whole story is the fact that we are comfortable attributing beliefs to languageless creatures; and it does not seem that this activity relies on some confidence on our part that the creatures in question engage in some activity of inner judgment.

Thus de Sousa writes:

> We attribute beliefs to dumb animals in the course of informally explaining their behavior. And this explanatory role of belief is underplayed by the theory that focusses primarily on "mental acts".[9]

Another, fairly obvious reason, is that we are all familiar with cases where a human being's action manifests a belief that P even though she is not disposed to judge that P nor even to judge some Q that straightforwardly entails P.[10]

What sorts of states are we purporting to pick out when we use folk psychological language in this sort of way? Well, (and here I follow de Sousa), such states are first appropriate to describe in terms of degree rather than the on-off switch model. It is here that the idea of a degree of belief finds its home. Second, our very conception of those states is intimately tied to that of action. Indeed, it seems that our sense of what such states are is bound up with the role we give them in explaining action. Third — and a little more controversially — it seems that the appropriate attitudinal objects for the purposes of such explanations is something as course-grained as a set of worlds. On the standard conception, degrees of confidence and utility assignments operate upon carvings-up of the possibility space. When we conceptualize judgment and assertion, meanwhile, a rather more fine-grained conception of content is called for. I want at this stage to flag the second and third of these points as ones that are going to be especially important for us.

In sum, there seems to be two sorts of related though distinct phenomena that our concept of belief gestures towards — on the one hand, judgments and commitments acquired through judgments and/or dispositions to judge. On the other, degrees of confidence whose role in folk psychology is usefully systematised by some sort of decision-theory. Corresponding to these sets of phenomena, I would suggest that there are two layers of explanation of action — on the one hand, explanations that appeal to occurrent states of the on-off variety. On the other, explanations of a more decision-theoretic sort. Does this make our concept of our belief a grab-bag? Well, that depends upon what sort of unified story one can tell about how degrees of confidence relate to judgment. One would certainly hope that while the phenomena of judgment and degree of confidence are distinct, some systematic account of their relationship could be provided.[11]

2.2

Let us now focus on the states that we are, prima facie, willing to describe as ethical judgments. If the anti-truth-aptness proponent is to make any headway at all here, he will have to convince us of something like the following: 'So-called ethical judgments bear a very different relation to action that judgments in the unproblematic category. It is constitutive of so-called judgments that a certain state of affairs is good that that one finds actions that one thinks will bring about that state of affairs compelling and hence, ceteris paribus, one will be disposed to bring about that state of affairs.

So far, we have been trying to look at the prospects for anti-truth-aptness arguments in as favorable light as possible. Unfortunately, my optimism can go no further. Suppose someone made good on all the optimistic goals I just set for the anti-truthaptness arguer in ethics. Ought this to shake our confidence in the claim that we make ethical judgments or lead us to concede that it is only in a loose and idiomatic sense that we do so? Now without a systematic account of the relationship of judgments to decision-theoretic explanations, it may be felt impossible to choose between these responses in a satisfying way. Such pessimism is unwarranted, however. I offer two sorts of considerations to the effect that we have excellent reason in the envisaged scenario to affirm that so-called ethical judgments are genuine judgments.

First, as should be clear, it seems to me that the view that so-called ethical judgments are not really judgments has been motivated largely by a mistaken identification of the state of high degree of confidence with judgment. The states of confidence that play a decision theoretic role with regard to action are neither states of judgment nor dispositions to judge. Given the distinctness of these states, it seems that no obvious violation of commonsense platitudes are going to be achieved by endorsing the first response. The relation of occurrent judgments to action is, in general, conceptually rather more tenuous than is the case for decision-theoretic states. We don't get our handle on the concept of judgment by giving it a constitutive connection to action. Thus if we were to discover that one particular class of so-called judgments — in this case ethical ones — have a distinctive relation to action of some kind, that is not going to violate some general platitude about how a judgment, qua judgment, connects with action. Meanwhile, common sense is clearly inclined to say that we indulge in ethical judgments. After all, there are all sorts of features that so-called ethical judgments have in common with the standard case: They are phenomenologically just like paradigm judgments; the game of giving and asking of reasons that we play with ethical judgments and their linguistic expression has a great deal in common with critique of ourself and others in other domains; we happily use the language of judgment when describing our own attitudes to ethical claims; and so on. It thus seems clear to me that if our aim is to stay within a conservative framework that strives to respect common sense platitudes, unless given decisive reason not to, and that aims to construct a psychological theory within that framework, then we should maintain that there are ethical judgments no matter what interesting discoveries we make concerning how they relate to action. At this point, it should also be clear that similar considerations will apply to other disputed domains. If our aim is to stay within a conservative framework, then we should recognize that so-called religious, mathematical, aesthetic and modal judgments are all genuine judgments.

Second, while moral judgments may have a somewhat desire-like relation to action, there is good reason not to see the same point as applying to putative high degrees of confidence in ethical propositions. The reason has to do with the widely maintained view, mentioned earlier, that it is sets of worlds that form the proper objects for degrees of confidence. Take any set of worlds. A high degree of confidence that this world belongs to that set need not be essentially motivating. To see this, one need only bear in mind that any set of worlds can be described in many different ways: Take the set of worlds where Joe is good. That may be the very same set of worlds where Joe has God's favorite property, the very same set where Joe a certain incredibly long disjunction of physical descriptions is true of Joe and the very same set of worlds where various metalinguistic claims are true (most obviously, where Joe has the property denoted by the English word 'good'). Only relative to a certain description does any set of worlds have even a prima facie case to be essentially motivating. Now if the widely maintained view about the proper form for decision-theoretic explanations is right, degrees of confidence ascriptions will be insensitive to various ways of describing the sets that are the objects of confidence. But if that is so, one wouldn't expect that a high degree of confidence in the set that Joe is good has any conceptual connection to action. After all, that is also a high degree of confidence that Joe has God's favorite property and so on. Hence the only conceptual connection that such a probability assignment would have with action is of the normal decision theoretic sort — one that works with utility assignments to explain action. Thus even if so-called judgments about goodness conceptually involve the agent's seeing certain ways of acting as compelling, that need not at all require one to maintain any such conceptual ties between degrees of confidence in ethical propositions and action.

In connection with all this, let us reflect on a well-known debate, namely, whether ethical thoughts can explain actions without implicitly relying on certain desires. The answer may depend upon which layer of explanation one has in mind. While at the decision-theoretic level, one's desires — utility assigmnents — will be called on to explain action, it may be that for the purposes of causally explaining action in terms of occurrent mental states, ethical judgments need not be coupled with desires. In this way, one might happily use ethical discourse both to describe the objects of judgment and the objects of probability assignments, without fearing that a constitutive connection of judgment to action will inevitably vitiate one's decision-theoretic assignment of confidence in ethical propositions.

In sum, then, there are two clusters of states that come under the broad rubric 'belief' that are no doubt related in interesting ways and yet must be kept distinct: judgment on the one hand, degrees of confidence on the other. The former is most intimately related to the concept of assertion and the latter to the concept of action. Focusing on the case of ethics, I have tried to show that once one distinguishes between these two clusters, it will become rather clear that the standard anti-truthaptness argument grounded in the philosophy of belief is unlikely to succeed. For there don't seem to be psychological data that ought to shake our conviction that we make judgments in the problematic areas. Meanwhile, the connection between judgment and truth-aptness is a conceptually strong one. If one is to somehow undermine the claim to truth-aptness of these problematic areas, it seems unlikely that reflection on belief will provide the argumentative resources.

Before concluding this section, there are two further lines of thought that I would like to pursue very briefly.

2.3

First, let me rehearse an objection to our main truth-aptness argument that doesn't trade on the bifurcation I have claimed between judgment and confidence. Let us recall how, say, the emotivist in ethics is likely to respond to the query, 'Do you mean all belief ordinary ethical belief ascriptions are false?': 'No because, strictly speaking, such ascriptions do not attribute beliefs. People are merely speaking loosely when they say something of the form "I believe (some ethical P)"'. On this account, then, we have to distinguish a liberal use of 'believes' according to which people have ethical beliefs from a strict use of 'believes' according to which they do not. The following issue now arises. Which usage of 'belief' figures in the platitudes? Should we understand 'belief' as it occurs in 'Belief in some P involves commitment to its truth-aptness' strictly or loosely? Is the folk connection between truth-aptness and belief confined to the strict use of belief? Or is it rather that the connection with truth-aptness holds for even beliefs that are loosely so called?

These questions are, of course, extremely pertinent to the whole dialectic. Suppose we convinced ourselves that in a strict and literal sense, belief requires a certain connection to action but that in a loose sense we can call other states 'beliefs', perhaps by analogy to cases that, in the strict sense, count as beliefs. If the platitudes about truth-aptness concern both strict and loose beliefs, so to speak, then one will not thereby have made any progress whatsoever to undermining the truth-aptness of a domain on the basis of psychological reflection. (The problems will of course deepen further if we begin to distinguish strict and loose senses of 'true'.)

Let us explore a simple analogy. It might be felt that in a strict and philosophical sense, no physical object is flat on account of the bumpiness endemic to matter while conceding that, loosely speaking, plenty of things of flat. Clearly, however, folk platitudes about flatness should be read as having the liberal sense in mind. When someone says 'If something is flat, it has no bumps' they are not thinking of the antecedent as something that couldn't be satisfied by any physical object at our world.

What good evidence is there, then, for supposing that the folk platitudes about belief and truth-aptness concern the strict use of 'belief'? It seems to me that if there is any good reason at all here, it will be grounded on metaphysical reflection. If one were antecedently convinced that there were no facts for 'loose beliefs' to correspond to, then one might charitably read the platitude as concerning belief in the strict sense.[12] There is thus no straightforward route from folk platitudes about truth-aptness and belief coupled with the claim that, strictly speaking, belief requires a certain relation to action, to the conclusion that, say, ethics is not truth-apt. Again, it seems that metaphysics will have to call the tune.

There is a general lesson that is to be learned here, I think. Once one begins to make systematic use of the distinction between loose ways of talking on the one hand, strict and philosophical ways of talking on the other, then there may be far less to be gleaned from common sense than one might initially think; for in the light of that distinction, the issue as to how to translate the words of common sense into the words used to express certain philosophical theses may be extremely problematic.

## 2.4

One who found something like the above remarks about judgment persuasive might try to reconstruct a rather different sort of argument concerning the truth-aptness of ethics. Persuaded that a being radically unfit to act in the world would be perfectly capable of conscious judgments, one might find the following line of thought compelling: 'There is no reason in principle why such a being might not know everything we know. Any objective truth is one that is in principle graspable by an actionless being of this sort. If so-called ethical truths and falsehoods cannot be grasped by an actionless being, then they are not truth apt after all.'

I actually find this argument rather more interesting that the main argument that we have been addressing. But it is still rather indecisive. The argument requires a fine-grained individuation of truths — if the truth that Joe has God's favorite property is the same truth that Joe is good, then the argument will break down for reasons that should be obvious. The crucial issue, then, is whether we have good reason to think that every mode of presentation of every state of affairs would be in principle available to an actionless being. Lacking the space to discuss this matter fully, I merely wish to remark that I see no good reason why some modes of presentation cannot be constituted in part by some motion from the mind to action, especially when the truths in question are, arguably, neither observable nor explanatory of what is observable. (Another relevant point is that some modes of presentation may well have to do with experiencing various possibilities of one's own practical involvement in the world. I have been told that this sort of theme is important in both Husserl and Heidegger.)

Before turning to protruth-aptness arguments, I should stress that the point of this section has not been to suggest that all anti-truth-aptness arguments are bound to fail. Rather, it has been to suggest that such arguments are likely to gain little impetus from reflections about the nature of belief.While I have focussed on the case of ethics, the obstacles raised are not peculiar to that case.

Perhaps there are no sets of worlds that constitute the truth conditions of, say, ethical discourse and which can serve as the objects of probability assignments or of judgments. But it oughtn't be philosophical reflection upon belief that convinces us of this.

### *3. The Pro-truth-aptness Argument.*

3.1

The most interesting version of a pro-truth-aptness argument will not proceed by merely arguing that, up to this point, we have had beliefs in some problematic domain. For the philosopher may quite properly respond: 'Well that goes to show that in the past I have improperly committed myself to the truth-aptness of that domain. I am now going to revise my doxastic practices and with good justification, since I now have good reason for thinking that the domain is not truth-apt.'

The interesting case, I suggest, is not simply one where we feel confident that people have had beliefs in some domain, but where we have excellent psychological evidence that this is not a practice that they are about to give up. Stronger still, we might have excellent psychological evidence that this is not a practice that any healthy human being — philosopher or not — is about to give up. Due to the complexities concerning 'belief' that I raised in the last section, lets focus on judgment in particular. We might have excellent psychological evidence in the case of, say, ethics, that no healthy human being is about to refrain from ethical judgments for good.

To many of us it will seem that such a situation already obtains for many clusters of concepts. Suggestions about, say, displacing folk psychological or ethical concepts by some philosophically sanctioned surrogates, do seem to manifest a heady, almost comic, optimism about the ability of philosophy to intervene in human history, to tinker with the fabric of our conceptual scheme.

One is reminded here of Hume, whose psychological investigations convinced him that the mind is so constituted that abstruse philosophical reasoning of even the most flawless kind will have little effect on our beliefs.

I dine, I play a game of back-gammon, I converse, and am merry with my friends; and when after three of four hour's amusement, I would return to these speculations, they appear so cold, and strained, and ridiculous, that I cannot find in my heart to enter into them any further (*A Treatise of Human Nature* p. 269).

Presaging Goethe's contention that what won't appeal to the senses will have little weight with the intellect, Hume insists that the 'effort of thought' required for metaphysics 'disturbs the operation of our sentiments on which the belief depends' (*A Treatise of Human Nature*, p. 185). Another Humean theme is that it is for our own good that our minds are largely immune to philosophical speculation. Thus he writes of our belief in the external world that 'Nature has not left this to his choice, and has doubtless esteem'd it an affair of too great importance to be trusted to our uncertain reasonings and speculations' (*A Treatise of Human Nature*, p. 187). Even those of us that don't think with Hume than unfettered philosophy would lead us to self-destruct through sceptical doubt may concur that the fixity of certain basic beliefs in the face of philosophical argument evolutionary beneficial, if only because we're not all that good at philosophy.

Let us now ask what would follow were we convinced of the fixity of certain doxastic practices. How would a philosophical view that denied truth-aptess to one of the relevant domains of belief fare in that situation? One might certainly worry that this philosophical enterprise was importantly futile in this connection. This will no doubt bother some metaphysicians and not others. Ought they to be bothered? This brings us to the issue of rationality, and it is that issue that I wish to focus on here. Here we need only recall the consistency argument raised in section one. Isn't it a dictate of rationality that one espouse views that are consistent with other judgments that one makes? Suppose we had reason to think that we are, by nature, bound to indulge in ethical judgments. How can one rationally embrace philosophical convictions that conflict with the presuppositions of the ethical judgments themselves? Isn't this just an ordinary case of embracing a view that is inconsistent with views that one already holds and hence an act for which epistemic criticism is due? Viewed this way, psychological data about the inevitabilities of our doxastic practices seem capable of rendering certain philosophical enterprises both pointless and irrational.

Can metaphysics really be at the mercy of psychological investigation in this sort of way? The question is a delicate one. On the one hand, it seems almost distasteful, when doing metaphysics, to dirty our hands with psychological observations about what sorts of judgments are inevitable for us. But on the other, we don't want to violate our own canons of rationality. Let me briefly explore two strategies that the denier of truth-aptness might follow here.

3.2

First, he might be comforted by the fact that there do seem to be cases where one can rationally embrace a view that is inconsistent with other views one holds. For example, I can rationally embrace the belief that at least one of my beliefs is false. Perhaps a metaphysic that conflicts with certain fixed beliefs can be assimilated to this sort of case. Or so a philosopher might hope. In this connection, van Inwagen imagines (though is not concerned to endorse) the following speech delivered by one who conceded that a human's deliberating manifested a belief in free will and yet, on the basis of philosophical argument, wished to maintain that there was no free will:

> I have inconsistent beliefs, but that does not trouble me. The only way for me to achieve consistency would be for me to give up my assent to the obviously true proposition that we lack free will. This is too large a price to pay for consistency, the only virtue of which, after all, is that it does not logically necessitate having a false belief. But, apparently, I must have a false belief whatever I do, so this feature of consistency, though a virtue in the abstract, would not benefit me (1983, p. 160).

Perhaps a more interesting strategy is exemplified by Quine, who faces the sort of dilemma that we are discussing here with regard to intentional idioms. For while Quine, qua philosopher, wishes to deny that such idioms are truth-apt, he seems to concede that intentional talk is something which, as people, we are inevitably committed to.

With regard to the first point, he writes:

> If we are limning the true and ultimate structure of reality, the canonical scheme for us is the austere scheme that knows no quotation but direct quotation and no propositional attitudes but only the physical constitution and behavior of organisms.But if our use of canonical notion is meant only to dissolve verbal perplexities or faciliate logical deductions, we are often well advised to tolerate the idioms of propositional attitude (1960, p. 221).

And with regard to the second point:

> Brentano was right about the irreducibility of intensional discourse.
>
> But there is no dismissing it. It implements vital communication and harbors indispensible lore about human activity and motivation, past and expected. Its irreducibility is all the more reason for treasuring it: we have no substitute. At the same time there is good reason not to try to weave it into our scientific theory of the world to make a more comprehensive system (1990, p. 71).

It is as well that Quine makes this last concession, since his own work is litered with the intentional idiom. Consider for example: ' In the same spirit even the mathematician, realist *ex officio*, is always glad to find that some particular mathematical results that had been thought to depend on functions or classes of number . . . ' (1960, p. 270).

There is a natural reading of the pair of passages indented above: We use some discourses with the intent of saying what is true, others with the intent of peforming some other task — dissolving verbal perplexities and so on. The trouble is that this can't be Quine's considered view. That way of describing things relies on a distinction between intending to describe how things truly are and intending to do something else; yet the distinction between those two sorts of intentions can't really have a place in his system. After all, an irrealist about intensional discourse would surely be in trouble if his irrealism were grounded upon a distinction — between two sorts of intending — that requires intensional discourse in order to be made out. Quine, qua philosopher, recognizes only dispositions to assent to sentences. Lacking no notion of 'quasi-assent' , he will inevitably concede that we all assent to sentences employing the intentional idiom. Further, he concedes that any sort of theory to which we assent can usefully enjoy a truth predicate — the basic function of a truth predicate being the modest one of generalizing 'along a dimension that cannot be swept out by a general term' (1990, p. 80) as when we say that every sentence in a certain book is true. He ought thus to concede that not only are we going to continue assenting to intensional discourse, but moreover that it is appropriate to invoke a truth predicate in that context. So what is going on? Isn't Quine now guilty of the doubletalk that he has inveighed against elsewhere, assenting to the truth of psychological talk in the common affairs of life and then retracting it in his philosophy books?

I have just described one important point at which Quine's philosophical views totter on the brink of incoherence. Yet I do think that he has some sort of story to tell that does not rely on an important distinction between intentions to limn the true and ultimate structure of reality and other intentions. Let me then sketch what I take to be Quine's considered position on these matters. While I haven't the space to defend the interpretation here, it does suggest another general strategy by way of which the philosopher can respond to the pro-truth-aptness argument we are considering.

Put informally, Quine's strategy is one of oscillation. One can usefully deploy here the Wittgensteinian idiom of switching from one language game to another, an idiom that Quine himself deploys in his recent *Pursuit of Truth.*[13] He sees himself as oscillating from an austere scientific theory where the aim of the game is prediction and where an austere physicalistic language is appropriate, to a more liberal language game, better suited to other human tasks such as communication. When speaking from with the language of austere physics, Quine recognizes truth-aptness as being restricted to such language. When speaking from within the more liberal language game, Quine assigns truth-aptness to, inter alia, psychological discourse. One particular Quinean view about truth fits in nicely here. Truth, Quine tells us, is immanent. When we embrace a theory, our truth predicate ranges not over all possible sentences but over the sentences that figure in the language of that theory. His austere physicalism will deploy a theory T1 with a truth predicate — 'True in T1' — that ranges over the sentences in the language of T1. Common sense intensional discourse will not figure in the language of T. Hence, when one is speaking T1, one will not reckon common sense intensional sentences as truth-apt. When engaged in the language of ordinary affairs, Quine will, in effect, speak a different language, T2, with a different truth predicate 'True in T2' that ranges over the sentences of that langauge. He then oscillates between T1 and T2. When speaking the language of T1, he will not say, 'It is true or false that John believes that snow is white' but when speaking the language of T2, he will assent to talk of that sort. He thus oscillates from an austere view of the world, which in philosophical moments suits him. to a more generous view of the world which his humanness forces upon him. Interestingly, this sort of oscillation is rather like the sort of oscillation that Hume, that other great empiricist, described in himself. While in philosophical moments, he describes himself as capable of suspending belief in an enduring self, he concedes that 'we cannot long sustain our philosophy, or take off this biass from the imagination. Our last resource is to yield to it, and boldy assert that these different related objects are in effect the same, however interupted and variable' (*A Treatise of Human Nature*, p. 254).

If I may be permitted to deploy the notions of judgment and commitment, the picture we get is that in philosophical moments we are indeed capable of suspending our disposition to make certain sorts of judgments and hence our commitment to sentences in some domain. The psychological fixity of our commitments consists in this: we cannot suspend these commitments for long — in the ordinary affairs of life, such commitments will come flooding back.[14]

Suppose now a philosopher oscillates from commitment to ethical discourse to commitment to its non-truthaptness. What are we to make of this? Is it a sort of doubletalk? Clearly, it is, in a way. But returning to our original question, is it rational to oscillate in this way? I shall not pretend to be able to offer a considered answer here. Such questions about what can only be called the epistemology of metaphysics require a more articulate conception of the normative proprieties of judgment than I have at my disposal. Only against the background of some remarkably well developed account of this sort can such difficult cases as those of our two metaphysicians — the inconsistent and the oscillating — be properly judged.

Certainly — as van Fraassen notes in *Laws and Symmetry* (1989, ch. 7) — Bayesian accounts of rational belief change, as well as other accounts of a fairly formal nature are likely to be ill-suited to provide accounts of the rationality of radical belief change of the sort that the oscillator indulges in. Perhaps such a shift is fundamentally arational on account of the fact that our concept of rationality is ill-fit to apply to such a case. Perhaps instead we might pursue van Fraassen's idea that judgments about rationality in this sort of case will be guided by rather looser considerations pertaining to our broad sense of cognitive virtues and vices — recklessness, foolishness, courage, modesty and so on. Perhaps it is even appropriate here to draw on broader considerations of value, where the question of whether we ought to pursue this brand of metaphysics will acquire at once an epistemic and moral flavour .

At least this seems clear enough, however: if synthetic a priori knowledge of the sort that metaphysicians have historically yearned really were possible, then one could hardly convinct a philosopher of irrationality who suspended ordinary convinctions on the basis of a synthetic a priori vision, even knowing that she were so constituted that she would inevitably lapse back into those convictions. Were such synthetic a priori knowledge a pipedream, (as Quine and others insist), the situation will be less clear. Yet one might still find reason to admire the metaphysician's insistence on broad theoretical virtues and her concern to apply more exacting standards to her theory than ordinary folk do to theirs. The point is not that we should convict ordinary folk of irrationality. The question is whether we are well-placed to convict the metaphysician of irrationality.

There is a lot more to be said here, obviously, about the value and rationality of metaphysics. Suffice it to say that psychological reflection about belief is not by itself going to leave us feeling rationally compelled, as metaphysicians, to endorse the truth-aptness of a domain. What is needed in addition, is a well-developed conception of the normative proprieties of metaphysical theorizing.

*Conclusion*

Some philosophers clearly hope that reflections about belief will offer decisive reason for affirming or denying the truth-aptness of a domain. The results of this paper lend themselves, I hope, to a fairly pessimistic view of that project. In this respect at least, the project of limning the ultimate structure of the world by way of reflection on how the world is represented is likely to fall flat.

*The Universtiy of New South Wales*

NOTES

[1] For helpful discussion, I am indebted to Mark Lance, Michael Smith, Diane O'Leary-Hawthorne and especially Michaelis Michael.
[2] This last feature figures prominently in Brandom, this volume.
[3] I am grateful here to conversations with Richard Holton and Michael Smith.
[4] They may not. One might try to answer these issues by formulating a theory of truthaptness. But one might instead be more Moorean. Thus one might say that it is just obvious that certain claims of ethics, mathematics and so on, are true and that any philosophical argument against those certainties will have premises less obvious than the conclusion that attacks. One who proceeded this way might feel no need to formulate general criteria of truth-aptness, minimalist or non-minimalist.
[5] See Simon Blackburn (1984, ch. 6) for a variant of this argument.
[6] Or at least this is how the conservative will characterise the radical. The radical might be less sure of the factuality of the distinction between what belongs to common sense and what doesn't, what is platitudinous and what isn't, what is analytic and what isn't and so on. Tricky metaphilosophical issues abound in this area.
[7] We should be careful not to place too much ontological weight on 'genuine property' here, however. The nominalist ought not to be excluded from these debates. Even the nominalist may want to insist that some words that occupy the grammatical place of a predicate are unfit to be true or false of anything, thus distinguishing sense from nonsense.
[8] The last sentence introduced the notion of commitment. In connection with someone's judgments we might happily speak of what she is committed to: this might include those things she has judged and not subsequently retracted or a little more liberally, those propositions she is disposed to judge were she to entertain them or more liberally still, those propositions that follow from what she are disposed to judge whether or not she recognizes those entailments. Robert Brandom (1983 and this volume) is helpful here.
[9] De Sousa, op. cit., p. 54.
[10] I myself find the second rather than the first reason ultimately the more persuasive, since I am

not sure of how to give a fully coherent account of even decision-theoretic states where inner and outer judgment are altogether absent. I discuss this sort of issue in O'Leary-Hawthorne (1994).

[11] To see what an account might look like here, the reader should look to de Sousa (1971). De Sousa's central idea is to think of judgment as an action that can be explained by decision-theoretic states.

[12] Even here, though, there are difficulties. One might instead say that the platitude uses a liberal sense of 'believes' together with a liberal sense of 'truth', raising the question 'What, if anything, does common sense have to say about what, strictly speaking, is true?'.

[13] Now of course one might try to use such lack of explanatory value as a sign in itself of non-truthaptness. This line of thought certainly needs addressing, but since it belongs to the metaphysical approach, I shall not be addressing it here.

[14] 'But when I cite predictions as the checkpoints of science, I do not see that as normative. I see it as defining a particular language game, in Wittgenstein's phrase: the game of science, in contrast to other good language games such as fiction and poetry' (1990, p. 20).

[15] How to describe such a scenario in decision theoretic terms, I will not worry about here, though interesting questions arise in that connection. In particular, one may reasonably wonder whether one can even temporarily cease to assign degrees of confidence (high or low) to those claims belonging to the various problematic areas of discourse under dispute.

## REFERENCES


Blackburn, S. (1984) *Spreading the Word* (Oxford: Clarendon Press).

Brandom, R. (1983), 'Asserting', *Nous* 17.

Davidson, D. (1984), 'Radical Interpretation' in *Inquiries into Truth and Interpretation* (Oxford: Clarendon Press).

Dennett, D. (1978), *Brainstorms* (Cambridge, Mass: Bradford Books).

Dummett, M. (1976), What is a Theory of Meaning? (II)' in G. Evans and J. McDowell, eds., *Truth and Meaning* (Oxford: Oxford University Press).

Ginnane, W.J. (1960), 'Thoughts', *Mind.*

Grice, P. (1957), 'Meaning', *The Philosophical Review* (July).

Hume, D. *A Treatise of Human Nature*, ed. Selby-Bigge, revised by Nidditch (Oxford: Clarendon Press).

Jackson, F., Smith, M. and Oppy, G. (1994), 'Minimalism and Truthaptness', *Mind*, (July).

Kant, I. *Critique of Pure Reason*, translated by Norman Kemp Smith, Macmillan Press.

Mackie, J.L. (1977), *Ethics - Inventing Right and Wrong* (Harmondsworth: Penguin).

Malcolm, N. (1972) 'Thoughtless Brutes', APA Presidential Address.

O'Leary-Hawthorne, J. (1994), 'Belief and Behavior', *Mind and Language.*

Peacocke, C. (1992), *A Study of Concepts* (Cambridge, Mass.: MIT Press).

Price, H. ( 1989 ), *Facts and the Function of Truth* (Oxford: Blackwell).

Quine, W.V. (1960), *Word and Object*, Cambridge, Mass.: MIT Press.

Quine, W.V. (1990), *Pursuit of Truth* (Cambridge, Mass.: Harvard University Press).

Russell, B. (1905), 'On Denoting', *Mind*, xiv.

de Sousa, R. (1971), 'How to give a piece of your mind: or, the logic of belief and assent', Review of Metaphysics.

Strawson, P. (1950), 'On Referring' in *Mind.*

Van Fraassen, B. (1989), *Laws and Symmetry* (Clarendon Press: Oxford).

Van Inwagen, P. (1983), *An Essay on Free Will* (Oxford: Clarendon Press).

Wright, C. (1992), *Truth and Objectivity* (Cambridge, Mass.: Harvard University Press).

MICHAELIS MICHAEL

# CUBISM, PERSPECTIVE, BELIEF[1]

Artists and even school children learn the techniques of representing in perspective, of creating representations involving points of view. As such it might seem that, as notions go, the notion of perspective is fairly straightforward; the sort of notion which is capturable within a simple teachable theory. Perspective, its various techniques and theories, can be seen as a relatively late development in the pictorial arts. It can be seen as a development which was eventually displaced from the centrality it had achieved during the Renaissance in the late nineteenth century. And in this frame of mind we can discuss the various techniques for achieving 'illusions' of perspective. We might, as ancillary to this discussion, engage in the psychological inquiry of uncovering the processes exploited by these various techniques and we might reasonably hope to achieve some considerable success in our endeavours. An important step towards this goal was taken by Ernst Gombrich [1962]. But there is a frame of mind, which is not so very hard to achieve, in which the notion of perspective is as peculiar and interesting as the mind itself. In this frame of mind the notion of perspective lies within a constellation of notions which together form a prototheory of agency in the world. To be an agent is *inter alia* but quite centrally to have a perspective on the world. In this frame of mind there is something in the very notion of a perspective which threatens irreducibility of the notion.

In this essay I will be trying to make some suggestions about the illumination that the notion of perspective can bring to the notion of agency, or intentionality. I will begin by examining the possibility laid before us by the Cubists of a synthesis of perspectives of a certain sort. The Cubism I have in mind is that which sometimes gets called 'Synthetic Cubism' and which dates from around 1910. This part of the Cubist project was governed by a desire for a kind of 'view from everywhere'. I will then examine a different but related desire harboured in the hearts of many philosophers for an objective sort of perspective which deserves the name Thomas Nagel gave: 'the view from nowhere'. I will be arguing that mistakes in disparate areas of philosophy are attributable to this desire. In particular, I will suggest that implicated are mistaken views about objectivity, and about the relation between semantics and psychology.

*Michaelis Michael and John O'Leary-Hawthorne (eds.), Philosophy in Mind, pp.243-276.*

*I Cubism and the view from everywhere*

When Braque and Picasso began trying to overcome what they saw as problems inherent in representational painting as it had been done, they embarked on the project of representing a number of perspectives at once.[2] The way they conceived of this project brought about the creation of cubism as an ideology in art. The way they conceived of this project also contains within it the impossibility of the successful resolution of the project itself. However, and in a real sense, cubism stands apart from the dominant trends in modernist art which by their self-consciousness of the creative act bring into question the very notion of representation and its limitations. Cubism was a confident and brash attempt to create works which not only were representations but represented all aspects of that which is being represented in a kind of synthesis. The successful cubist work would be a complete synthetic representation in a way that no previous art work could claim to be.

Cubists sought to represent not just the objects in the world, but rather the very apprehension of those objects by perceivers. Michaelangelo's omniperspectival sculpture, the first designed to be observed from any position, may well have been a kind of model for what they had in mind. But even with Michaelangelo there were limitations. His sculpture could not be seen from more than one point of view at once. A viewer would have to physically move themselves around the sculpture to attain the multiplicity of perspectives captured by the artist. Moreover, the work in itself represents the object in the world not the relations of that object to the perceiver. How can the cubist artist succeed in capturing the multiplicity of perspectives which Michaelangelo might be thought not to have?

One thing the cubists made explicit in their works was something which can be seen in the techniques of Velásquez and Cézanne, the way the perceiver makes a significant contribution to the perception. The technique of partial or fragmented perspectives achieve this end in a very elegant manner. A very nice example of this is the work of Jean Metzinger called 'Tea Time' painted in 1911 and which includes a teacup and saucer.[3] The work contains no full representation of the teacup. Rather, we are treated to a series of fragmentary, overlapping, and almost transparent glimpses or perspectives of the teacup. We find ourselves filling out the various partial glimpses to give a series of perspectives on the scene. We see the teacup from above and from the front and these filled out perspectives are distinct views of the same scene. The project of cubism is to create a work of art that captures the multiplicity of perspectives associated with the scene, but not simply as distinct perspectives. The perspectives are to be synthesised, interrelated and yet not collapsed into a unitary perspective.

Whether or not the sort of phenomenalisms which were in vogue at the same time in positivist metaphysics had a direct bearing on cubism or not has not, to my knowledge, been uncovered.[4] Even assuming such a phenomenalism, the project of cubism is a rather surprising one. If a representation were a representation of a teacup, then, in order to be complete, it would seem to have to be able to represent itself, as well as the other representational perspectives, otherwise the representation would not be complete in the way it was alleged to be. But then maybe the representation does not present a distinctive perspective on the teacup at all. That would be odd. The threat is that from the phenomenalist point of view, a complete collection of representations is not at all distinct from the object of representation. While this is rather like Bishop Berkeley's identification of objects with congeries of ideas, obviously what the artist creates is not the represented object itself.

The problem is that striving after completeness in the cubist sense seems to introduce the threat of paradox into the very project. Perhaps it is better to restrict ambitions by restricting the sorts of representations incorporated in the work. But even if we ignore completeness, and focus on the achievement of the synthesis of perspectives problematic residues remain. Even restricted in some way, the project retains its paradoxical tensions. How does one synthesize a multiplicity of perspectives without their collapsing into a single perspective? What does it mean to incorporate this multiplicity? Are the perspectives 'still there' or have they been 'transcended'? I suggest that this question is of central importance if we are to understand what it is to have a point of view.

At least in the theoretical writings of Georges Braque and Jean Metzinger, the project of cubism was representational.[5] The idea of achieving fidelity by assiduous falsification has been a trend in the so-called representational arts for hundreds of years. According to Michael Frede [1990], the development of Diego Velásquez's style is a discovery of the lie that reveals the truth. Indeed this view echoes José Ortega y Gasset's earlier remark that 'On a canvas of Velásquez, each section contains only vague and monstrous forms' [1968: 115]. The truth of Velásquez's canvas is not to be sought in the correspondence of each part of the canvas to a part of the world, but rather the relation of the whole to the whole experience of seeing. The seeing, not the seen (scene), is the object of the representation. Cubism is a continuation of this idea that fidelity demands judicious lying. It stands as heir to the tradition which led from Velásquez up to Cézanne, a tradition which involved the application of highly philosophical doctrines to the study of the phenomenon of seeing. These doctrines did not prove their truth through the aesthetic power of the works they brought about. For example, the works of Georges Seurat represent, amongst other things, the aesthetic power even of utterly false doctrines of the phenomenon of seeing. In much the same way, utterly false doctrines can be philosophically powerful. The claims of the cubists to be presenting the world as encountered, in fractured and essentially plural perspectives, might be thought either to fall into the mistakes of Seurat or, alternatively, to develop the insights of Velásquez.

So what did the cubists attempt to capture in their work? It would be a mistake to think that cubism, or any but the most single-minded of projects, is characterizable in a straightforward manner.[6] It remains nevertheless true that *a* motivating concern, at least at one point, was the problem of how to portray a synthesized multiplicity of perspectives associated with a given object.

Velásquez figured in this curious problem of synthesizing multiple perspectives when Kenneth Clark [1957; 9] once wrote of how he tried endlessly to discover the way Velásquez achieved some effect but succeeded only in seeing dabs of paint and brush strokes, or alternatively the full illusory effect in its representational glory. There seemed to be no way of seeing how the effect was achieved by the dabs and strokes. Clark could not solve the problem of synthesizing these disparate perspectives.

This is no straightforward problem. To get a sense of this problem let us see what would not count as a successful solution. We already saw that the sculpture of Michaelangelo designed to be viewed from a multiplicity of perspectives was not a solution. Moreover we could not solve the problem simply by representing the object from a multiplicity of perspectives, much in the way that architects will display a building from both plan and elevation. This is certainly a way of representing a multiplicity of perspectives but without the requisite synthesis. We can imagine the fortuitous circumstance in which the different perspectives on an object are represented in the same way, say in the case of a perfectly round ball or in the case of a perfect cube. In these cases, an elevation view and a plan view would look alike. Now we could imagine an artist drawing one of these views with the aim of achieving the goals outlined by cubists, and then claiming that since the very same drawing would suffice as a representation of the other views as well, that this work achieved the goal of synthesizing a number of views. But this claim would be mistaken. The drawings may be synthesized but the perspectives are not.

The idea of parts of a representation carrying with them their own point of view is clearly implicated in the work of the cubists, and reaches another high point in the work of M.C.Escher but it is quite likely a plausible way to view the early work by painters who preceded the Renaissance. Giotto's solid figures fit this characterization well. When we look at one of Giotto's frescoes we find ourselves drawn into the work at numerous points, discovering many independent parts of the whole with their own centre of attention, with their own point of view. It is almost as though the work as a whole comprises the various scenes that captured the artist's attention. In that case then the whole work is not to be thought of as something which presents the artist's experience at any time, but rather a collage of smaller elements which are created more or less independently. That is not to say that when we look at Judas kissing Jesus, that this part of the larger work will form a part which carries with it a unique point of view. Indeed it is quite remarkable that what is said of the parts in relation to the whole can be said of many of the parts of the parts themselves. The sense in which these paintings fail to be perspectival is that they suffer from too many perspectives rather than too few. Of course this multiplicity cannot extend indefinitely: we see Judas in profile. We see his left eye but not his right eye. So in some imperfect way Giotto's work carries with it the location of the viewer. The multiplicity which is imperfection in the work of Giotto is perhaps most completely exploited by Escher.

Shrugging their shoulders, the bemused artists might then move to another medium to attempt the achievement of the synthesis. Using the exciting possibilities of stereoscopic photographs and the appropriate viewing apparatuses artists can achieve something which they might claim to be what the cubists were looking for. After all each of the photographs presents a distinct perspective on a given subject matter. When they are put together in the appropriate way by the apparatus, the result is a synthesis of these distinct perspectives. Indeed the artist might reflect for a moment and realise that the stereoscopic photographs are just a way of achieving what we ordinary binocular viewers achieve all day long. This surely, they might think is the greatest evidence that cubism is a form of representational art. Again, such an artist would be mistaken. The conclusion they draw is true enough, cubism is a form of representational art (at least ideologically) but the premiss used to reach this conclusion is false. It would be wrong to think that our ordinary vision is an exemplar of the sort of synthesis of different points of view the cubists sought. With our ordinary vision, and by natural extension stereoscopic photographs we get a unity of perspective that the cubists sought to escape. It was important for them that the perspectives associated with an object always remain a plurality.

Holograms are representations which present different images to the viewer depending on the viewer's position in relation to the hologram itself. The effect is that the viewer can get radically different views of the object depicted, all somehow captured in the few lines scratched in the polarizing film. We could imagine an artist who discovered this medium and excitedly worked to carry off the successful completion of the cubist project. After all, the few scratches in the film capture in a synthetic manner the various different perspectives, even though the scratches do not unify the perspectives in the way that ordinary vision does. We do not end up with one perspective as we do with ordinary vision. But if that attempt to achieve a synthesis were successful, why is ordinary vision not successful in attaining the synthesis the cubists sought? After all, the different perspectives are not seen at once in a hologram. The hologram in this respect is no better (and usually worse) than Michaelangelo's sculpture at achieving the goals sought. Only one perspective is seen at any one time and different perspectives are seen when there is a change in the relative position of the viewer and the polarizing film. But this sort of multiplicity of perspective is true just in the same way as in ordinary vision. When I change my position relative to an object, I gain different perspectives of the object. That fact does not turn every object or piece of sculpture into a cubist work of art. This would not solve the puzzle of why the cubists thought that they needed a genuinely new way of representing objects. They sought a way of representing objects which stood apart from the traditional ways of representing objects by making explicit something that the traditional ways of representing do not, namely, that there are a number of perspectives associated with any object, that there is (as some would put it) no dignified perspective over and above the multiplicity of perspectives.

But what are we to say? Is there a way of achieving this? Is there, in other words, conceptual space to manoeuvre between the multiplicity of perspectives as presented by an architect and the unity of perspective that we have when we look out onto the world? Is there conceptual space for the synthesis of perspectives which is not a collapse of perspectives and not a mere multiplicity either? I hope the pessimistic claim that there is not and that therefore the cubist project was one which was doomed looks plausible after considering the cases above. What I am most interested in is what we can learn about points of view from the failure of the cubist project.

That nebulous notion of perspective, or of point of view, can be treated in many ways. There is clearly an ambiguity over whether bodily movement I undergo in and of itself induces a change in my point of view. On one way of using the term say when we talk about a person's point of view, meaning roughly their fundamental beliefs about the world, bodily movement would not change my point of view. On another, perhaps more literal, understanding of the notion of point of view it would. I aim to stay closer to the latter, more literal, understanding. The point of the discussion about cubism is to suggest that the notion of point of view can be partially defined by this curious feature: there is no conceptual room for a synthesis of points of view which lies between the pluralism of a number of points of view, and the monism of our ordinary binocular vision. Let me put the point this way. When I look at something with both eyes open, I have a point of view on the thing. I could then shut one eye. I again have a point of view on the thing. I could shut that eye and open the other. Yet another point of view. Now about these three ways of seeing the thing, is the first a product of the other two points of view? The lesson of the earlier remarks about cubism seem to be that it is not — that when I see in using both eyes, the resulting point of view is not the result of putting together two points of view each associated with a different eye. That is not to say that the physical stimulation reaching my eyes when viewing with two eyes is not the product of the stimulation I get when viewing each of the two other ways, nor that my point of view is not a function of the totality of the physical stimulation I receive. For, though my binocular point of view is the function of a certain totality of sensory stimulation, there is no reason to think that it can be decomposed into constituent points of view each of which is the function of a certain physical stimulation.

It might be said that if we understand the notion of a point of view in the literal sense, we are committed to saying that my ordinary vision with both eyes open involves two points of view since with one eye only open, there is a difference in my point of view depending upon which eye I have open. But this is a bad argument. To see that it is a bad argument consider an analogous argument: When we performed a split brain experiment by severing the *corpus callosum* in a patient, the patient showed every indication that there were two selves in the one body. We presented a question aurally and got no answer. We presented the same question visually and got a correct written answer. But even at that point when the question was represented aurally we got no response. *Et cetera*. It might be contentious but let it be granted that we have two selves after the procedure. Now someone might say that this shows that there were two selves before the procedure. But this is clearly a bad argument. Suppose there were other procedures to which we could have subjected our poor patient. Suppose that we could have cut through the brain at an angle orthogonal to the direction we did in fact cut and that if we were to so cut we would have got analogous results but that the two selves then would have been different in character to the two selves resulting from the actual operation. What should we say now before any of the operations? That there are two selves or four or just one? If we modify the example so that we cannot go forth with both procedures at once, so that we can only ever get to two selves we still cannot conclude that there were two selves before the procedure since we might imagine that the different procedures produce different pairs of selves. So the answer that there are two selves before the procedure is unjustified. The only answer that retains any plausibility in light of such possibilities is that before the procedures have been performed there is one self. Why should we give the analogous suggestion any more credence in the case of points of view?

Perhaps we can begin to spell out what is involved in the notion of a point of view and also address some of these worries by suggesting that the notion of a point of view is tied to the notion of my location. In particular my semantic location. If it is true, and it seems true, that whenever I say 'I am here' that I utter a truth then the location associated with that use of 'here' may as well be called my semantic location.

Adding an eye to the back of my head won't mean that my perspective has more than one location. It would however flesh out my perspective in the same way as having binocular vision does, in so far as that does not involve my having two points of view. But if this way of trying to multiply my points of view won't work, what about something analogous to Daniel Dennett's [1981] example of the de-brained body? In this example Dennett imagines that his brain is removed from his body because the radioactive environment he is going into to defuse a bomb is dangerous to his brain alone. His brain is hooked up via electronics to his body and manages to control his body in much the way that it does ordinarily. Although his brain is in a lab somewhere it seems to him that he is where the rest of his body is located. Indeed it seems plausible that when his body utters the word 'here' the semantic content of that word varies with the location of his body and not his brain. Now consider some variations on the theme. Suppose that the brain was connected by electronics not to one whole body, but rather to two bodies and that each body is adequately controlled by the half of the brain it is hooked up to. Perhaps the on/off switch is determined for the body by whether it has its left eye open in the one case and whether it has its right eye open in the other. As long as only one of these is open we seem to be wholly somewhere. Now, switching between shutting your left eye and your right eye can have an interesting consequence. Perhaps it has the effect of shifting your perspective from one side of the table to the other. What is the effect of having both eyes open? Can we imagine it? Perhaps it is something like having superimposed photographs. Perhaps like watching two televisions at once. Perhaps not. Simple cases that seem analogous might not be. We can use the stereoscopic apparatus mentioned above to introduce not the pair of photos which the apparatus was designed for but rather two unrelated photos.[7] What happens then is that the brain does not synthesize the information received in the usual manner. Rather what the two eyes see is not superimposed in any sense and we can quite easily pay attention to what is seen by just one eye at will. Now it might be that this is indeed what it is like to have the one brain controlling two bodies, I now pay attention to how it is in this body, ignoring how it is in the other, and now I switch my attention. Possibly, but not necessarily.

What more would need to be added to the story to get the sort of unity which would be analogous to the way we can all imagine the eye at the back of your head being incorporated into a more fleshed out but unitary point of view? Is the unity of the point of view constituted by the unity of your body? Or is it the other way around? If both bodies, x and y, push a car up a hill, is there, considering the two bodies together, the sort of unity of action involved when I now push a car up a hill? After all, I now use two legs to push the car, but it is not apt to say that the legs each act separately, even when I have to force myself to take each slogging step. We can imagine that the situation goes two different ways. On the one hand that in some real sense the two bodies become one compound object. Perhaps in the sense in which when I type at the keyboard, my left and right hand become a kind of compound object. My conscious intention to type a certain word, for example 'that', does not divide that typing up into subintentions for each hand to type the letters that they do. In this sense I am certainly where both of my hands are. On the other hand it might be that I am not where the all the parts of the compound object are. When I am typing and I am not paying any attention to my left knee, I am not where my left knee is, so in this case, depending on where my attention is, there will I be.

The notions of the unity of action, of the unity of body, and of the unity of point of view do seem interrelated. It may well be unprofitable to wonder which underlies which. It remains a significant worry that the sorts of thought experiment we have been considering, split brains and such, might be as misleading about the way we actually work as the considerations which led to mistaken view that 'two persons after the operation so two persons before'. As personhood is underwritten by and gives content to a kind of unity, so it is with points of view. Because of the unity of a point of view, the sort of view from everywhere which the cubists sought, is not to be found anywhere.[8]

*II. Perspective in Philosophy: The Desire for Objectivity*

Some philosophers have felt a desire to know the world as it is in itself and have felt that the ordinary ways of understanding things are merely the apprehension of appearances. Not to say, for these philosophers, that relinquishing the ordinary ways of understanding the world in favour of science would enable them to apprehend the world in itself. Science, according to this story is merely the systematization of more and more appearances. In doing science we would just be compounding appearances without encountering reality itself. Objective reality is conceived on this account as not any combination of appearances, not even all the appearances there are. In some moods Plato fits this picture but so do many more recent thinkers from Kant to Bradley. The idea that objectivity is not to be understood as a multiplicity of appearances has led philosophers to wonder how to understand the notion of objectivity and how it is that we can gain knowledge of objective reality.

If the objectivity of a representation and, by extension, judgement is not to be found in its incorporation of all or at least a multiplicity of points of view, how are we to characterise objectivity? A move made by some philosophers is captured in Nagel's [1986] apt phrase 'The view from nowhere' and Williams' [1978] 'Absolute Conception of the World'. The notion of an Absolute Conception of the World, and especially the characterisation of the view as a view from nowhere, builds in the idea that subjectivity is tied to the notion of a perspective, but moves beyond that to assert that certain views of the world are more objective than others — they are not as perspectival. The dependence of something on a perspective seems the important notion.

The contrast between perspectival and non-perspectival ways of conceiving seems to have been a powerful motivation behind much philosophical reflection about our minds and their relations to the world. To pick an example, almost at random, we can consider Henri Bergson. He contrasted what he called 'Intuition' from ordinary ways of conceiving as follows:

> If we compare the various ways of defining metaphysics and of conceiving of the absolute, we shall find, despite apparent discrepancies, that philosophers agree in making a deep distinction between two ways of knowing a thing. The first implies going all around it, the second entering into it. The first depends on the viewpoint chosen and the symbols employed, while the second is taken from no viewpoint and rests on no symbol. Of the first kind of knowledge we shall say that it stops at the *relative*; of the second that, wherever possible, it attains the *absolute*.[9]

My own view is that if this ancient and influential picture were substantially correct, it would be hard to take seriously the very project of metaphysics. The motivation for an alternative mode of conceiving the world comes from the recognition, or seeming recognition, of contradiction. The presence of contradiction at the ordinary, merely relative, level indicates another level, a level of true being at which these apparent contradictions are reconciled. This much Kant accepted. He very famously did not think that such curious modes of awareness were available to allow us to judge significantly about the things-in-themselves.[10] For the metaphysicians however this is not a place that they can rest. If ordinary belief and all the modes of knowledge associated with ordinary belief are not properly speaking ontological, it would become rather problematic how can we make the leap to an ontological mode of conceiving the world. It is in response to this problem that the view that there is a special sort of knowledge or mode of knowing which characterises philosophical knowledge makes its presence felt. In these modes of knowledge, the suggestion goes, is a way of conceiving the world which is of such a nature as to be ahistorical, non-perspectival and revelatory of the true nature of the world. Given this sort of account of philosophical knowledge, ideas of modes of access under the rubric of the light of reason, or intuition, or dialectics, or reflection, find their motivation, for these are modes of access which are supposed to allow us to get beyond our ordinary beliefs, to achieve a level of critical distance without which the claims to objectivity would seem presumptuous and preposterous. Attendant to the distinguished modes of knowledge are also distinctive theoretical concerns which lead us to philosophical reflection.[11]

In this section I shall argue that the contrast that we find in Bergson, and which is implicit, as he says, in many philosophers before and after him, is mistaken. Rather than focussing upon Bergson however, I shall look in more detail at two philosophers who do use such a distinction to their detriment, one implicitly, Rene Descartes, and one explicitly, Bertrand Russell. I will also say a little about how the distinction involved here resurfaces in contemporary debates about *de re* attitudes, and in particular *de re* belief. I want to try to separate issues which I think have unfortunately been run together and show how once we do make this separation the issue of the nature and significance of semantics is thrown, if not into question then, at least, into sharp relief.

It might with some justice be argued that Leibniz is the perspectival philosopher *par excellence*.[12] However, Descartes may have suffered most notably for not paying enough attention to the notion of a point of view and the fact that our judgements are made from points of view. Descartes, in his second Meditation, voices for his interlocutor a remarkable question. The question is remarkable because it seems to raise a rather modern problem for Descartes, and a question which is all the more interesting because it is far from obvious what Descartes thought his reply to the question should be.[13] I will argue that the question is a deep reason for worrying about the metaphysics exemplified in the Meditations, that in effect there is no solution to this problem if our theory of the mind retains the Cartesian notion of an intentional object.

And yet may it not perhaps be the case that these very things which I am supposing to be nothing, because they are unknown to me, are in reality identical with the 'I' of which I am aware?[14]

The apparent interpretation of this passage is that the aspect of the object presented by reflection is different from the aspect presented by sensory perception but the objects thus differently presented are nevertheless the same object. The problems raised for Descartes by this objection are clear. How does Descartes know that the mind and the body are distinct?[15] He goes on to say that he does not know how to answer this worry at this point but that he will go on talking only about that of which he is sure. Of course there is a tradition of interpretation which has it that this is just what he should say here because although at this point he does have a clear and distinct idea of the thinking self by virtue of the *cogito* reflexion, he has yet to form a clear and distinct idea of the body. This would explain why he goes on to talk about the wax. So one might think that he is led to this discussion by the need to give an account of the essence of body in general. The wax passage is of course difficult to interpret since it starts by discussing the ontological relationship of the substance which is the wax and the properties we perceive but ends with a claim that it is by virtue of the understanding that we have an idea of the wax as a persisting thing. This seems a claim about how we come to have a certain idea and not a claim which will answer the ontological worries which seemed to give rise to the discussion earlier. Thus it is far from clear what Descartes thought his reply to the objection about aspects to be.

When we take the worries about aspects seriously, does thinking become a necessary and eternal property of the mind as Descartes suggests, or is it that thought is a property *a priori* linked to the aspect of the self presented via reflexion? The latter connection is *a priori* since whenever I reflect on myself I find that it is involved in thought. That is hardly surprising for what is reflexion but a mode of thought. Such a conclusion will obviously not support the claim to eternal thought since there may be aspects of the same self, such as those presented to others while we sleep, which do not entail that we are thinking. This point, too, was raised by Arnauld when he wrote:

Someone may also make the point that since I infer my existence from the fact that I am thinking, it is certainly no surprise if the idea that I form by thinking of myself in this way represents to my mind nothing other than myself as a thinking thing. For this idea was derived entirely from my thought. Hence it seems that this idea cannot provide any evidence that nothing belongs to my essence beyond what is contained in the idea.[16]

There is a reply which Descartes makes to this objection. A sensible reply, but one which misses the crucial point of the objection. There are two aspects to the objection. One has to do with the completeness of the idea we have of ourselves from reflection, the other as we might say, with the perspectival nature of the idea. The completeness problem is this: How can I know that I lack a property, even just as a part of my essence, just because when I reflectively consider myself, the idea that I form does not include that property? After all, couldn't it be that that idea I form is incomplete? Well, to this Descartes has a reply: It doesn't matter if the ideas are not complete. If the idea of body includes the idea of divisibility and the idea of the body includes the idea of indivisibility, we can infer that these ideas, incomplete though they be, are ideas of distinct things, for nothing can have contradictory properties.

But this leads to the second and I think more pressing problem: If, when conceiving myself as thinking thing, I form the idea of a thinking thing in which I do not include the idea of extension, and when I consider my body I do include the idea of extension, and not the idea of a thinking thing, might it not be that these are ideas of the very same thing? Might it not be that the ideas that I have are, as we would be tempted to put it, perspectival? Then the fact that one idea had from one perspective, includes the idea of extension and the other idea, had from another does not, is not a conclusive reason to think that they are ideas of distinct things. Now, in some senses we can find passages in Descartes where he does pull back a bit and seems to hedge his bets as to what sort of argument he has provided. In particular, I am thinking about his remarks about the role of God's granting revelation in our coming to recognize that a certain idea is adequate.

A created intellect, by contrast [with God's intellect], though it may in fact possess adequate knowledge of many things, can never know it has such knowledge unless God grants it a special revelation the fact. [CSM, II, 155]

However, it seems that he never really comes to terms with the worry we can find in Arnauld's pressing objection. Descartes makes the reply that he never really needed that the ideas we had were adequate nor known to be so, but rather that they be complete.

> ...for there to be a real distinction between a number of things, each of them must be understood as 'an entity in its own right which is different from everything else'. [CSM, II, 156]

But this response just raises anew the same worry. How is it that Descartes can be confident that his understanding of the mind and the body is such that he understands them as entities in their own right and as different from everything else?

In the end, the whole question turns on the notion of clear and distinct ideas. It must be remembered that Descartes distinguishes imagination from pure understanding. Recall his discussion in the Sixth Meditation of the chiliogon and the adequacy of his idea of a chiliagon: though his imagination cannot produce an idea, in the sense of an image it can distinguish from many other polygons, nevertheless his understanding has an idea which is adequate and suffices to distinguish that polygon from any other. So the claim to a clear and distinct idea of the self arising from the *cogito* reflexion cannot be thought of as an image in the mind of the reflecter. But it is equally clear that ideas must be thought of as having content. We need to know more about clear and distinct ideas. In particular, we need to know whether there could be two people with clear and distinct ideas of the very same thing who differ as to the content of their ideas. Second, could clear and distinct ideas fail to determine the nature of that of which they are the ideas? It seems we can go two ways here. We can say that clear and distinct ideas can fall short of the requirement for distinguishing their object from any other object, or we can insist on that requirement as a criterion of clear and distinct ideas.

If we refuse to countenance non-perspectival ideas whether complete or incomplete, at least for finite beings, this does not mean that the objects of the ideas are themselves in any way conditioned by these perspectives. This seems to have been a problem some have found with Kant.[17] In a sense it is a grammatical mistake. As Aristotle remarks when we talk of Bloggs qua philosopher being imaginative, 'qua philosopher' modifies the predicate term and not the subject term.[18] To think otherwise is to think that there is in the world apart from Bloggs herself, a curious sort of intensional object, Bloggs qua philosopher. There is obviously more to say about this but notice the analogy with the present case. The chair from this point of view is soft. The chair from that point of view is hard. Are these two sentences about the same chair, or have we two separate chairs? Only the former which will avoid various puzzles and maintain ontological economy. There is but one chair but we buy our ability to say this at the cost of increasing our ontological burden with respect to properties. Now we have as well as the property of being hard, the relational property of being hard with respect to a certain point of view. There is nothing contradictory about our ascriptions to that chair. The aptness of the challenge raised by Arnauld's comments is evident, since clear and distinct ideas got going in the proof of the independence of mind and body by being essentially non-perspectival. The challenge to Descartes has the form of a dilemma. If we have clear and distinct ideas and they are essentially perspectival, the premisses cannot establish the truth of the conclusion of the argument for the real distinction. They fail to establish the conclusion since, like the chair and its hardness in the earlier example, there is no contradiction in the idea of the self having extension when considered in one way, and not having extension when considered in another way. On the other hand, if the ideas are not perspectival then we need an explanation of how it is that we could come to have such ideas. Certainly these non-perspectival ideas are unlike the sorts of ideas we have of the world when we look out at it.

I suspect that Descartes would have chosen the latter horn of the dilemma since his argument depends on it. But nowhere do we find an account of how it is that we can have ideas of this form. Indeed he thought that some of our ideas really did represent the world as it is, for instance, those ideas we have of the primary properties. But these are perspectival ideas as much as any others. We lack an account of how it is that we attain the sorts of ideas Descartes needs in order to respond to the objections we can find in Arnauld's comments. I think these objections are successful at undermining Descartes' claims to certainty and/or having provided demonstrations.[19]

Let me restate the questions which let us see how Descartes was presupposing a non-perspectival kind of idea. One way of putting what might seem to be the crucial question is whether, if we have two true ideas one attributing F to x and another attributing not-F to y, we can infer that x is not y. This is clearly the form of argument in his argument for the real distinction between mind and body. But there is a substantial assumption in the way this question is stated. The mode of argument seems to rest securely on Leibniz's Law, that a thing cannot differ from itself. This is so for any true ideas, perspectival or not. All the work has been done for us when we say that the idea is the idea that x is F or that it is the idea that y is not F. In the case of the argument for the real distinction the work is done in making clear that it is extension and not extension-when-considered-as-body which is the property contained in the idea. In that case we are in a position to assess Descartes' argument, and the diagnosis just follows what Arnauld said in his Objection. As this seems to suggest, there is a problematic relation between ideas and their content. In following sections we will see that there is something that we can say about the notion of content which may be helpful here.

*III. Non-perspectival Belief in Recent Analyses*

H.L.A.Hart [1949] discussing Russell's notion of knowledge by acquaintance describes Russell's notion as a 'part of the *damnosa hereditas* left by Plato to his philosophical posterity.' The disfavour shown by Hart to Russell's notion is severe. It extends to the very principle of knowledge of the sort Russell envisaged, not merely to the extent Russell thought such knowledge might be available to us. I believe Hart was right to align Russell's account of knowledge by acquaintance with Plato. Presumably he had in mind Plato's account of knowledge as a grasping of being in the *Theaetetus*.[20] In view of his critical spirit, I side with Hart against Russell. However, it must be said that our view is a minority view. Knowledge by acquaintance comes as a package with Russell's more famous doctrine of descriptions. It is this doctrine which Russell thought would enable him to construct surrogates for propositions with which he was not acquainted. Russell explicitly aligned his notion of knowledge by acquaintance with singular propositions. Moreover it was Russell who again made the point that this sort of knowledge was at the bottom of our semantic capacities, that at the basis of each case of significant 'knowing that' was a case of 'knowing who'. He said such knowledge 'brings the object itself before the mind'. In more recent times this sort of knowledge has come to be discussed as knowledge *de re* and workers in the area have explicitly paid due homage to Russell's early pioneering work. In this section I will argue that this area of concern is a bit of a mess but that at the heart of it is a conception of non-perspectival belief which is deeply mistaken and is underpinned by an almost comic attempt to avoid scepticism.

Although they are typically identified, the notions of knowledge *de re*, knowledge by acquaintance, and knowing who must be separated and indeed separated from the issue of *de re* attribution of knowledge. I will suggest that there is no significant relation that deserves the name of knowledge *de re*, nor belief *de re*. However there is a genuine practice of *de re* attribution of intentional states, but that this is not the attribution of a distinctive sort of intentional state. Moreover I will argue that once we see that the epistemic guarantees Russell alledged to flow from his notion of knowledge by acquaintance are overblown we will also see that the connection between semantics and the explanation of behaviour may not be as immediate as he thought it would be. At the basis I will be affirming the perspectival nature of our beliefs without thereby committing myself to the dubious idea that the objects of our beliefs are thereby perspectival. I hope to facilitate a *rapprochement* between a Fregean approach to beliefs and a direct reference theory of the semantics of names by emphasizing the differences between a semantics of beliefs and the semantics of belief attribution.[21]

There is a familiar motivation for Frege's notion of sense. For Frege it is just because we can be perfectly rationally ideal and find ourselves in the situation in which we have two beliefs, as it turns out, about the same thing, under two different modes of presentation that we need to introduce an notion of sense. Sense is an aspect of meaning which is correlative to knowledge and logical relations.[22] To call sense an aspect of meaning is a rather ambiguous thing. Usually a theory which posits intermediary objects between the linguistic expressions and their referents as does Frege's has been taken to be inconsistent with a direct realist understanding of the semantics of proper names. However following a lead by Kripke, we can distinguish a theory of meaning from a theory of reference. My contention is that Fregean senses have a role in a theory of meaning even if not in a theory of reference. Moreover, this tells us that where semantic properties of the whole depend only on the referents of the parts this amalgam theory will produce the same predictions as the view that names only contribute their referent to the semantic value of the whole. That is not to say that in so called opaque contexts that we need to follow Frege in identifying a name's reference with its customary sense. The role of the notion of sense is tied more closely to logical relations among objects of knowledge than it is to their semantics. On a direct realist account the only contribution of a proper name to the semantic value of an atomic sentence in which it occurs is to provide an object to talk about. This is sometimes equivalently put by saying that sentences including names have as their content singular propositions. These can be variously construed. One way of construing singular propositions is as sets of possible worlds involving the object in question. Another way is as a structured object which contains that object. This seems to have been Russell's view.

To be in a state properly indexed by a singular propositions has been thought to be rather like possessing non-perspectival ideas. Sometimes these are called the contents of *de re* attitudes. The very name 'direct reference' suggests a kind of immediate latching onto the object in question. Curiously this way of thinking of singular propositions has a long history in analytical philosophy. It goes back to Russell's distinction between knowledge by acquaintance and knowledge by description. Here is Russell introducing the topic:

The object of the following paper is to consider what it is that we know in cases where we know propositions about 'the so-and-so' without knowing who or what the so-and-so is. For example, I know that the candidate who gets most votes will be elected, though I do not know who is the candidate who will get the most votes. [1918; p.209]

...It would seem that, when we make a statement about something only known by description, we often *intend* to make our statement, not in the form involving the description, but about the actual thing described. That is to say, when we say anything about Bismark, we should like, if we could, to make the judgment of which he himself is a constituent. [1918; p.218]

...in judging, the actual objects concerning which we judge, rather than any supposed purely mental entities, are constituents of the complex which is the judgement. [1918; p.222]

Notice that we are tempted to say that here Russell has confused the object of the judgement, with the judgement itself. The complex, we want to say, is the object. However behind this identification lies Russell's

... fundamental epistemological principle in the analysis of propositions containing descriptions... : *Every proposition which we can understand must be composed wholly of constituents with which we are acquainted.* [1918; p.219]

It is instructive that Russell here characterises his principle as epistemic, while it is really a semantic principle, a principle about what can be in the content of the meanings of sentences we use meaningfully. The point is not centrally about what we can know, though it will have consequences for this. It is most directly about what we can mean. The reason Russell gives for his principle, the only reason as far as I can tell is that

it seems scarcely possible to believe that we can make a judgement or entertain a supposition without knowing what it is that we are judging or supposing about. [1918; p.219]

Now this seems right but then again it seems in danger of becoming trivial if it is not false. What can he mean by our knowing what we are judging about? When I make a judgement about Julius Caesar, and someone asks me whom I am judging about, can't I simply reply 'Julius Caesar'? They may want more, but why should I always be able to give more? There seems to be no real reason. It might be that the judgement I make expresses all my beliefs about Julius Caesar. Still the question can be asked, but unless I repeat myself I cannot answer. Russell seems to have imported his notion of acquaintance into his argument for this very notion at this point. He seems to have thought that you could always use the demonstrative 'this' to indicate what it is that you are supposing about, that in other words you are immediately acquainted with the objects of your supposition. So it would follow that 'Julius Caesar' is not a logically proper name and that I am not acquainted with Julius Caesar. But I take it that this argument is unsatisfactory.[23] So there is no really good reason to think that the class of thoughts you can have is limited in the way Russell thought. But there is another epistemic consequence of the way Russell thought acquaintance 'brings the object itself before the mind'. Namely, we are guaranteed at least that the object exists. Whenever we are thinking about something, that something is guaranteed to exist. Now this means pretty quickly that the only things we can really think about are ourselves, our sense data and universals.

> We have acquaintance with sense-data, with many universals, and possibly with ourselves, but not with physical objects or other minds. We have *descriptive* knowledge of an object when we know that it *is the* object having some property or properties with which we are acquainted; that is to say, when we know that the property or properties in question belong to one object and no more, we are said to have knowledge of that one object by description, whether or not we are acquainted with the object. [1918; p.231]

It is odd that, according to Russell, we can think about universals in a way we cannot think about the objects having the universals. A universal apart from the object it inheres in is an odd thing to have before your mind. If we cannot be acquainted with physical objects, how is it that we can be acquainted with their properties? Recall that it is the universal itself that is supposed to be before your mind not an idea of it or anything mental like that. Perhaps Russell meant to restrict the sorts of universals involved here to those which inhere in the objects with which you are acquainted, yourself and your sense data. But this is not clear from what he says and anyway would either lead us to a curious phenomenalism as we construct other universals, like the universal had by all and only the Kings out of properties of sense data, or it would falsify what Russell says when he says

> We have *descriptive* knowledge of an object when we know that it is *the* object having some property or properties with which we are acquainted. [ibid]

Thus according to Russell the very same property is such that we can be acquainted with it and it can be had by objects (presumably even physical objects) which we cannot be acquainted with. This is a very hard condition to satisfy. It looks as though one of the these two conditions is going to have to go. After all if we can be acquainted with an objects properties, what more is needed before we are acquainted with the object itself?

One of the things you cannot be deluded about is the identity of those things you are acquainted with. Moreover, if you are acquainted with x and you judge that it is F, and you are acquainted with y and you judge that it is not F, you can infer that x is not y. The interest of Russell's semantic picture accrues from this sort of epistemic consequence. We can summarise Russell as saying that we can only have singular thoughts about things we are acquainted with and that we are acquainted with things only when we stand in a certain privileged epistemic position with respect to those objects.

Russell is at least consistent, but not very plausible. For a number of reasons to be found in Kripke's *Naming and Necessity* it seems that singular propositions are far more extensively available than Russell's notion of acquaintance would allow. Nevertheless some sort of notion of acquaintance seems attractive. But once we start to weaken the idea of acquaintance from the strong epistemic access that Russell invoked, we lose the idea that in using a sentence which expresses a singular thought we have gained any particular epistemic guarantee. For we can be using such sentences and fail to refer to anything. But when we lose this epistemic dimension of the semantics, we lose an important connection between the semantic story and the story we tell about the explanation of behaviour.

The epistemic aspects of Russell's notion of acquaintance gets revived in some of the recent work done under the rubric of belief *de re*. Here the issue is the question of knowing who, or as Burge [1977] puts it, belief *de re*. I will concentrate the rest of this section on that very notion. According to the account offered by Burge, *de dicto* beliefs are fully conceptualized (1977; p.345) whereas *de re* beliefs are those 'whose correct ascription places the believer in an appropriate non-conceptual, contextual relation to objects the belief is about.' (1977; p.346) The idea that belief *de re* is co-extensive with knowing who can only be made plausible if we are willing to ditch the connection between belief attributions which are *de re* and belief *de re* itself. I do not see the point of that. We may as well say that all there is is the peculiar kind of attribution, and no peculiar kind of belief at all. Still there is the question when a person can be said to know who they are talking about. My own view on the matter is that there is no good answer to that question, but that the significance attributed to the notion of knowing who, is utterly misplaced. Behind it is the bad old mistake, enshrined in Kaplan's notion of a vivid name. Recall that *a represents x to* Ralph (symbolized 'R(*a*, *x*, Ralph)') if and only if (i) *a* denotes *x*, (ii) *a* is a name *of x* for Ralph, and (iii) *a* is (sufficiently) vivid.' [Kaplan, 1969: 363]

A vivid name is one which has certain rich internal connections for the user of the name. Once we have vivid name at the ready we can express singular propositions. Accordingly singular propositions are only available to us for object-name pairs which have an appropriately rich set of internal connections for the user. Singular propositions are simply about the thing in question, unmediated by any descriptional content. The temptation is to conceive of these contents in the same sort of way Russell did as containing not anything analogous to Fregean senses but rather the objects themselves. Into this semantic picture is built the non-conceptual aspect of belief *de re* by way of their immediacy. The question which was raised about Russell and Descartes arises here with just the same pertinence. Is it possible for someone to be acquainted with an object by way of two vivid names and not realise that they are names of the same thing? Clearly that is a possibility. But then the contents assigned to the sentences the speaker will assent to using these names will not capture the logical relationships among this speakers beliefs. It is not a rational failing that they do not realise that these two names are names of the same thing. But as the contents are assigned, there is nothing to mark the lack of logical connection between the vehicles of the contents. The temptation would seem to be to make the richness of the connections such that this sort of case is ruled out. But then it would seem that the sorts of things we could have singular thoughts about would be similarly restricted. They would seem to have to be limited to the sorts of occurent momentary objects Russell thought we were acquainted.

As it is, the richness of the connections associated by a speaker with a name is a matter of degree but anyway it is odd that we should require the richness.If the idea is supposed to be that a singular proposition can only be assigned to someone who has and uses a vivid name for the objects of the proposition then the idea is mistaken. A sentence which expresses a singular proposition can be used by someone who is not in a position to know who the sentence is about. If I say'Laurie Daley plays for the Canberra Raiders' I use this sentence to express a singular proposition about Laurie Daley. The sentence in my mouth has the usual meaning even though the only connection I have to Laurie Daly is that I heard this on the radio the other day. I have no idea what Laurie Daley looks like, though I would expect him to be a rather large (perhaps fearsome) person, maybe with a nose which has been broken. I can be said to believe of Laurie Daley that he plays for the Canberra Raiders. That is, exportation seems appropriate in this case, even when the richness of the information I have about Laurie Daley and so the vividness of the name is rather thin. It might well be that he is also the only player in the Kangaroos with the initials 'L.D.'. Someone who knew that (not me!) could truly say about me that I believed of the only player in the Kangaroos with the initials 'L.D.' that he plays for the Canberra Raiders. That belief attribution *de re* is perfectly true even though I have not the faintest belief either way whether Laurie Daley is the only player in the Kangaroos with the initials 'L.D.' and even though it would be wrong to think that I know who he is. Certainly I believe that he is the only player in the Kangaroos with the name 'Laurie Daley', but apart from that description there is not much I know which would serve to individuate him from other Rugby League players. Thus it is wrong to think that before one can use a belief attribution *de re,* the believer must have an ability to distinguish the object of the belief from all others. Still the belief that I do have, that he plays for the Canberra Raiders, is indexed by a singular proposition. It is true if and only if that person has the property of playing for the Canberra Raiders. I might have had other beliefs which were not singular beliefs involving that person. So I might have heard on the radio that the only player on the national team with the initials 'L.D.' plays for the Canberra Raiders. Hearing this and being particularly trusting of the media, I come to believe it. I think it is still acceptable for someone to say about me that I believe of Laurie Daley (now a name I have not heard) that he plays for the Canberra Raiders. There is no need to limit the *de re* attribution to believers whose beliefs express singular propositions. We do not ordinarily do so. That being so, there cannot be a close connection between the appropriateness of a *de re* attribution and the nature of the content of a belief. Moreover, as we have seen *de re* attribution is largely independent of the issue of knowing who.

The mistake seems to have been imported by Russell when he explicitly aligned the notion of knowing who with singular propositions. The purported cognitive significance of the distinction between beliefs indexed by singular propositions and those indexed by non-singular propositions is a mistake. Indeed in this context we can understand what might lead to the mistake: the fact that the distinction between the two dimensions of content have not been adequately distinguished.

The question we have been worrying about, the indexing of beliefs by propositions is on the whole too simplistic. Beliefs have at least two different dimensions of content. One dimension we might call their metaphysical dimension, it is the class of possible worlds in which what was expressed is true. Another dimension is given by epistemic consequence relations. Kripke drew attention to the way these dimensions were independent with his work on the contingent *a priori* and the necessary *a posteriori*. It is clear that two beliefs may have the same metaphysical content without having the same *a priori* consequences. We can define a notion of content by taking the epistemic content of a belief to be all those beliefs which follow *a priori* from that belief.[24] We need to sort out which of the two dimensions we might be interested in. I have been talking in the manner of the other writers I have commented on, in effect using the notion of metaphysical content. The notion of what the sentence in our mouths says. There are other notions, in particular, the notion of what contexts the sentence is true in and about.

We obviously have to think hard about the work belief attribution sentences are doing. If belief is a relation to a proposition, and a proposition is just a way the world is or could be, then any equivalent way of reporting the way the world is should be equally acceptable as a belief attribution. But clearly this is not the case. Failure of substitution of co-referential names or necessarily co-extensive predicates is manifestly a feature of beliefs. So belief attributions are not merely reporting the presence of the relation of the believer to the proposition.[25] What work could a belief attribution be doing other than reporting the fact that a relation stands between the believer and the proposition which indexes the sentence used in the attribution? Well in this sort of story we have left the epistemic dimension of the content out of the picture. But that is the dimension of content which is relevant to deliberation since it is precisely that dimension of content which is accessible to the agent. It is not for nothing that we do hold people responsible for failing to recognise the *a priori* consequences of their beliefs. If belief attribution is indicating more than the metaphysical content of the belief, by indicating the epistemic content as well, we will have an explanation of why belief attributions do not admit of substitution of co-referential names in general. In this account we can see how we can countenance the semantic story offered by the direct reference theorists for sentences involving proper names without committing ourselves to the substitution principle for names in all contexts. By doing so we distinguish the metaphysical content of a belief from the impact the belief makes on the epistemic state of the believer as a whole.

I have been using Kaplan's 'Quantifying In' as a locus of a certain mistake. He has not stayed with the doctrine I found so mistaken. In a later work [1975: 326] he says:

> All this familiarity with demonstratives has led me to believe that I was mistaken in 'Quantifying In' in thinking that the most fundamental cases of what I might now describe as a person having a propositional attitude (believing, asserting, etc.) toward a singular proposition required that person to be *en rapport* with the subject of the proposition. It is now clear that I can assert *of* the

**first child to be born in the twenty-first century that *he* will be bald, simply by assertively uttering,**
**(29) Dthat['the first child to be born in the twenty-first century'] will be bald.**
**I do not now see exactly how the requirement of being *en rapport* with the subject of a singular proposition fits in. Are there two different sorts of singular propositions? or are there just two different ways to know them?**

What are we to make of the retractions made by Kaplan? It seems in light of our discussions above that one thing he says, namely that there is no need to be *en rapport* with the subject of a singular proposition before being able to take up propositional attitudes to that proposition, is right. I can be in a position to use many names in the appropriate manner without the any claim to being *en rapport* with the bearers of the names. Even the name 'David Kaplan' is one of those. As far as I know, I have never met or even seen David Kaplan, and yet I can say all sorts of things about him. And some of what I say, might even be true. But false or not, it is still about *him*. That is, the propositions will often be singular. So the positive aspect of the doctrine in 'Quantifying In' is false, being *en rapport* with the subject of a singular proposition is no part of the precondition for appropriate relations to that proposition. Now in this failure we see immediately that one aspect of the epistemological role Russell carved out for his knowledge by acquaintance, in so far as it is associated with attitudes to singular propositions is misplaced. Though I utter remarks and have beliefs which depend on how things are with him for their truth, I am in no special position to know that David Kaplan exists. For all I know, the use I follow trustingly is like the practice emanating out of Canberra a little while ago of thanking a certain Bruce Toohey for help with streamlining proofs and discussions in logic papers. This was an elaborate playful hoax. Poor trusting soul that I am, how am I to know that the case of David Kaplan is any different? And yet, if this is not so, if the world is as I think it is, I am talking about David Kaplan. However things are with David Kaplan the *a priori* consequences of my beliefs would be the same. On the other hand, the metaphysical content of my beliefs would change radically if I were situated in a different sort of world, but one which presented itself to me in much the way that this one has up to now.

Should we then be encouraged to endorse the second of the options offered to us by Kaplan? Should we say that there are singular propositions and at least two very distinct ways of knowing them, depending on whether you are *en rapport* with the object of the singular proposition or not? No, not if we think that the notion of being *en rapport* is best thought of as your displaying a kind of practical knowledge, a knowing who. In that case singular propositions enter into the story in an inessential manner. We might as well say that there are singular propositions and two very distinct ways of knowing them, depending on whether you have had dinner with the object of the singular proposition or not. There is a distinction between being *en rapport* with an object and having a belief which is to be indexed by a singular proposition. However, they are as irrelevant to each other as the distinction between having dinner with someone and having a belief which is to be indexed by a singular proposition involving that object.

We saw that we can have beliefs indexed by a singular proposition even though we are not *en rapport* with the objects of those propositions. As well, there are cases of being *en rapport* or knowing who in which singular propositions are not involved. This occurs, for example, when a description is known to you to distinguish an individual from others. An example given by Gareth Evans involves the description

'The unique inventor of the zipper'.

If we are told by a reliable informant that one and only one person invented the zipper, we are in a position to use the description in a way that singles them out. We might for example entertain the thought that the unique inventor of the zipper might have died a sad person. Of course we could introduce a name on the back of such a description, in the way Gareth Evans [1979] discusses in his elaborations of Kripke's work on this topic. That name would enable us to express thoughts not expressible before the name was brought into the language, unless we muddy the waters with modal operators and the like within the individuating description.[26] But the central point remains, individuating descriptions need not be more than actually individuating (and usually need not even be that) before we accept that the agent is *en rapport* with the object in question. Certain sorts of practical ability seem vitally important before we will accept that the agent knows which object is in question. Pointing and displaying seem paramount. However that cannot be all that is involved.

What was the lesson I suggested we learn from the independence of issues surrounding belief *de re*, singular belief, and being *en rapport* or knowing who? In summary, there are no beliefs which in and of themselves are beliefs *de re*. We do not attribute *de re* beliefs, rather we use *de re* attributions of belief. Now the issue of being *en rapport* with an object is utterly distinct from that discussion. I can have a belief attributed to me *de dicto*, about an object I am *en rapport* with. For example, I am *en rapport* with my cat if I am *en rapport* with anyone. Yet it is perfectly acceptable to say of me that I believe that my cat has black hair. It is just as acceptable to say of me that I believe that Dragon has black hair. The latter but not the former gets indexed by a singular proposition, but in both cases I am *en rapport* with the cat the belief is about. So what is it to be *en rapport*? It is to be in certain actual relations with the object; to be able to display a certain degree of practical recognitional expertise. But one thing is not involved: it should be no part of the notion that you can always recognise the object you are *en rapport* with. This is easily seen in the case of numbers.[27] I may well be *en rapport* with the numbers less than 1000. But there are many sentences about particular such numbers about which it is true that I do not know which number it the sentence is about. This is directly analogous to the masked man paradox.

Recall that one of the crucial issues which got this whole thing going was over the difference between

Ralph believes that someone is a spy,

and

Someone is such that Ralph believes that he is a spy.

We were supposed to feel that there was a great difference here. Well I agree. The first reports a belief by assigning a propositional content to the belief. The second does not. It is a *de re* attribution of belief. That is not to say that it is an attribution of a *de re* belief. There is no such thing. I simply deny that ordinary speakers uncorrupted by the hypersensitivities of philosophers will restrict *de re* attribution of attitudes to cases where the subject of the attribution has a special sort of attitude. Consider the pair of sentences:

Ralph believes that John Denver is a spy,

and

John Denver is such that Ralph believes that he is a spy.

Now it is clear that the former indexes Ralph's belief by a sentence taking a singular proposition. In the latter we do not have the complete proposition provided for us. The attributor of the belief may well be using a name that Ralph has never heard. He may know of John Denver under the code name 'The Dutch Kid'. Indeed Ralph may not even be using a name himself. Ralph may be using a description which though not uniquely satisfied by John Denver leads *us* to think that Ralph thinks of John Denver that he satisfies the description. Perhaps he has heard about a suspiciously innocent looking blond-haired country-rock star who used to sing about Rocky Mountains. Ralph jumps to the conclusion that the person he just heard of is a spy. Now there may not be just one person meeting the content of the description, but salience comes to the fore and we attribute a belief about John Denver to Ralph. Changing the example just a little, Ralph may decide to call the person he has just heard about 'Mr X', in which case he may then express his belief by uttering the sentence

Mr X is a spy.

What proposition do we index this sentence with? I do not think this question has a single determinate answer. But not because of any intrinsic difficulty with this sentence. An acceptable answer would see this sentence take a singular proposition as its index. But the problem about indeterminacy also arises with the sentence

John Denver is a spy.

The notion of the content of a sentence or of a belief is to some degree capable of context sensitivity. Even when we bear this vagueness in mind, it is quite appropriate to assign the same propositions to both of these sentences. Not that this is obligatory, but it is not ruled out by conditions of appropriateness.

*De re* attribution of belief is attribution of a belief about a *res*. That means that the attributor of the belief commits themselves to the existence of the *res*. In this respect *de re* attributions support exportation: as far as the attributor is concerned the object of the belief exists concurrently with the belief being false. Even here there is reason to be cautious. What I have just said is strictly speaking false. It is false because we use names ordinarily which we know not to refer. While it might be thought that the utterance of

The Greeks worshipped Zeus

does not commit the utterer to the existence of Zeus because of the opacity induced by 'worshipped', this explanation is at best partial and indeed seems wrong in any case. No such explanation will apply to an utterance of

Othello killed Desdemona.

Worshipping might be an intensional relationship but killing is not. It is clear that this raises the possibility of attributions which are *de re* but where the attributor does not hold that such a *res* exists. In fact it seems that these cases are few and that we use markers to indicate what is going on. As in

> Of their alleged gods, the Greeks believed that they and they alone could partake of nectar.

Notice however that 'alleged gods' is not to be understood as 'things alleged to be gods' for that is ontologically committing in a way that the sentence is clearly not meant to be. Thus the warning stands: *de re* attributions can be appropriate even when there is no conversational implication or presupposition of the existence of the *res* in question.[28]

If we accept what I have been saying about belief *de re*, what are we to say about knowing who, or knowing which? For surely we do not want to throw out that notion. That notion is a part, and a crucial part of our view of ourselves as situated in the world. The fact that I know who Dragon is, reflects more about the rich sort of interaction I have had with my cat than does the fact that I know that Dragon enjoys the morning sun. But we do not necessarily lose that, after all I am not suggesting that we do not make distinctions between contexts in which a *de re* attribution of belief is appropriate and those in which they are not. Clearly we do make such distinctions. But utilizing such a distinction is not a way of indicating a feature of the content of the beliefs, nor of their associated dispositions, nor of their idealized associated dispositions. Though it may be a way of sometimes indicating the sort of interacting web of beliefs this particular belief is located within. That is what guarantees that even if I am *en rapport* with two entities I cannot be sure that they are not one entity. This just is the mark we saw earlier of perspective. We are always dealing with the world from perspectives.

It might be thought that the right way to do the semantics for beliefs, when we take into account the fact that the beliefs are had from perspectives is to treat the beliefs as though they were implicitly relational. The suggestion has some attraction but it needs to be spelt out, for the immediate question is: in relation to what? One suggestion which comes to mind is point of view. But that won't work since one object can be presented twice over within the one point of view in a manner which will not make the failure to recognise the object as the same object a rational failing, to any degree. But then the sceptical question might seem appropriate: why is any failure to recognise an object a rational failing. Perhaps there is no really good answer to this. Consider the argument:

All men are mortal.
Socrates is a man.
Therefore,
Socrates is mortal.

Someone might refuse to accept the conclusion as following from the premises because they did not know whether we meant to refer to the same thing with the second occurrence of the name 'Socrates'. If we had not, obviously this argument is subject to the fallacy of equivocation. But surely we do not need this as an additional premise? The actual argument is an instance of a number of forms, some of which are validity preserving and some of which are not.[29]

In this section I have argued that there are no *de re* beliefs. Lest I be misunderstood let me reiterate and reemphasize, while there are no *de re* beliefs, we do sometimes use attributions of the form in question, and that these are particularly useful ways of attributing a belief to someone and giving an indication of the metaphysical content while abstracting away from the epistemic content. Moreover, we are sometimes said to know who someone is, but this is no particularly special ability. In many contexts providing a name will suffice. One such context is if you are asked if you know who the Consul killed on the Ides of March was. We contrast that sort of ability with knowing the person, but even when we know a person, we need not have the ability to distinguish that person from all possible others, or even all actual others.

*Conclusion*

The Cubists attempted something impossible: to create a kind of multiple perspective which turned out to be inaccessible to us. Their failure stands as a puzzling cairn. Many have sought to understand the difference between archaic and modern thinkers as built around the notion of a point of view. The notion of a point of view, the notion which is so important to the cubists is importantly a notion of a unity. The attempts to develop a integrated multiplicity of perspectives foundered. But this led us to wonder about the ground of the unity.

Though we have seen numerous attempts by philosophers to deploy a kind of non-perspectival belief or way of apprehending the world, these have not succeeded and certainly not the attempt to characterise belief *de re*. Our beliefs are perspectival. We see the world from our perspectives and not even our most general theories can avoid the fact that they are our theories. All is not left in a perspectival state though, by appropriately distinguishing between metaphysical and epistemic dimensions of content we leave open the possibility that we have singular thoughts. These are, however, without the epistemic guarantees Russell and others have claimed for them. We can still make sense of the fact that there are facts in the world, that the world does not depend on our perspective. It is after all a perspective on the world. Though our judgements about objects are all made from perspectives, we should not think that this means that the objects are themselves perspectival. As Leibniz said in his *New Essays* [p.141]

> Witty thoughts must at least appear to be grounded in reason, but they should not be scrutinized too minutely, just as we ought not to look at a painting from too close.

It is instructive that Leibniz himself gives the example of the painting.[30] Some people might think that the facts that Leibniz adverts to show that parts of the world at least are perspectival. I think that this is a mistake of the same order as the one Aristotle warned us about. Paintings and witty remarks are alike in this respect: they are parts of the world which can play the role they do in our lives only if we access them from certain perspectives. This should no more be counted against their objectivity than the need for microscopes to observe flagellates counts against them. I do not think that in this paper I have really tried to refute the concerted perspectivist – the person who holds that things only exist from perspectives and that there is nothing behind the perspectives. The drive to perspectivism can be motivated by making the grammatical mistake of which Aristotle told us. Kant stands accused, though certainly not as yet convicted, of making just that mistake. There is a tendency to make this mistake when translating Leibniz's law which says that a thing cannot differ from itself, from a material principle into a formal principle. This translation is not so hard for a formal language when we can simply stipulate that the problem does not arise; in formal languages names are logically proper names. In natural language the situation is much harder; stipulation is impossible for already existing languages. We must be aware of the tendency to slip into the mistake of thinking of natural language as if it were a formal language.

*University of New South Wales*

## NOTES

[1] This paper would have been worse but for some timely comments from Catriona Mackenzie, Lee Overton and John O'Leary-Hawthorne. My original motivation for in particular the first two sections came from a series of chats with Michael Janover.

[2] By calling cubism a representational art form I am aligning it on the depictional side of the hard-to-explicate depict/describe distinction. Scepticism about this distinction comes from many sources, not the least from the conception of description in Wittgenstein's *Tractatus* which makes all description at some level depiction. On the other hand there is the radical scepticism about similarity which tends to make all depiction forms of description to be found in Goodman's *The Languages of Art*. While acknowledging that some powerful minds have contended this distinction makes no sense, I would not be satisfied unless such a distinction could be made to work,

[3] This painting is in the Arensberg Collection of the Philadelphia Museum of Art.

[4] This sort of philosophical approach was however widespread in modernist works. See for example, Robert Musil's novel, *The Man without Qualities*.

[5] Jean Metzinger's article 'Note on Painting' appeared in 1910 in *Pan*. He says of Picasso: *Picasso reveals to us the very face of painting. Rejecting every ornamental, anecdotal, or symbolic intention, he achieves a painterly purity hitherto unknown. I am aware of no paintings from the past, even the finest, that belong to painting as clearly as his. Picasso does not deny the object, he illuminates it with his intelligence and feeling. Cézanne showed us forms living in the reality of light, Picasso brings the material account of their real life in the mind—he lays out a free mobile perspective.* quoted from pp.369–70 William Rubin [1989].

[6] In the evolution of cubism much more is involved than the attempt to create synthesized perspectives, more that owes much to Picasso's borrowings from African art and to the expressionists, and also as part of the project of creating representations from everywhere, the attempt to have the artist disappear which led to the many unsigned and unnamed paintings in this period produced by Picasso and Braque.

[7] This suggestion stems from a remark made by David Braddon-Mitchell.

[8] The description of the cubists' aim as the view from everywhere, as contrasted with Nagel's *View from Nowhere*, was suggested to me by Gilbert Harman.

[9] From an excerpt of *Introduction à la Métaphysique* translated in Runes [1962].

[10] It is significant that the claims Kant makes in the First Critique about things-in-themselves are largely negative. It seems that for Kant saying how things-in-themselves are not is not saying how they are. The exception to this rule might seem to be his discussion about the freedom of the transcendental self, but this is itself explained in terms of the transcendental self not being subject to the principle of causality

[11] Aristotle suggested that we are led to philosophy because of our wonder at the world. 'All men by nature desire to know. An indication of this is the delight we take in our senses; for even apart from their usefulness they are loved for themselves.' *Metaphysics* Bk 1 chapter 1, 980ᵃ22-24

[12] See in particular his remarks on the nature of the singular substances at §9 of his *Discourse on Metaphysics*.

[13] Though I have characterised the problem as modern, it is true that the problem has a long history and is intimately related to the paradox of the masked man, a paradox known in Hellenistic philosophy. It would be interesting to trace discussions of this "paradox" in Descartes' reading but I have found no sign of it in his own work.

[14] *Meditations on First Philosophy*, in Cottingham *et al*, *The Philosophical Writings of Rene Descartes*, Cambridge Univ. Press, Cambridge, page 18.

[15] This talk of aspects is reminiscent of Spinoza's 'dual aspect' theory of the mind–body relation. My hope is to investigate that at some later time.

[16] Arnauld, Fourth Set of Objections to Descartes' *Meditations*, in CSM page 143.

[17] I think that the desire to avoid things-in-themselves such as we find in Fichte and others is

based on this mistake. If things-in-themselves are never given in experience then what need have we for them? This question would only make sense if things-as-they-appear are things distinct from things-in-themselves. If that is not the case, if the things that are things-in-themselves are the things-that-appear, though obviously in a different guise, such drive for ontological economy is confused. For an interpretation that is amenable to this construal see Allison [1981]

[18] *Prior Analytics* Bk I, ch 38.

[19] Incidentally, there is a curious aspect to this objection, it mirrors the structure of two modes of Aenisedemus, an ancient sceptic; in particular the eighth mode mentioned by Sextus Empiricus as the argument from relativity. Descartes certainly knew the work of the ancient sceptics so it is strange that he did not seem to see the objection coming.

[20] See in particular the passages 186ᵉ ff. I must note in passing that it is far from clear that we ought to saddle Plato with this view as his considered judgement. It must be said that the passages about grasping the birds are less than obvious in their import. It is, for us, quite natural to treat the birds not as objects about which we have beliefs but rather the contents of the beliefs themselves. In the latter case we would no longer be left with a basic model of knowledge as grasping being. However, even starting to sort these issues out is beyond me here.

[21] Here all I want to do is indicate the plausibility of a *rapprochement*. I hope to argue for the amalgam theory elsewhere.

[22] They are just the questions raised by taking seriously the structural analogy between ideas and Fregean senses. It might be thought that I am making Descartes a rather anachronistic and simple-minded Fregean by collapsing the Fregean distinction between a sense and the associated idea (in Frege's use of the term). I think that's not quite right. First once we distinguish as Descartes does between ideas considered objectively and images produced by the imagination we get some way toward the distinction that Frege has in mind. Perhaps even so, it would be anachronistic to identify their distinctions since Frege is clearly influenced by Kant's rather different picture of objectivity. See H.Sluga [1980] for evidence of this influence. The aims of Descartes and Frege are rather different. The two focussed their attention on different intensional media, one on thought and the other on language. Now for certain (and perhaps quite sound) theoretical reasons we might regard the two media as related as Descartes certainly did. Davidson follows Descartes quite explicitly on this, though he has reasons which make the connection of thought to language-use even closer than Descartes wanted it. Whereas for Descartes language use is a good sign of thought Davidson, with his philosophical behaviourism, makes language use constitutive of thought.

[23] Regardless of its seeming endorsement by Gareth Evans [1982].

[24] If we could make sense of the notion of immediate consequence for beliefs we could define a more fine-grained notion of epistemic content by looking just to immediate consequences. It is not obvious whether such a notion is defendable.

[25] Some might ask whether this is to be counted as part of their semantics or their pragmatics. After all, might it not be that the feature of failure of substitution is to be explained by conventional implicature, a non-cancellable feature part of meaning but not part of the truth conditions. I am in general sceptical about the presence of a conventional implicature which is taken to distinguish the attribution of truth from the appropriateness of utterance, when that implicature cannot be cancelled. For this is a form of pragmatic implicature which is invariant upon the conditions pragmatic factors are sensitive to, namely, contexts of utterance. It is hard to imagine that we could come to predict the presence of such pragmatic factors.

[26] But note that the description in question has none of these devices.

[27] The case of what have been called *de re* attitudes to numbers was discussed by Saul Kripke in lectures at Princeton in the Fall of 1990. I think he would disagree with my comments here, certainly his views as expressed then are inconsistent with what I say here. My interest in the topic of *de re* attitudes was sparked by his remarks both in those lectures and in his discussion of the contingent *a priori* in *Naming and Necessity*.

[28] A trickier case is what happens to the conversational presuppositions when there is no wide-

spread assumption in the non-existence of the *res* in question. That, it seems, would mean that the *de re* attribution will bring into the pragmatic presuppositions one concerning the existence of the *res*.

[29] A crucial thing to keep in mind is that these are not symmetrical properties. If an argument is of the form of any validity preserving inference it is valid. However if the argument is of the form of any non-validity preserving inference, it does not follow that it is invalid. The issue with the argument in question is whether it is of the form of a validity preserving inference. That would seem to require that the two occurrences of the name 'Socrates' are tokens of the same type. That is the important point. They could be co-referential without the argument being of a valid form, for example, if they were tokens of two distinct types of name, two types which seemed on the surface the same. How might they be different? Well, imagine that there are two linguistic communities in which Socrates hangs out, neither knows much about the other. In particular, someone from one community when bumping into someone from the other community who tokens the name 'Socrates' does not know whether that person is referring to the same person as he would refer to using that name. Now such a person, might get themselves in a muddle, start an argument with 'Socrates' as it is used in his own community and end with the conclusion using 'Socrates' as it is used in the other person's community. The argument would then be invalid but contrary to surface indications, not of the form of the valid argument. To see that, suppose that when using the name with different referential intentions the speaker speaks with a distinct accent. So says the name as English speakers say it, when speaking one way, and says it as a Greek speaker would say it, when speaking the other way.

[30] This is reminiscent of the passage discussed earlier about Kenneth Clark 'stalking' an illusion by Velásquez

## REFERENCES


Allison, H. (1983), *Kant's Transcendental Idealism* (Yale Univ. Press).

Arnauld, A. 'Fourth Set of Objections to Descartes', *Meditations*', in CSM.

Bergson, *H. Introduction à la Métaphysique* translated in part in *Classics in Logic*, ed. D.D.Runes 1962.

Burge, T. (1977), 'Belief *De Re*', *Journal of Philosophy* 74:338–62.

Clark, K. (1957) 'Six Great Pictures, 3: 'Las Meninas' by Velázquez' *London Sunday Times* June 2nd. Referred to in Gombrich (1961).

Dennett, D. (1981), 'Where am I?', *Brainstorms*, pp.310–23, (Cambridge, Mass.: MIT Press).

Descartes, *R. Meditations on First Philosophy*, in Cottingham *et al* (editors and translators), *The Philosophical Writings of Rene Descartes* (Cambridge Univ. Press: Cambridge) (referred to as CSM).

Evans, G. (1979), 'Reference and contingency' *Monist*, 52:161–8.

Evans, G. (1982), *Varieties of Reference*, (Oxford Univ. Press: Oxford).

Frede, M. (1990), 'Fictions of Truth: Velásquez at the Metropolitan Museum' *Arts Magazine*, (April).

Gombrich, E. (1961), *Art and Illusion* (Princeton: Princeton Univ. Press).

Goodman, N. (1969), *The Languages of Art*, (London).

Hart, H.L.A. (1949) 'Is there knowledge by acquaintance?', *Aristotelian Society Supplementary Vol.* XXIII.

Kaplan, D. (1969), 'Quantifying In', in Martinich (1985).

Kaplan, D. (1975), 'Dthat' originally published in *Syntax and Semantics Vol. 9,* ed. P.Cole. Reprinted in Martinich (1985); quote is from p.326 in reprinted version.

Kaplan, D. 'Afterthoughts' in Almog and Wettstein edd. *Themes from Kaplan* (Oxford Univ Press: New York).

Kripke, S. (1980), *Naming and Necessity* (Blackwell: Oxford).

Leibniz, G.W. (1980), *New Essays on Human Understanding* trans and edd P.Remnant & J.Bennett, (Cambridge Univ. Press, Cambridge) 1981.

Martinich, A. (1985), *Philosophy of Language* (Oxford Univ. Press: Oxford).

Metzinger, J. (1910) 'Note on Painting' *Pan.*

Musil, *R. The Man without Qualities,* published in three volumes by Picador.

Nagel, T. (1986), *The View from Nowhere* (Oxford Univ. Press: New York).

Ortega y Gasset, J. (1968), *The Dehumanization of Art,* (Princeton University Press: Princeton).

Rubin, W. (1989), *Picasso and Braque: Pioneering Cubism.* (Museum of Modern Art: New York).

Russell, B. (1910-11), 'Knowledge by Acquaintance and Knowledge by Description' *Aristotelian Society Proceedings,* republished in Russell(1918).

Russell, B. (1918), *Mysticism and Logic; and other essays* (Longmans: Green and Co.; London).

Sluga, H. (1980), *Gottlob Frege* (Routledge: London).

Williams, B. (1978) *Descartes: The Project of Pure Inquiry,* (Pelican: Hammonsworth)

Wittgenstein, L. *Tractatus Logico-Philosophicus* trans Pears and McGuinness, (Blackwell: Oxford).

Wright, C. (1984) ed. *Frege:Tradition and Influence* (Blackwell: Oxford).

GIDEON ROSEN

# OBJECTIVITY AND MODERN IDEALISM: WHAT IS THE QUESTION?[1]

*I think many persons now see all or part of what I shall say: but not all do, and there is a tendency to forget it, or to get it slightly wrong. In so far as I am merely flogging the converted, I apologize to them.*

J. L. Austin, ' The Meaning of a Word'

## I. Introduction

If you're going to call a volume 'Philosophy in Mind' you should eventually point out that in one crucial sense, the mind cannot possibly matter as much to philosophy today as it has in the recent past. As David Stove has recently reminded us (Stove 1991), most of the good philosophers writing in the 19th century took it for granted that *the world as a whole* was in some sense psychic — penetrated through with thought or mentality — and hence that the study of Mind was the proper foundation for the study of absolutely everything. These days, of course, we can hardly take the idea seriously. Metaphysical idealism of the old German sort strikes us as simply incredible. And while the facts surrounding the eclipse of idealism are no doubt complex, it's not very hard to say what it is about the way we think now that places the view beyond the pale of serious possibility.

The Mind of the idealists was, after all, a very peculiar thing by our lights: an entity not quite identical with anything we encounter in the natural world — and this includes the 'subject' of empirical psychology — which nonetheless somehow constitutes or conditions that world. And the trouble with idealism is that we just can't bring ourselves to believe in this Mind anymore. A flexible and relatively undemanding naturalism functions for us as an unofficial axiom of philosophical common sense. This naturalism is so vague and inchoate that any simple formulation will sound either empty or false. But it is a real constraint: and one of its implications is that if we believe in minds at all, they are the embodied minds of human beings and other animals. Our most basic assumptions thus leave no room for the trans-empirical Subject whose relation to nature was the urgent problem of post-Kantian metaphysics. And since it is just plain obvious that empirical, embodied minds

*Michaelis Michael and John O'Leary-Hawthorne (eds.), Philosophy in Mind, pp.277-319.*

do not actively constitute the bulk of inanimate nature, the idea that the world as a whole is in some sense mental can only strike us as an incredible fantasy.

So understood, the problem of idealism is about as dead for us as a philosophical problem can get. Still, to judge from the jargon of recent metaphysics one may legitimately suspect that the ghost has not been entirely laid. We may not discuss ' idealism' any more. But we do discuss — at great length, and with surprising intensity — something called 'realism' and the various 'antirealisms' to which it is opposed. Now the debates that unfold under these colorless terms are terrifically various; and the emergence of a daunting technical apparatus can give the impression that in each case some relatively precise thesis in logic or semantics is at stake. But of course these logical and meaning-theoretic theses are rarely so urgent in themselves. Rather they derive their urgency (along with their unity as a class) from their supposed connection to something much vaguer — a metaphysical stance or picture, deriving directly from the concerns of post-Kantian philosophy, the cogency of which is supposed to turn on these seemingly more tractable, technical questions.

Like the older debates over realism and idealism, the modern discussions are concerned *au fond* with our right to certain imagery. Whether the topic is moral realism, realism about causation and modality, platonism in the philosophy of mathematics, or Metaphysical Realism about the world as a whole, what the realist mainly claims is the right to say things like this:

Our discourse about X concerns a domain of fact that is *out there*. These facts obtain *anyway*, regardless of what we may think. When all goes well, inquiry in the disputed area *discovers* what is *already* there, rather than *constituting* or *constructing* its object. Successful thought amounts to the *detection* of something real, as opposed to a *projection* onto the real of our own peculiar or subjective perspective ...

And so on down the list of familiar words and pictures. The antirealist denies our right to this imagery, just as the post-Kantian idealist would have done. The link between the old problem of realism and the various new ones is thus forged at the level of rhetoric. And the most basic question one can ask in this area is then simply this: once one has rejected the post-Kantian metaphysics of experience with its commitment to a trans-empirical Subject, what can this denial possibly mean?

We can epitomize the realist's basic commitment by saying that for the realist as against his opponents, *the target discourse describes a domain of genuine, objective fact*. The basic foundational question is then: What is objectivity in the relevant sense, and what are the alternatives? Can we find a definite and debatable thesis upon whose truth the legitimacy of the rhetoric of objectivity depends? If anything has emerged from the long and noisy discussion of these issues, it is the pointlessness of trying to engage them without first attaining a sharper account of what the familiar imagery is all about. This is the question I would like to discuss here.

I should say from the start that I am pessimistic about the prospects for an answer. So far as I can see, it adds nothing to the claim that a certain state of affairs obtains to say that it obtains objectively. To be sure, we do have 'intuitions' of a sort about when the rhetoric of objectivity is appropriate and when it isn't. But these intuitions are fragile, and every effort I know to find the principle that underlies them collapses. We *sense* that there is a heady metaphysical thesis at stake in these debates over realism— a question on a par in point of depth with the issues Kant first raised about the status of nature. But after a point, when every attempt to say just what the issue is has come up empty, we have no real choice but to conclude that despite all the wonderful, suggestive imagery, there is ultimately nothing in the neighborhood to discuss.

I won't have time to defend this pessimism here. I'll spend most of the paper distinguishing this 'fugitive' question from other perfectly good ones with which it is sometimes confused. I shall then consider only a small handful of the extant proposals for framing a definite question of objectivity. The proposals I want to discuss take the traditional distinction between primary and secondary qualities as the starting point for describing a more general contrast between features of the objective world — the world as it is in itself — and features of the world as it is for us. I will suggest that thcsc proposals fail — not because they fail to draw an interesting line, but because the line they draw fails to correlate in any interesting way with our intuitions about the appropriateness of the realist's rhetoric of objectivity. In a later paper I hope to extend the critique to a range of distinct proposals. For now, however, my pessimism about the genuineness of the issue plays the role of a working hypothesis, nothing more.

I should guard against one misconstrual, however. Some philosophers will be tempted to conclude that to say that there is no clear statement of the fugitive question is to say, in effect, that the realist wins on the grounds that his opponent has failed to stake out a genuine thesis. My own view, by contrast, is that it would be more accurate to say that in that case neither side wins. If it makes no good sense to deny the realist's characteristic claims, then it makes no good sense to affirm them either. Quietism, as the view is sometimes called, is not a species of realism. It is rather a rejection of the question to which 'realism' was supposed to be the answer.

*II. Realism*[2]

Now in fact this overstates my worry. What I have called the realist's central contention — the thesis that the disputed facts are objective facts — naturally factors into two parts: the thesis that there really are facts to discuss in the relevant area, and the further claim that those facts are in some sense objective. The first claim is clear enough, I think, and insofar as the discussion of realism turns on it, that discussion makes excellent sense. My doubts are confined to the second claim.

Let us follow what seems to be an emerging consensus by dividing the first thesis in turn into two more basic components.

The realist's most basic commitment is to the view that

(1) Declarative statements in the disputed area are genuinely assertoric: they normally express beliefs rather than some other state of mind, and are therefore properly evaluated in terms of truth and falsity.

To reject this part of the realist's view is to embrace a species of 'non-factualism' or 'non-cognitivism' in the disputed area. The paradigms here are emotivism and prescriptivism in ethics, though the view has lately received much wider application at the hands of Simon Blackburn and others (Blackburn 1984, Gibbard 1990). Obviously, if a class of sentences resembles the typical linguistic manifestation of a command or a question in not being apt for truth and falsity, it would be a mistake to regard it as correctly representing a region of reality or a domain of genuine fact. So any view that sustains the realist's rhetoric will incorporate something like (1).

But equally obviously, (1) is not enough. The atheist grants that talk about the Godhead is genuinely assertoric — that theology is in the market for truth and falsity — but he is no theological realist. So beyond (1), we should say that realism also demands a commitment to something like (2):

(2) Our current doctrine in the disputed area is not massively mistaken. Some of our core commitments are true, or at least 'on the right track' , and some of our most confident posits exist.

This is vague, but still clearly substantial. To reject this aspect of the realist's commitment while retaining the first is to leave it open that our view may be infected with wholesale error. This sort of non-realism comes in varying degrees of strength, corresponding to the varieties of atheism and agnosticism in the theological case. Bas van Fraassen's constructive empiricism in the philosophy of science is a species of 'agnostic' non-realism (van Fraassen 1980). It is a sort of local skepticism which consists in the resolution to suspend judgment on all matters concerning the unobservable causes of observable things. J. L. Mackie's moral skepticism, on the other hand, is a species of 'atheistic' non-realism (Mackie 1977). On Mackie's view, ordinary ethical thought is committed to the existence of 'objectively prescriptive' states of affairs: facts which by their very existence provide all rational beings with reasons for action. Since he also holds that no such facts exist, Mackie charges everyday morality with a massive ontological error. Error theorists in various areas differ about what the upshot of their non-realism should be. Some, like Mackie, propose that we retain the discourse, at least in its surface aspect, while revising our interpretation of it. Others, like Hartry Field in the mathematical case, argue for retaining the discourse with its standard interpretation while withholding our assent from its false commitments (Field 1989). And still others, like the Churchlands in the case of discourse about the mind (and most atheists in the theological case) propose that we simply abandon the error-ridden discourse in favor of something entirely different (Churchland 1984). In each case, however, the basic metaphysical stance is the same. The error theorist refuses to accept the basic claims of the disputed discourse. What the realist calls a fact, the error theorist calls a fiction. The only question then is what to do with it: use it in good conscience, change it, or reject it altogether.

Say that a philosopher who accepts both (1) and (2) for a given region counts as a *minimal realist* in that area. Minimal realism is just the view that our core beliefs about the disputed subject matter are more or less true. Or to put the point somewhat differently, the minimal realist holds that in a thin and metaphysically unambitious sense, our doctrine *correctly represents* or *corresponds to a genuine domain of fact*: at least some of the objects the discourse posits really exist, and the corresponding singular terms refer; some of the predicates pick out genuine properties, and these are really instantiated, etc.[3] The irrealist, the agnostic and the error theorist in one way or another reject this, and so refuse to allow that anything in the world is faithfully described by the characteristic judgments of the disputed area.

The issues surrounding minimal realism in a given area are perfectly clear, up to a point, even if it is often quite unclear how best to resolve them. There is nothing distressingly metaphorical about (1) and (2) — and insofar as our gloss on the view imports the imagery of correspondence between representations in the mind or language and certain worldly items — objects, properties, states of affairs — we know how to dispense with that imagery in favor of relatively colorless talk about whether certain declarative judgments are true; an idiom which gives way in any particular case to first-order discourse in the material mode about whether certain objects exist and what they are like. So to illustrate, the minimal realist about ethics holds that moral judgments are genuine assertions, capable of truth and falsity, and also that our received moral view is not massively mistaken. We can put this by saying that for the minimal moral realist, our moral opinions correctly describe some of the moral facts. But to say this is just to say that some of our moral opinions are true. Or we can formulate the view as the claim that some of the things we call 'good' and 'right' do in fact possess the properties of goodness and rightness. But to say this is just to say that some of the things we call 'good' really are good, and so on. This casual way with the idiom of facts and properties is not uncontroversial; but it is tremendously convenient for approaching the issues that interest us. I propose to acquiesce in it for now. We shall have occasion to reconsider this concession later on.

Some writers have suggested that minimal realism is really the only realism we understand: that the only genuine issues in the neighborhood are the question of truth-aptitude and the question of truth. If this were right, it would plainly be a mistake to construe the modern debate as an interesting transformation of the traditional concerns of post-Kantian metaphysics. To reject minimal realism is obviously to forgo all right to the realist's imagery. If you don't believe that moral or modal judgments are ever true, then you shouldn't say that they describe a domain of fact that it is out there, independently of us, waiting to be detected, etc. *And yet it would be just as misleading to say that there are moral or modal facts of some other, less-than-fully-objective sort, somehow constituted by us or our practices.* On a view like this the realist's rich imagery conveys nothing of substance, and so neither does its denial. Our sense that these debates engage a subtle metaphysical issue about the *status* of certain facts or states of affairs is just a mistake.

The persistence of the Kantian imagery suggests, however, that for many writers there is a residual dispute — one that arises only after minimal realism has been accepted. The residual issue concerns not the existence of the objects, properties and facts described by the disputed discourse, but rather what we have called their objectivity. The challenge, then, is to indicate a line within the world (considered as the totality of facts, or the totality of things together with their properties and relations) between the objective items and the rest.[4] A solution need not actually say which things fall where. It will be enough to say, in relatively non-metaphorical terms, what it is for an item to fall on one side or the other. The line will correlate, in a rough and ready way, with whatever intuitions we possess about the legitimacy of the rhetoric of objectivity.

Until the challenge is met, our sense that there is a genuine issue for realists and their rivals to debate over and above the 'flat' ontological issues associated with the debate over minimal realism can represent at best the pious hope that one of the most compelling problems of traditional philosophy is a real problem after all, and not just a rhetorical illusion.

### *III. Some Senses of 'Objectivity'*

First, however, a word is in order about certain *other* senses of 'objectivity' and related terms. Just as there are a number of reasonably straightforward issues about realism that have nothing to do with objectivity in the sense that interests us, there are a number of reasonably clear questions about objectivity that must be carefully prised apart from the ones that divide realists from their antirealist rivals.

The dominant use of 'objectivity' these days has nothing to do with metaphysics or ontology, but belongs rather to epistemology or methodology. Methodological objectivity, as we may call it, is primarily a feature of inquiries or methods of inquiry, and derivatively of the people who conduct inquiries and the judgments they form as a result. To a first approximation, we call an inquiry 'objective' when its trajectory is unaffected in relevant ways by the peculiar biases, preferences, ideological commitments, prejudices, personal loyalties, ambitions, and the like of the people who conduct it. When the Bulgarian judge falls in love with the Romanian gymnast and gives her clumsy routine on the balance beam a 9.8, we may suspect that his objectivity was less than ideal. When the Church investigates Galileo's astronomy and finds it unsupported by the available evidence, we may doubt the conclusion was driven by an entirely unbiased or objective appraisal of the data. And so on.

These paradigm cases are clear enough. Still, it is not easy to say what unifies them and the others we might supply. On one conception, the idea of a perfectly objective inquiry is the idea of one whose outcome is determined solely by the way the world is together with the fact that the inquirers in question are rational. An objective view would then be one which any purely rational inquirer who turned his attention to the matter would eventually hold.[5] Against this background we may recognize it as a falling away from ideal objectivity whenever we find that an inquiry was sensitive to the peculiar, extra-rational features of the inquirers — their sensory endowment, their powers of concentration or computation, their histories, their cultures, their training, their senses of salience and interest, etc.

It is a post-modern commonplace that methodological objectivity so conceived is a chimera. Indeed, it is far from clear that an inquiry with no substantial starting point and no substantial constraints apart from the formal constraints of rationality makes sense at all, even in areas like mathematics where it has historically had the most play.[6] Of course this does not prevent the idea from functioning as a sort of regulative ideal for actual practice. This is to suggest that even though perfect methodological objectivity is impossible, whenever we find ourselves falling away from it we have reason to correct our bias. From this point of view, pointing up surprising failures of methodological objectivity is the central task of a critical philosophy.[7] But of course this perspective implies a prior judgment to the effect that failures of objectivity are ipso facto undesirable. And this is doubtful. The fact is that sometimes the disclosure of imperfect objectivity debunks an inquiry or undermines its legitimacy and sometimes it doesn't. And the really interesting philosophical problem consists, it seems to me, in reflection on how to draw this line between pernicious and benign lapses from the ideal. [8]

These are tremendously important issues, and they are not unrelated to the problem of realism. One standard non-realist tactic consists in pointing out that in some potentially surprising way, the processes by which we fix opinions in areas like science and ethics are crucially sensitive to our starting point, or to various contingent cultural and political forces that have nothing to do with the facts. This observation is supposed to undermine our confidence that our way is (even approximately) the right way, as against the various routes we might have taken if not but for fortune. The suggestion is that failures in objectivity should push us towards skeptical suspense of judgment— and hence to a rejection of minimal realism. Against this the realist may argue either that even given the failure of objectivity in the near term, in the long run these alien pressures are diluted to the point of vanishing under pressure from the facts themselves, or instead that a certain solidarity with one's actual (albeit contingent) non-rational inheritance is rationally permissible even in the absence of a compelling proof that our way is the best way. The first sort of defense is most prominent, for obvious reasons, in the philosophy of science (Boyd 1984). The second is associated with the realism of Richard Rorty, among others (Rorty 1989).

This is obviously a central line of debate. But it is a debate over minimal realism, whereas our present quarry is a debate that allegedly remains even after that one has been settled. I mention the issue only to point out that on the face of it, whether an inquiry is objective in the methodological sense has nothing to do with whether the facts it aims at are objective in the sense that interests us. Let it be granted that due to the inevitable complications of transference and countertransference every psychoanalytic investigation of neurosis is subtly influenced by the unconscious affinities of analyst and analysand, and so counts as less than fully objective. This recognition is nonetheless compatible with saying that the facts about why Dora refuses to speak German are entirely objective or *out there*, and that psychoanalysis constitutes a flawed but valuable procedure for *discovering* what is *already there*. The methodological conception of objectivity may have something subtle to do with the metaphysical notion that interests us. We will return to the suggestion later. But they are clearly not the same idea, so it is important to keep them distinct.[9]

We should also mention a second conception of objectivity — this time a more 'metaphysical' one — that has just as little to do with the concept we're after. I have in mind the sense in which the 'objective' world is sometimes opposed to the 'inner' or 'subjective' world, a contrast central to the Cartesian tradition in the philosophy of mind, and anathema to post-modern thought. According to the traditional conception, some items exist in an external, public space — rocks, stars, human bodies, etc. — to which no thinking subject has any privileged relation, while others exist in a private world, non-inferentially accessible to only a single subject. To the extent that we can be acquainted with things like rocks and stars, any number of people can be acquainted with any such item; but there is supposed to be a sense in which only I can be acquainted with my pains and other objects of consciousness; the extent of this special sort of acquaintance is the extent of inner space.

Now this conception takes a number of forms, and I wouldn't want to defend any of them.[10] Nonetheless, I think we can agree that even if subjective phenomenal states or objects in the traditional sense do exist, the facts about them would not be less-than-fully objective in the sense that we are concerned to capture. When I consider the spider and ask (perhaps idly), 'What is it like to be him?,' even if I can't frame an adequate answer, the fact that does answer my question (or would if I could only 'grasp' it) is conceived as an entirely objective state of affairs. If anything its claim to objectivity is driven home by my admitted incapacity to form any conception of it. For according to the rhetoric that surrounds our topic, one of the hallmarks of the objective is precisely its independence of our theories or conceptions. Likewise, if Jones is essentially alone with his pain, still the fact that he feels a pain of a certain perfectly determinate sort is conceived in the tradition as an entirely objective fact — a fact God would know, but about which any ordinary external observer might be radically mistaken. So without troubling to make this contrast precise, we may say that as a conceptual matter there is a clear distinction between the subjective/objective contrast as it derives from Descartes and the contrast between the objective and the less-than-fully objective that is at stake in debates over realism. Cartesian experience may be private. But the Cartesian is a robust realist about it. The facts about subjective experience are — in our sense —objective facts, even though they are obviously, in another sense, 'subjective'.

It is not too much of an exaggeration to suggest that what is sometimes 'post-modern' philosophy is constituted by its denial of two allegedly Cartesian principles: the claim that any good inquiry must be objective in the methodological sense, and the claim that the proper starting point for such inquiry is the 'inner' or the 'subjective' world. So in one sense the post-modern philosopher sets himself against objectivity, and in another he sets himself for it. Obviously this sort of posturing is not the sort of thing one can argue about. I mention it only to point out that our fugitive issue — the question of 'objectivity' at stake in disputes over realism — is neither of these.

### *IV. Mind-dependence*

Unlike objectivity in the methodological sense, the objectivity that interests us is a feature of worldly items and not just of judgments or representations. I have spoken fairly loosely of 'facts' or 'states of affairs', though we might just as well have spoken of objective properties and relations, and in some cases even of objective objects. It is important for present purposes that we not be distracted by the various detailed metaphysical proposals about how these worldly items are to be construed. We should not presuppose any particular construction of facts or states of affairs, or any particular theory of properties. The best way to formulate the issue is to acquiesce in what is sometimes called minimalism about these entities. For the minimalist, whenever we have a true sentence *p* we may harmlessly invoke an entity, *the fact that p*; and whenever we have a meaningful predicate *F* we may harmlessly invoke *the property of being F*. The slogan sometimes associated with the view is that these worldly items are mere shadows of the linguistic objects to which they correspond. But this image suggests a certain insubstantiality or dependence on language that has no proper place in the minimal conception. It is open to the minimalist to maintain that the fact that snow is white, or the property of being a tree would have 'obtained' or 'existed' even if there had been no language at all.[11] For him, this is just a way of saying that even if there had been no language, snow would still have been white and there would still have been trees. The minimal view has problems, of course. It aims to be deliberately naive. But paradox threatens if we unreservedly endorse the transition from '*X* is *F*' to '*X* has the property of being *F*'; and if nothing else, the theorizing that responds to the paradox will involve some distinction between meaningful predicates that pick out properties and those that don't.[12] Let us set these genuine concerns to one side, however. From the present perspective, the line between the objective and the rest is a line drawn within the world of facts so conceived. If the world is the totality of facts, then we may distinguish (at least notionally) the objective world — the totality of objective facts — from the world as whole. *P* is an objective property if it is an objective fact whether an object possesses it: and an object *x* is objective if the fact that *x* exists is an objective fact.

Given this way of speaking, the question about the objectivity of morality — if it is not a question about the methodological objectivity of moral judgment — is to be understood as a question about whether properties like goodness and rightness are features of the objective world which in the most favorable cases we discover, or whether their distribution is somehow determined by our moral attitudes, institutions and practices. The question about the objectivity of mathematics (which may or may not be the question Kreisel called *the* question) is similarly: Are the mathematical facts — the states of affairs that correlate with the truths of mathematics — entirely independent of our mathematical thinking, or are they rather somehow constructed by it? And so on.

The objectivity that interests us is thus evidently to be contrasted with a sort of mind-dependence — or, to use a more modern idiom, a dependence on our linguistic and social 'practices'. What is less-than-fully objective owes what reality it possesses to our thinking, and is to that extent something mental. This way of talking makes vivid the supposed point of contact between the modern debate and the post-Kantian metaphysical problematic. But it can also be quite misleading. A few plodding words about the varieties of mind-dependence are therefore in order.

Tables, frescoes, political institutions and baseball games all owe their existence to thought in one obvious sense. They are artifacts. They come into being in part because someone intends that they should. The depletion of the ozone layer, global warming and the middle class are also mind-dependent, though in a different sense. They are the unintended consequences of intentional social activity. Let us say that an item depends *causally* on the mind iff it is caused to exist or sustained in existence in part by some collection of everyday empirical mental events or states. Mind-dependence of this sort is perfectly familiar. And no one supposes that the observation that an item is causally mind-dependent in any way undermines its claim to objectivity in the sense that interests us. The most full blooded realist about frescoes and social classes can allow that these things are artifacts in a generalized sense. If the Kantian imagery is to be made sense of, it must therefore be possible at the very least to sketch a species of mind-dependence distinct from this sort of causal dependence.[13]

And as soon as this is pointed out, it becomes equally clear that a range of other not-quite-causal conceptions of mind-dependence are also distinct from our target. Mental activity is itself mind-dependent in the trivial sense that it would not exist if there were no minds. The fact that Jones is in pain would not obtain if Jones were not a thinking thing. This sort of mind-dependence need not be causal. The causal antecedents of Jones's pain may be strictly physiological. So we need a more general term. Let us say that an item depends *existentially* on the mind just in case it could not exist or obtain if there were no empirical minds or mental activity. Other examples include states of affairs which, though not exclusively psychological, nonetheless essentially involve the psychological, like the fact that New Zealand is a democracy or the fact that baseball is more popular than sled dog racing. In general, the subject matters of history, psychology, anthropology and the other human sciences all essentially involve the mental in this way, and so count as existentially mind dependent. But once again, this observation has no tendency to impugn the objectivity of these regions of fact all by itself. We can grant that these disciplines concern the mind without forgoing our right to the rhetoric of objectivity, discovery and detection. Of course there are subtle doctrines in the philosophy of the social sciences that are supposed to undermine this imagery. For now my point is only that there is no simple connection between possessing a psychological subject matter and failing of objectivity in the sense relevant to the debates over realism.

Objectivity, if the notion makes sense at all, is thus opposed to some sort of non-causal, non-existential (for short: non-empirical) mind-dependence. And it is precisely this that links the present debate with the post-Kantian problematic. For Kant the empirical world is 'conditioned' by the structure of the mind. This claim is not transparent. But one thing we know about it is that the mind that does this conditioning is not itself a part of the empirical world. It is not the object of empirical psychology or personal introspection. And this at least points in the direction of a thesis: for Kant, the structure of something that is not quite part of nature somehow determines the structure of nature itself. And whatever the nature of this determination is, it is obviously distinct from the various relations of dependency that may obtain among empirical subjects and the world they inhabit. Now as we said at the start, the modern discussion is defined, in part, by its rejection of this transcendental standpoint. The modern idealist — the antirealist who acquiesces in minimal realism — holds that some states of affairs depend non-empirically on *us* or *our practices*. But given the undemanding naturalism that conditions the debate, there is nothing for this plural subject to be except a part of the natural world itself. This means that the problem of staking out a debatable thesis in this general vicinity is if anything harder now than it was before. To put the point bluntly, the modern antirealist must apparently advocate an idealism that is *neither empirical nor transcendental*: a view according to which parts of the everyday world are non-empirically determined or constituted by other parts of that very world, namely ourselves. And the question, once again, is: What on earth could this possibly mean?[14]

*V. Objectivity as Response-Independence*

It would be pointless to attempt a survey of every effort to flesh out a distinction between the fully objective facts and the rest, not to mention the presuppositions of every philosophy that takes such a contrast for granted. Instead, by way of an illustration of just how hard the problem can be, I want to consider a family of important proposals, all of which take the distinction between primary and secondary qualities as a starting point. Here is why this should seem promising. On the traditional conception, it can be true to call this tomato 'red', just as it can be true to call it 'round'.[15] So in our preferred, metaphysically light-weight sense, the conception admits facts about the distribution of primary and secondary qualities alike. The difference is that a thing's being red, though an intrinsic feature of the object, is nonetheless supposed to have something to do with us and our sensibility, whereas a thing's being round is supposed to have nothing to do with us. And yet this 'subjectivity' or 'mind-involvingness' of the secondary is supposed to be compatible with the common-sense view that colors, like shapes, are neither causally nor existentially dependent on mental activity. Tomatoes are not literally tinted red by our perception of them; indeed they would have been red even if there had been no perceivers at all. The traditional conception therefore promises to provide a paradigm for a species of mind-dependence that is not an empirical dependence. What's more, the only minds in the picture here are the embodied minds of ordinary human beings. The contrast between primary and secondary qualities, and its attendant version of the contrast between objective and subjective, predates the conception of the transcendental subject. If the contrast can be sharpened, defended and generalized, then, the issue between realists and their rivals about the objectivity of a certain class of facts might reduce to the question of whether the facts in that class concern only the distribution of primary qualities. And this might be a genuinely debatable issue.

This line of thought can be developed in more than one way, depending on which strand in the traditional distinction one regards as central. We begin with a relatively straightforward suggestion due to Mark Johnston.

Johnston's central suggestion is that the contrast between primary and secondary qualities is fruitfully viewed as a special case of a more general contrast between what he calls 'response-independent' and 'response-dependent' concepts. The general contrast is drawn as follows. First say that a concept *F* is *dispositional* just in case there is an identity of the form

(3) The concept *F* = the concept of the disposition to produce *R* in *S* under *C*,

where *R* is the manifestation of the disposition, *S* is the locus of the manifestation, and *C* is the condition of the manifestation. So the concept fragility is the concept of the disposition to break when struck, where the locus of the manifestation is the fragile thing itself.

Next, let us say then that the concept *F* is a *response-dispositional* concept when something of the form (3) is true and (i) the manifestation *R* is some response of subjects which essentially and intrinsically involves some mental process (responses like sweating and digesting are therefore excluded), (ii) the locus *S* of the manifestation is some subject or group of subjects, and (iii) the conditions *C* of manifestation are some specified conditions under which the specified subjects can respond in the specified manner. Moreover, we shall require (iv) that the relevant identity does not hold simply on trivializing 'whatever it takes' specifications of either *R* or *S* or *C*.

Finally,

> A concept is *response-dependent* just in case it is either a response-dispositional concept or a truth-functional or quantificational combination of concepts with at least one non-redundant element being a response-dispositional concept. Otherwise we shall say that the concept is a response-independent concept. (Johnston 1993, 103-4; cf. Johnston 1989, 145 ff.)

Johnston then goes on to suggest that the secondary qualities of traditional epistemology may be identified with the response dependent concepts for the special case in which the response is a sensory state or event.[16]

On this sort of view, to say that *red* is a response-dependent concept is to say that it is identical with some concept one of whose components is the concept of a disposition to produce a psychological response. More specific versions of the proposal would include

(4) The concept of being red = the concept of being disposed to look red to statistically normal human beings in broad daylight at sea level.

or

(5) The concept of being red = the concept of being disposed to look red to us as we actually are under conditions which we are actually disposed to regard as good conditions for the perception of colors.

Because a thing's looking red to a subject is a sensory state, either account would imply that red is a secondary quality concept. (For the record, Johnston himself is explicitly skeptical about the possibility of providing such equivalences for our ordinary color concepts. So for him, red is *not* a secondary quality, at least not as we ordinarily conceive it.)

The important point, however, is that the contrast has application beyond the standard lists of secondary qualities. Thus David Lewis has recently argued that

(6) x is a value iff we would be disposed to value x under conditions of fullest possible imaginative acquaintance with it.

and Hume may have held that

(7) c caused e iff we are disposed to imagine the occurrence of an e-like event upon perceiving or imaging a c-like event given sufficient exposure to the course of nature.

We may take these biconditionals to be underwritten by claims of conceptual identity.

(6*) The concept of being a value = the concept of being such as to be valued by us under conditions of fullest possible imaginative acquaintance.

(7*) The concept of causation = the concept of the relation that holds between events when we are disposed to imagine the second upon perceiving the first given sufficient exposure to the course of nature.

And if we do then we may say, as Johnston does, that these and similar proposals represent response-dependent conceptions of value and causation, even though the responses here are not sensory responses — valuing and imaging, while psychological, are not forms of sensation — and the concepts in question are therefore not secondary quality concepts *strictu sensu*.

Now there is some reason to doubt that the traditional contrast between primary and secondary quality concepts is properly drawn in these terms. For note that on the present view, whether a concept is response-dependent or not is always an a priori matter. It is a question of considering and evaluating claims of concept equivalence; and on most ways of understanding this task is always, at least in principle, a reflective, non-empirical enterprise. By contrast, the traditional conception treats the claim that colors and the rest are secondary qualities as an empirical discovery of the highest order. The scholastics who denied it with their doctrines of sensible species were not necessarily confused about the contents of their concepts, and no early modern scientist would have suggested that a priori reflection by itself should have changed their minds. The scholastic mistake was a mistake about the psychophysics of color perception. Their view was an explanatory hypothesis, subject in principle to experimental refutation.[17]

Still, there is an obvious affinity between Johnston's contrast and the traditional one. In Philip Pettit's phrase, response-dependent concepts 'implicate subjects' in a way in which response-independent concepts don't. (See Pettit 1992.) If *red* were response-dependent [18] then the thought that a certain tomato was red would be a thought about how the world is for subjects of a certain sort. And yet if the explanation of the concept features one or two judicious deployments of the rigidifier 'actual' (as in (4)), then the thought will not vary in truth value from world to world as the facts about subjective responses vary. Ripe tomatoes intrinsically like the ones in my garden are disposed to look red to us as we *actually* are even if they never get the chance, either because there are no perceivers around or because the only perceivers differ from us in their patterns of sensory response. Hence the prospect of securing a sort of mind-dependence — the concepts implicate subjects — that is at the same time neither causal nor existential.

A promising thought, at least on its face; so let us consider the following proposal.

(8) When the central concepts of a discourse are response-dependent, the true sentences within the discourse represent a range of subjective or mind-dependent facts. A fact is genuinely objective, then, when it is represented in a discourse whose central concepts are response-independent.

Johnston calls the theorist who regards the central concepts of a discourse as response-dependent a Descriptive Protagorean. Clearly, the Descriptive Protagorean can be a realist in our minimal sense. Against the non-factualist and the error theorist he may consistently hold that our central claims in the discourse are true, which is just to say that they represent a domain of genuine facts and properties. Things may be good or red; these properties may be instantiated by real things; there may be facts about their distribution, etc. The suggestion in (8) is that the Descriptive Protagorean is nonetheless something less than a *full-blooded* realist, since he holds that the subject matter of the discourse depends (though not empirically) on our patterns of subjective response.

So we must ask: Does this gloss on the notion of a less than fully objective state of affairs do the trick. That is, does the Realist's rhetoric of objectivity, already-thereness, discovery and detection really fail to cohere with the recognition that a range of facts is given to us through the employment of response-dependent concepts? There is something quite natural in supposing that when we use such concepts what we are describing is the world as it is for us, and not the world as it is considered in itself. And this suggests that debates about the extent of response-dependence are a genuine sharpening of much older metaphysical issues. Is this natural thought a good one? [19]

No, it isn't. Consider (9):

(9) The concept of being annoying to fox terriers = the concept of being disposed to annoy statistically normal fox terriers under ordinary conditions.

Annoyance is a mental state, albeit a rather primitive one. So the concept is response-dependent in Johnston's sense. We may add that it is plainly instantiated, e.g., by certain pullings of tails and pokings of eyes. So there exists a range of facts about which things are annoying to fox terriers and which are not, and the discourse we use to represent these facts trades in a response-dependent concept. We may therefore ask whether there is any reason to treat these facts as anything but entirely objective. In calling the concept response-dependent are we in any way abrogating the right to think of these facts as robustly real constituents of a mind-independent order?

Well, unless we have concerns about the status of dispositions in general or about the status of psychological states like annoyance — legitimate worries, perhaps, but no part of Johnston's view — then it is hard to see why the facts about which sorts of proddings annoy a certain breed of dog should count as anything short of robustly real. The point is the obvious one: dispositions to bring about mental responses would seem to be on a par, metaphysically speaking, with dispositions to produce merely physical responses in inanimate things: the qualities Locke calls mere 'active Powers'. Absent a reason to construe mentality itself as less than fully real, the facts about the annoying, the embarrassing and the rest are no different from facts about the poisonous or the corrosive. If we have no reason to withhold the rhetoric of objectivity in the latter sort of case, we have no reason to withhold it in the former either.

Could it be that the plausible core of the view emerges when we restrict (8) to cover only the concepts of dispositions to produce mental responses in *us*? There are two ways to take the suggestion. *F* might be the concept of a disposition to produce mental responses in a group G to which we all happen to belong. And for concepts of this sort the proposal is no better now than it was before. The facts about which things annoy *human beings* or *late twentieth-century bourgeois intellectuals* are not materially different from the corresponding facts about fox terriers. On the other hand, *F* might be given as an *essentially first-personal response-dependent concept*:

The concept of being F = the concept of being such as to produce psychological response *R* in *us* (as we actually are) in *C*.

I take it that a proper theory of concepts may well distinguish between indexical concepts of this sort and the corresponding concepts where a non-indexical phrase is used to pick out the same group of subjects. The plural pronoun brings with it a certain vagueness, of course. 'Us' might mean the species, the culture, or some smaller group, or perhaps even a larger group like those who share our 'form of life'. But in any particular case some more or less definite restriction will be in place, and there will still be a conceptual difference between an indexical specification of a group as the one which bears a certain relation to the subject and a non-indexical specification of the same group. So we can ask: is there any reason to think that a discourse whose central concepts are first personal response-dependent concepts is concerned with a less than fully objective subject matter?

The thought is no doubt encouraged by the fact that when we employ such concepts there is now a definite sense in which our topic is the world as it is for us. This has a Kantian ring. And yet it seems to me that no matter how apt the phrase may be in this context, it is not usefully opposed to 'the world as it really is', or 'the world as it is in itself' as the rhetoric of objectivity demands.

Consider value as Lewis conceives it: (6*). So understood, the concept of value is a first-person response dependent concept. So we ask: is there any reason to think that the facts about which things are valuable in this sense are less than fully real or objective?

One might argue that there is as follows. Suppose I assert '*x* is a value' and someone else from another group — not part of my 'we' — asserts '*x* is not a value'. The analysis implies that we could both be right. Both utterances could be true. But at the same time, we obviously disagree, in the sense that our judgments cannot be coherently conjoined. So it must be that our judgments somehow concern different domains: his the world as it is for *him*; mine the world as it is for *me*. His true judgment picks out one evaluative fact; mine picks out another. These facts are incompatible. They can't sit together in a single world. So they must sit in different worlds. By symmetry, neither of these worlds can be the real world — the single objective world that is the same for everyone. Hence value judgments must concern features of not-quite-objective 'projected' worlds, distinct from the world as it is in itself.

I won't try to pin this line of thought on anyone in particular. It is clearly tempting, and it brings out a connection between a species of metaphysical relativism and the rejection of realism which a number of writers have plainly felt.[20] All the same, it is really a gross confusion.

Consider: I assert 'That is my foot' and someone else from another group asserts (with the same object in view) 'That is not my foot'. Clearly we could both be right. But at the same time we disagree, since our judgments resist conjunction. So it must be that our true judgments concern facts that inhabit different worlds. But symmetry precludes our calling one of them 'real' to the exclusion of the other. The facts about which feet are my feet must therefore reside in projected worlds, distinct from the world as it is in itself. So the fact that these feet are mine is not an objective fact.

This is just silly, of course. The mistake obviously lies in the suggestion that the two judgments disagree. In the relevant sense, judgments of the form fa disagree when they attribute incompatible *properties* to a single object. In the absence of indexicals, ambiguity and other related phenomena we may indeed move reliably from the fact that $\neg fa$ is the syntactic negation of $fa$ to the claim that they disagree in this way. But when indexicals are on the scene, the inference obviously fails. Tokens of a single indexical predicate can pick out different properties on different occasions of use. So the property attributed to a by the use of $f$ in the first sentence need not be incompatible with the property attributed to it by the use of $\neg f$ in the second. This is clearly what happens when the predicate is '... is my foot'. And it is only because the indexical is not right there on the surface in the case of Lewis-style value attributions that the corresponding argument is even remotely compelling. Run it again with the officially equivalent idiom 'we are disposed to value $x$...' and the two cases plainly converge.[21]

It may be objected that this diagnosis implies a controversial view about the individuation of properties. I have assumed that indexical predicates like '...is my foot' pick out different properties on different occasions of use. And yet throughout the discussion I have intended my talk of facts, properties, and the like to register only the most minimal commitment. Our heuristic slogan was: facts are the ontic shadows of sentences; properties the shadows of predicates, etc. And it might well be asked: with what right do I then suppose that two uses of the same predicate in the same language in the absence of material lexical ambiguity pick out distinct properties? Why not say: 'There is one property — the property of *being my foot*, as distinct from the property of *being Rosen's foot*. I attribute it to *x* and you don't. But both attributions are true. So it is a fact that *x* has this property, and also a fact that *x* lacks it. Surely these facts cannot cohabit. So the worlds must be many: and as before, a symmetry argument suggests that none of the worlds we describe using indexical concepts — including first-person response-dependent concepts — deserves to be called the 'real' world'?

The first thing to say about this line of thought is that even if it is granted, (8) is still off target. The line between the objective facts and the rest would then have nothing to do with dispositions to produce mental responses, since as we've seen, many of these can be described in a non-indexical idiom, and when they are we have no reason to regard them as less than fully objective. It has rather to do with the contrast between facts corresponding to sentences involving indexicals and the rest. Whenever a sentence (considered as a type) can vary in truth value from one context of use to another the argument will work to show that the facts it describes are not features of the real world. So when I truly say 'This ring contains 0.6 g of platinum', the fact that I describe is not quite objective, because you might utter these very same words and say something false. (You might be talking about a different ring.)

This is clearly preposterous, and at this stage any link with the traditional metaphysical debate over realism has been utterly severed. Still, it *is* a fair question for those of us who want to play fast and loose with a metaphysically light-weight notion of property and fact to say where exactly the mistake occurs. It is clear enough what we want to say. The proposal treats properties as shadows of orthographically individuated predicates — or better: as correlates of predicates that share the same Kaplanian character — whereas we want to treat them (roughly) as correlates of predicates that share the same Kaplanian content (Kaplan 1989). So if I say: '*x* is my foot' and you say, addressing me as part of the same conversation, '*x* is not *your* foot' or '*x* is not Rosen's foot', we do attribute incompatible properties to a single object. Still it is unclear what the motivation for this preference should be.

I propose to leave the issue unsettled. Clearly, if there is an interesting issue between realists and their rivals that goes beyond minimal realism and continues the metaphysical problematic deriving from Kant, it cannot be a question about whether the characteristic concepts in the relevant discourse are indexical or not. As everyone in the debate has always understood these matters, when we 'disagree' in the application of a predicate like '... is my foot' we are talking about a single world and attributing distinct and compatible relational properties to the indicated object. Perhaps this is an optional view. But it is hard to believe that weighty metaphysical issues of the sort that really have exercised the best minds of our generation could possibly hang on it.

### *VI. Objectivity as Judgment Independence*

We have rejected the suggestion that response-dependent discourse is always discourse about a less-than-fully objective world for a simple reason. The concepts 'implicate' us, to be sure. But when we consider the properties they pick out — the subject matter of the discourse that involves them — they turn out to be *merely anthropological.* A being from another group might discuss the very same properties without implicating himself at all. He would be talking about the dispositions of one part of the objective world to affect another part of that world in certain apparently objective respects. And setting to one side the closing conundrum of the previous section, we have seen no reason to think that a realism about such facts is anything but a full-blooded realism about them.

I've gone on at such length about this proposal because it permits a particularly sharp formulation of the point. But with this in the background, we can afford to be briefer in discussing some related suggestions. Crispin Wright's latest book (Wright 1992) contains the most detailed and systematic treatment of this problem so far. Wright's view is that there are in fact *several* debatable issues about objectivity that may be taken to divide realists from their opponents even after minimal realism has been taken on board. A full discussion of Wright's suggestions is clearly to the point; but there is no space for it here. For now we focus on a single proposal that bears evident affinity with the last.

Let us say that a concept *F* is *judgment-dependent* if and only if

(10) It is a priori that: *x* is *F* iff certain subjects *S* would judge that *x* is *F* under conditions *C*.

— where as before, the class of subjects *S* and the conditions *C* are specified substantially, i.e., in such a way that the embedded sentence is not just a trivial logical truth. This approximates the more Johnstonian formulation of the corresponding idea:

(10$_j$) The concept *F* = The concept of being an object disposed to elicit the judgment that it is *F* in *S* under *C*.

If the two were equivalent we could view the judgment dependent concepts as a species of the response-dependent concepts, where the response is restricted to judgments involving the very concept at issue. The equivalence fails, however, in part because talk of dispositions in (10$_j$) corresponds to a counterfactual in (10); but more importantly for our purposes, because claims of concept identity are (on the face of it) stronger than claims of a priori coextensiveness. One way to show this would be to defend the controversial claim that when a term f has its reference fixed by means of a description d, it may be a priori that *x* is f iff *x* is d even though the concepts associated with the terms are quite distinct.[22] But for our purposes it is sufficient to observe that, whenever *S* is some complicated mathematical truth, it can be a priori that *x* is f iff (*x* is d & *S*) without the concepts of *being f* and *being d & S* coinciding.

These differences between Wright's apparatus and Johnston's make a difference. (See Wright 1992, ch. 3 appendix and Johnston 1993, Appendix 3 for discussion). But it is unclear whether it bears upon our present concerns. Here I proceed in terms of Wright's preferred idiom — so-called ' basic equations' of the form (10).[23] Our remarks apply equally to a parallel proposal developed in Johnston's framework.

The proposal we need to consider then is this:

(11) When the central predicates of a discourse are judgment-dependent, the facts that discourse describes are less-than-fully objective.

Note that (11) offers only a sufficient condition for mitigated objectivity in keeping with Wright's suggestion that there are other ways for objectivity to be compromised. The question for us is: does the presence of pervasive judgment-dependence in a region of thought displace the rhetoric of objectivity and its attendant imagery? Are we any the less entitled in these circumstances to think of the facts we describe as facts that are already there: facts which we aim to describe or detect, and not to invent or construct?

The case for (11) can be made to sound quite compelling. As we observed in passing above, one familiar gloss on the notion of objectivity that interests us has it that the objective is that which obtains independently of our opinions about it. Now consider the judgment *in*dependent concept *F*. To say that *F* is judgment-independent is to say, in effect, that for any specification of conditions *C* we have no guarantee a priori that judgments about the extension of *F* formed in *C* are must be true. And this is as it should be if the facts about what is *F* are genuinely independent of our opinions about them. When two things are genuinely independent, after all, it is always an empirical matter which conditions favor their co-occurrence. So if the *F*-facts are out there independently of us, we should have no insight a priori into when certain psychological events — our judgments that *x* is *F* — should co-occur with the distinct facts about which things are *F*. But by the same token, when *F* is judgment-dependent and we *do* possess a priori insight into the conditions under which the facts about *F* and our *F*-involving judgments coincide, we seem precluded from thinking of the facts and our opinions about them as two genuinely independent realities. It comes much more naturally to say that the facts about which things are *F* — the *F*-world — are *constituted* by our practice of judging things *F*. Or perhaps, if the directionality of this metaphor seems inappropriate, to say that rather than the world's being independent of the mind, in these respects at least, the mind and the world together make up the mind and the world.[24] Thus we seem to have a promising way of making out the sort of non-empirical connection with the mind (our practices) characteristic of the denials of realism whose content has so far eluded us.

Some of the most ambitious global rejections of realism can be seen as affirmations of global judgment dependence. Consider, for example, the Peircean suggestion that the truth is somehow constituted by the opinion we would hold in the limit of inquiry. This is widely regarded as something less than a full blooded realism; Peirce himself sometimes called it 'idealism'. The slogan under which such views currently move is that realism implies that 'truth is not epistemically constrained' , whereas antirealism implies the opposite. We can think of these views as global endorsements of the schema:

> It is a priori that: *x* is *F* iff we would judge that *x* is *F* were we to investigate the matter so thoroughly and clear headedly that a stable rational consensus were achieved.

Alternatively, we can view them as endorsements of the judgment dependence of the concept of truth:

> It is a priori that: *S* is true iff we would judge *S* true at the limit of an ideal rational inquiry.

We call such views 'antirealist' because, on the face of it, they imply that the facts in general (the world) are *constituted or at least constrained by our practices for fixing opinion*. These global proposals are obviously implausible — which is not to rule out the possibility that a philosophical argument could establish them. (See Johnston 1993 for extensive discussion). But local versions are possible and in many cases quite natural. Thus *interpretivism* in the philosophy mind:

> It is a priori that *S* is in mental state *M* iff an interpreter employing our actual canons of mental state attribution who was fully informed about the non-intentional facts would attribute *M* to *S*;

or *constructivism* in political philosophy

> It is a priori that a social arrangement is just iff free and rational beings would judge it just under conditions of full information and reflective equilibrium.

In each case, this sort of picture is supposed to be compatible with minimal realism about the disputed area while revealing the facts in question to be so internally connected with our practices for describing them that the enterprise of description is not properly regarded as aimed at an independent reality. In these cases, we are supposed to find the full rhetoric of objectivity out of place. But is this right? Is the demonstration of significant judgment dependence in an area of discourse enough to displace the realist's imagery?

Now I think the air of progress here is really illusory, and there are several ways to bring this out. Notice first that as we have formulated the view so far, the notion of judgment-dependence does not yet presuppose that the subjects in question are *us*. So (to adapt an example of Johnston's) let us suppose that

(12) It is a priori that: A U.S. law is constitutional (at t) iff the majority of the US supreme court, after informed and unbiased deliberation, would judge it constitutional (at t).[25]

On this assumption, the concept of constitutionality is judgment-dependent. But now ask: are the facts about which laws are constitutional somehow less than fully objective for that? I don't see why they should be. So far we have been given no reason to think that the facts about what a certain group of people would think after a certain sort of investigation are anything but robustly objective. The facts about how the court would rule are facts of modal sociology. These may be very hard to discover, and the idiom that describes them maybe vague (which means that there may be truth value gaps in our discourse about constitutionality); but on the face of it they possess the same status as the facts about what any other collection of animals would do if prompted with certain stimuli, or set a certain problem.[26] The facts about what the court would do with a given case — the facts given on the right hand side of (12)— are thus, for all we've said, features of the objective world. And if the facts given on the left hand side just are these very facts, then (12) gives us no special grounds for thinking of them as less than entirely real.

But is it fair to *identify* the facts about constitutionality with the facts about what the court would judge? The identification would be perfectly appropriate if instead of (12) we had ($12_j$):

($12_j$) The concept of being a constitutional law = the concept of being a law that the court is disposed to judge constitutional.

But as we have remarked, (12) is significantly weaker than ($12_j$). So there is space for the suggestion that, while the facts about how the court is disposed to rule are objective, the facts about constitutionality which supervene upon them do not deserve this epithet.

In the absence of a real theory about facts and their individuation it is hard to know quite how to respond. My own view, for what it's worth, is that *intuitively*, if the facts in the contested class can simply be read off in a mechanical way from the facts in an uncontroversially objective class, then there can be no grounds for denying the same status to facts in the contested area. After all, if we have a class of facts which by universal consent the mind plays no role in constructing; and if it is also by universal consent an entirely *analytic* or *conceptual* truth that when those facts obtain, certain 'other' facts obtain, then it is entirely unclear where the mind is supposed to do its 'constructive' or 'constitutive' work. Think of it this way: an anthropologist studying the court might determine which laws are constitutional by theorizing about which laws the court would ratify. He thinks of the latter study as a matter of charting some modal facts that are *already* in place. His own way of thinking in no way *constitutes* these facts. But more importantly, the only sense in which anyone's thinking constitutes them is the sense in which they just are facts about the thinking of the members of the court; and we have already seen that this is not incompatible with complete objectivity. Now once these objective facts are known, he can deduce facts about the distribution of *constitutionality* by employing an entirely analytic principle — a conceptual truth. And the trouble is that even if we agree to call these facts 'distinct', it's still hard to see why they deserve to be called mind-dependent in any special sense. For the anthropologist they are every bit as 'out there' as the facts that allegedly constitute them: the facts about what certain people would think.

Of course the cases that interest Wright are all cases where the subjects in question are given as *us*. But given our previous discussion, I think we can see that this cannot make a material difference. Suppose that

It is a priori that: $x$ is funny iff we would judge $x$ funny under conditions of full information about $x$'s relevant extra-comedic features.

This allows that we may be wrong about what is funny in the familiar ways. A joke that strikes us as funny at first hearing may have meant and received as a vicious insult, or it may have had its origin as an in-joke among the guards at Dachau; and in these cases we should retract our initial judgment. 'It seemed funny at the time, but I was wrong. It wasn't.' The suggestion is just that the facts about what is really funny are fully determined by what we would clearheadedly take to be funny if we were not mistaken in any of these material ways.

Now the question is whether the proposal (11) gets any more plausible when restricted to such first-personal judgment-dependent concepts. And as before, we can convince ourselves that the move to the first person cannot make a metaphysical difference by considering the anthropologist's perspective. The anthropologist is studying us, and he has gotten to the point where he can reliably determine which jokes we will judge funny under conditions of full relevant information. Perhaps he has achieved this through rigorous inductive social science; or perhaps he has engaged in an exercise in sympathetic immersion in our way of life — not to the point of conversion, but rather to the point of *Verstehen*. He may not call such jokes 'funny' himself. This might suggest that he finds them funny, which may not be the case. But suppose he asks himself about the property *we* attribute to objects when *we* call them 'funny' . This may not be the property he refers to when he uses this word. But he can still identify it, perhaps as 'the property denoted by *their* use of 'funny' — 'F', for short' . And then he can ask: 'Which things are *F*?' . And he will have no trouble answering. He has translated enough of our language to know that things have *F* when we are disposed to judge them funny. And since he can track such dispositions by anthropological means, he can determine which things have *F* with reasonable accuracy. And the important point is that from his point of view, the facts about the distribution of *F* are 'mind-dependent' only in the sense that they supervene directly on facts about our minds. But again, this has no tendency to undermine their objectivity.

Assuming the anthropologist's perspective seems to me just the right tactic for addressing claims to the effect that a certain class of facts is interestingly constituted by our practices of judgment.[27] From this perspective, the only sense in which this is true is the sense in which the relevant facts turn out to *be* facts about our practices — or at least facts analytically supervenient on facts of this sort. To recognize a class of concepts as first-person judgment-dependent is in effect to recognize that a discourse that employs them is tacitly autobiographical. When we employ these predicates that seemingly attach only to extra-mental items like jokes or laws, what we are really talking about is what we would think of these things under certain specifiable conditions. But if we regard facts about what other tribes would think as robustly objective, we seem compelled to a similar view about ourselves. From a metaphysical point of view, biography and autobiography are on a par. The only difference lies in who is doing the talking. But that hardly converts into a difference in the facts described, unless one makes the peculiar error addressed in the previous section.

Still, there is an important difference between the anthropologist's stance and our own participant stance that may seem to cut against this verdict. Suppose that the *C*-conditions mentioned in the account of some first-person judgment-dependent concept *G* are met, and also that it is common knowledge that they are met. When the alien anthropologist considers whether a new object *x* possesses the property we pick out with *G*, the question presents itself to him as a substantive empirical question. He can wait to see what we in fact judge, or he can use his theory or his informed understanding to predict our judgment. Either way, he views his own inquiry as aimed at a fully objective and independent fact, identical with or supervenient upon facts about our dispositions to deploy certain concepts. But now consider our own ' inquiry' into whether *x* is *G*. We know that we are in conditions *C*, and we know a priori that whatever *we* judge in these conditions is automatically correct. Well, then it's hard to see how we can think of ourselves as *trying to discover* some independently constituted fact. If it's a question of something as brute as comedy, our 'inquiry' may consist in considering the joke and waiting to see how we 'respond' . We can of course try to theorize or speculate about our thinking before we actually do it. But this is just to take up the anthropologist's stance, temporarily treating ourselves as objects of a third-personal inquiry. When we are 'engaged' , on the other hand, we cannot think of our judgment as aiming to conform to anything apart from itself. Hence from this perspective it can seem very natural to say that the facts in question are 'up to us' . What we call a judgment can look more like an act of invention or decision (although these words suggest more deliberate activity on our part than is likely to be present). And of course, as soon as we start talking like this, the antirealist's rhetoric seems entirely rehabilitated.

The oddness is only exacerbated when we turn to more articulated practices of judgment, like the practice of deciding whether a law is constitutional, which in the nature of the case represent themselves as deliberative inquiries sensitive to reasons. If the concept of constitutionality were as we have described it, then from the anthropologist's point of view the facts about which laws are constitutional appear as thoroughly objective facts about what a certain group of people would think after due consideration. But consider the perspective of the sitting Supreme Court justice — and suppose, for the sake of vividness, that there is now only one sitting justice. When he tries to decide whether a law is constitutional, if he knows the analysis he knows that whatever he says will be right, provided only that he is thoughtful and clear-headed. And yet he must still think of himself as constrained to consider the balance of reasons, the force of precedent, and so on, on both sides of the issue. If he steps back from himself momentarily, he may see this process as simply a matter of providing input to a system whose output constitutes the relevant fact. He can theorize about what this output will be, and from this perspective, the facts about constitutionality will strike him as fully objective. But what is he to think when he is 'engaged' ? In a certain sense he must concede that the facts do not constrain his decision at all, but rather flow from it. He runs no risk at all of failing to weigh the evidence and arguments 'correctly' , and so failing to make the right decision. And yet, in another sense, when he is actively engaged in his deliberations, it seems to me that he cannot possibly think this. He must think of himself as trying to figure out where the arguments point: which decision they indicate as correct. He must think of himself as being led rather than leading. He must think of himself as aiming to conform his judgment to an independent fact in virtue of which it will be either correct or incorrect.

It seems to me a very interesting question what to say about the 'phenomenology' of the engaged perspective when the concepts employed are first-personal judgment dependent concepts and the conditions of deliberation are known to be 'ideal' . Does rational deliberation involve an inevitable illusion to the effect that one's verdict might fail to get things right? It is conceivable that a thought of this sort stands to theoretical reason — the practice of forming beliefs — much as the thought that one is metaphysically free stands to practical reason — the practice of deciding how to act: That is, it may be a thought that one cannot sustain on reflection, but which nonetheless forces itself upon us whenever we assume the relevant deliberative standpoint. Obviously, this requires further investigation. Nonetheless, I am convinced that it does not bear directly on the metaphysical question. It seems to me clear that

(a) The facts described by a discourse whose central concepts are first-personal judgment-dependent concepts are in principle describable from the anthropologist's stance in an entirely third-personal idiom,

and that

(b) The facts the anthropologist describes — the properties he attributes to objects — are, for all we have said, entirely objective in the sense that interests us.

And this seems to me enough to establish that the proposal presently on the table, however seductive, cannot be right.

Of course things would be otherwise if we had an independent argument to the effect that what I have called the 'anthropological' facts were somehow less than fully real. Recall that these are facts about what members of a certain population would judge under certain conditions. Now if one had independent grounds for thinking that either the modal facts or the facts about the content of judgment were less than entirely objective, these reasons might transmit to the facts which, according to the analysis, supervene upon them. More interestingly, since the 'conditions' in question will typically involve *inter alia* the requirement that the inquiry in question be 'rational' , 'clear-headed' , etc., it follows that if one took a less than fully objective view about the distribution of these properties, one might have grounds for denying objectivity to the facts represented by the left-hand side of the basic equation. Nothing I have said so far rules out these considerations. All we are entitled to conclude is that if one wishes to maintain that the facts given in a judgment-dependent discourse are less than fully objective for reasons like this, one must give an *independent* argument for the mitigated objectivity of facts about modality, semantic content, rationality or whatever. It will not suffice to show that these concepts themselves are judgment-dependent, as they well may be. For the point of the argument is to *show* that there is an interesting connection between judgment-dependence and failed objectivity; so it would be question-begging to take this for granted at the start.

## *VII. Objectivity and the Absolute Conception of the World*

I want to close with a discussion of a rather different family of proposals due largely to Bernard Williams. Like Wright, Williams is expressly concerned to draw a line between those aspects of our thought which represent an objective reality and those which merely represent the world as it is for us by generalizing the contrast between primary and secondary qualities. Williams understands this distinction rather differently, however, so the problems with his approach demand separate discussion.

Williams' leading idea is that an objective fact is one which receives representation in what he calls 'the absolute conception of the world' . Williams explains this suggestion in more than one way, and I doubt they are all equivalent. The general idea is nonetheless clear enough. To begin with, the absolute conception of the world is a *conception* — that is, a representation of the world. It is, moreover, a true representation. However it is not the whole truth. Some facts (true propositions) are left out. The chief exegetical question is how precisely Williams intends us to distinguish the absolute conception of the world — a catalog of elite truths — from the complete true story of the world — the catalog of truths tout court. The latter captures all the facts; the former only the objective ones.

Sometimes Williams seems to suggest that the absolute conception is distinguished by its vocabulary: the array of concepts that compose it (Williams 1978, pp. 244-5). For Williams, some concepts are parochial, in the sense that they are 'available' only to creatures of certain sorts, whereas others are available to any sufficiently sophisticated thinker regardless of his peculiar sensory endowment, history, culture or way of life. Williams offers no general theory of concepts or concept possession, so it's hard to know how to evaluate this sort of claim. But the examples convey the spirit of the view. Thus the usual secondary quality concepts are supposed to be available in the relevant sense only to creatures capable of the corresponding sense experience. The congenitally blind, no matter how fluent they may be in their use of the word 'red' , nonetheless lack the concept that sighted speakers of English typically express with that word. (Williams 1978, p. 241). Or, to take a rather different example, the 'thick' ethical concepts associated with traditional cultures — like the Homeric αιδωσ — are supposed to be available only to unreflective participants in such cultures and not to modern ironic sophisticates (Williams 1985, ch. 8). So no matter how good the sympathetic anthropologist gets at anticipating the native's application of a word, his sophistication or his failure to endorse the point of the practice in which the concept is embedded somehow prevent him from fully grasping it. By contrast, the traditional primary quality concepts along with the more advanced concepts of modern science are supposed to be available to creatures regardless of their parochial interests, sensory capacities, and so on. In one sense of the term, then, the absolute conception of the world is the truth about the world insofar is it is available to anyone at all, regardless of his peculiar way of life. In Nagel's phrase, it represents those aspects of the world accessible to the view from Nowhere.

So one proposal we might consider is this:

(13) A fact is less than fully objective if and only if it would not be represented in the absolute conception of the world, conceived as the true story insofar as it can be stated in an idiom none of whose concepts are parochial.

The intuition here is that the real is potentially accessible from indefinitely many distinct perspectives. It is there for everyone, regardless of his peculiar angle on the world. The less-than-real, on the other hand, manifests itself only from some restricted point of view; and so deserves to be called, in the Kantian sense, a mere appearance for that point of view of the real state of affairs that underlies it. This is suggestive, of course. But we have learned to be suspicious of suggestive verbiage. So let's take a closer look.

Note first the oddness of the suggestion that what is real is that which might be known by the least fully equipped inquirer. Our natural view, I suppose, is that blindness, deafness and the rest are defects in part because they represent obstacles to knowing what the world is really like. On the present view, however, this is not quite right. Blindness may be an obstacle to attaining certain concepts. But for just this reason those concepts are not essential for representing the objective world. In the limit the proposal would seem to imply that the only real features of the world are those which might in principle be represented by a disembodied intelligence with no senses whatsoever. And this, if not obviously wrong, is at least a very strange suggestion.

A deeper worry is that as presently formulated the proposal may well be vacuous. Suppose the Azande employ the thick ethical term 'blog' as a term of praise for actions that show a special sort of respect for the chief. It is a kind of piety, but not one for which we possess even an approximate counterpart notion. By hypothesis, we cannot possess the concept the Azande express by this word, since to possess it requires an immersion in their practice which is impossible for us. So when they apply the concept to an object correctly, the thought they express is one we cannot grasp. On the present view, this is supposed to show that the facts about the distribution of *blogness* are not features of the world as it is in itself, since no representation of them figures in the absolute conception. But this transition from a claim about thoughts to a claim about facts should make us suspicious. From the fact that one representation of the blogness facts fails to figure in the absolute conception, it does not follow that no such representation appears there. And indeed it is conceivable that even if the absolute conception does not include the claim that x is blog, it does include the claim that

x has the feature the Azande actually call 'blog' .

This claim employs no thick ethical concept. In fact, for all we have said so far it may employ no parochial concepts at all. But the property denoted by ' blog'. and the property denoted by the phrase 'the feature the Azande actually call 'blog'' are (arguably) one and the same. And if they are, the thick ethical *property* may well be an objective feature of the world, even though the thick ethical *concept* fails to figure in the absolute conception. In general, then, if the absolute conception provides the means for referring to the properties denoted by parochial concepts in this way, then (13) is vacuous. Whenever a parochial concept *C* applies to an object, there will be an anthropological attribution in the absolute conception that attributes the property it denotes — as the property denoted by 'C' — to that object. So here we have another example of the persistent failure in this part of philosophy to distinguish between concepts and the properties of which they are concepts.

This objection would lapse, of course, if we had independent reason for thinking that the absolute conception lacked the resources for anthropological attribution. The resources here include the capacity to refer to representations, say by quoting them, along with semantical terms like 'denotes' . Sometimes Williams talks as if the absolute conception were exhausted by the austere idiom of elementary particle physics; and if this were right, the present objection would not apply. As we shall see, however, Williams sometimes seems committed to a much richer conception, according to which the absolute conception includes the deliverances of a maximally detached anthropology or sociology. And if this is so, then pending clarification of the view, at least, I see no reason to think that any truths are excluded from the absolute conception.[28]

A second way of explaining what the absolute conception of the world amounts to points in a very different direction. On this approach, the absolute conception is defined as the limit of a certain procedure of abstraction. We begin with our naive, unreflective view of the world, where we take it for granted that the concepts with which we operate all reflect real features of things (occasional errors aside). We then notice that other people represent the world in different terms, not immediately commensurable with ours, and we set about trying to understand why our view differs from theirs. We step back for a moment from our naive representation and view it along with other representations as an item in the world. Our task is to understand how the variety of representations emerges from the interaction of variously constituted creatures with a single world. At the early stages of this procedure we may notice, for example, that the ways colored things appear to us depend in part on contingent features of our visual system, and that the best explanation for the variety of color appearances does not proceed by locating real colors in objects which some systems are better at 'receiving' than others. At more sophisticated stages we may realize that differences in ethical outlook or in judgments about the degree of confirmation attaching to a scientific hypothesis in light of certain evidence are similarly attributable to differences in us — in our ways of life, in our languages, in our peculiar historical circumstances, etc. In the limit, after we have taken into account all of the actual (and perhaps even all the conceivable) variations in outlook we arrive at the absolute conception of the world. This includes first, an empirical account of what the extra-mental world is like; second, an inventory of the various local representations of this world; and finally, an explanatory account of how those various local representations emerge from the interaction of interested, historically and biologically conditioned human beings with a world of that sort. The absolute conception so understood therefore includes not just the austere idiom of ideal physics, but a complete anthropology wherein the cognitive doings of variously situated groups — including the purveyors of the ideal conception itself — are displayed and explained.

Note how different this account of the absolute conception is from the last one. On the last proposal, the absolute conception did not employ parochial concepts, and was therefore in principle intelligible to creatures regardless of their peculiar interests. When it is conceived as the upshot of a maximal cognitive anthropology, however, it would seem that it must employ parochial concepts. If I am to explain why the Azande hold that a gift of turtles is certainly blog, there is a sense in which I have to *use* the concept *blog* in setting out the explanandum. To say that *S* believes *x* to be *F* is to use the concept of an *F*. To be sure, when I use the concept in this way — within the scope of a propositional attitude verb — I do not myself apply it to an object. But still, if I didn't grasp the concept — if I didn't understand the word 'F' as a word in my own idiolect — then it's hard to see how I could understand what was being said when it was used to characterize the content of someone else's thought.[29] The procedural account therefore seems to imply that the idiom of the absolute conception is not the austere language of non-parochial concepts, but rather the maximal idiom that includes every concept anyone actually possesses, parochial or otherwise. Indeed we can frame the tension between the two approaches to the absolute conception as a dilemma: Either there are no parochial concepts, or the absolute conception is radically unattainable. For if there is a parochial concept *C*, possession of which requires unreflective immersion in a certain local way of life, then it is not available from the detached perspective of the ideal anthropologist. This means that he will not be able to represent in his own language the contents of the representations of those who employ *C*: and what he cannot represent, he cannot explain. So any account of the origins of representations that he produces will leave out an account of the sources of C-involving thoughts, and will therefore fall short of the absolute conception conceived as a total anthropology of knowledge. [30]

This seems to me a real tension in William's view, but I will not press the point. The criterion implicit in the second approach to the absolute conception is a familiar one from the literature on realism. We can put it in a preliminary way as follows:

(16) A fact (property) is objective iff it must be mentioned (attributed to an object) in the best causal/explanatory account of the plurality of representations of the world.

The suggestion is that the primary qualities and the other properties described in the natural sciences will be attributed to things in an account of why people believe what they believe, whereas the secondary qualities won't. It is plausible that we can explain why people think that some things are red or delicious, at least in principle, without saying whether anything really is red or delicious; whereas, if I want to explain why people hold that some things are extended or massive, I may very well have to describe the world in terms of these very properties.

Now very often this sort of observation is made in the service of an fairly straightforward argument for the rejection of *minimal* realism. The nominalist argues, for example, that since abstract objects play no role in the causation of our beliefs (or of anything else, for that matter) we can have no reason to believe in them, and should therefore reject any platonist view which implies their existence (Field 1989; see also Armstrong 1989). The moral skeptic argues that we can explain the origins of our moral views without positing moral facts; but since this sort of explanatory role is the only reason we could have for believing in such unobservable states of affairs, we should reject the ontological claims of morality out of hand (Mackie 1977). Whatever one makes of such arguments, however, they must be clearly distinguished from the sort of thing Williams has in mind. Williams does not claim that discourse that fails the test described in (16) is false or that it is unreasonable to believe it. Indeed he holds, at least in the case of discourse involving thick ethical concepts, that it may sometimes even constitute knowledge. The test is not a test for mere existence; it is a test for *objective* existence. The real items that fail the test are features of the word — but they are not features of the world as it is *anyway*, independently of us, etc. [31]

What are we to make of this sort of criterion of objectivity? Unlike the others we have considered, it does not on the face of it presuppose the legitimacy of a transition from a distinction among concepts to a distinction among properties. So if we are to object, we must object in a different way. There is a great deal to be said. My remarks constitute only the main outlines of a response.

Note first that (16) threatens to be profoundly restrictive. It is conceivable, for example, that all talk of natural substances like gold and water is eliminable from the account of our beliefs about these things in favor of accounts that mention only subatomic particles. That is, it is conceivable that we should be able to explain why physicists believe that gold is a metal without explicitly mentioning gold, speaking instead only of the collisions between subatomic particles.[32] Or, to take a less fanciful example, it is conceivable that we should be able to explain why all human beings believe that the stars are quite far away without mentioning the property of being a star, conducting the whole explanation at a smaller scale. The account thus threatens to imply that only the most fine-grained features of the world are real or objective features. Now if 'objective' is taken to mean something like 'ultimate' or 'basic' , then this might not be so counterintuitive. But when it is contrasted with 'mind-dependent' or 'projected' , its plausibility diminishes considerably. Are we really prepared to infer, from the fact that our opinions about the composition of the stars can in principle be explained at the level of elementary particles, that the fact that stars contain hydrogen is somehow constituted by us or our practices; that it is something we somehow project onto the world instead of detecting it? I doubt it. There may very well be an interesting contrast between the ultimate explanatory basis — the smallest set of properties in terms of which the phenomena can be explained — and the superstructure of supervening or constituted properties. But on the face of it this distinction is at right angles to the purported contrast between the objective and the rest that is at stake in the debates over realism.

To be sure, one can imagine all sorts of refinements in the proposal. We could say that an objective feature is one which *supervenes* on the features which must be mentioned in any account of the origins of our ideas. But this is likely to be far more permissive than those who endorse this line of thought intend. If there are facts about what's funny, then they probably supervene (at least globally) on the microphysical facts. So on the present refinement we should count them as objective, whereas it is clearly in the spirit of this line of thought that we shouldn't. Alternatively, one may say that the objective features are the ones that can figure in *some sort* of causal explanation of why our representations are as they are, even if it is not the most fundamental or basic sort of scientific explanation. But again, it is unclear what this proposal would exclude. It can be perfectly appropriate to say that we laughed our heads off because the movie really was very funny (and not because we had been drugged or tricked or whatever...) (Wiggins 1976).

Crispin Wright has suggested an interesting variation on this line of thought. The striking thing about explanations involving the funny is that while it seems perfectly all right to invoke the property in certain explanations of our comic responses or of events in which these responses play a crucial role, it doesn't look like the facts about what's funny can explain anything else. That is, insofar as these facts play a causal or explanatory role, their influence is always mediated by some representation of them. By contrast, the properties one would like to call genuinely real seem capable of affecting all sorts of objects in all sorts of ways, many of which do not involve any intermediary thought about them.

Let us say that a property *F* possesses *wide cosmological role* iff

(14) Facts about which things are *F* figure in good causal explanations of a wide variety of phenomena, and some of these explanations do not involve as an intermediary any judgment about which things are *F*

and consider the proposal that

(15) A feature of objects is less than fully objective if it fails to possess wide cosmological role.

This is really quite compelling. Some properties make their presence felt only through our thought about them. Even if we do not literally make these properties, still their reality — their effect on things — is dependent on the mind in peculiar and crucial way. Others, by contrast, do their work independently of our conceptions, and so possess what deserves to be called a species of objectivity or mind-independence. What are we to make of this?

Well, the first thing to note is that (15) can't be quite right. It is possible to imagine a subtle physical property Q which, though intuitively thoroughly objective, is nonetheless nomically connected in the first instance only with brain state B — where this happens to be the belief that some things are Q. This peculiar discovery would not undermine our confidence that Q was an objective feature of things, as it should if (15) were true.[33] Another example is sensible species as conceived by scholastic Aristotelians. These were presumably objective properties of things whose only causal interactions were mediated by their representation in sensitive intellects of the appropriate sort. They failed to possess wide cosmological role, presumably as a matter of nomological necessity. But they were not for that matter conceived as being less than entirely real. To the contrary, detection of them was understood as the paradigm case of detecting what is objectively present in objects.

But this is just a minor glitch. The real difficulty with all these proposals is much more simpler. Consider speculative angelology. This is a discourse about angels, of course. It asks how many there are, what they think about, etc. But let us suppose that unlike ordinary angelologists, the speculative angelologists are all agreed that the angels have no causal commerce with us or perhaps even with one another. This being the case, they concede that it is very hard to know when you've got things right — hypotheses cannot be put to the test, as the skeptical angelologists are wont to point out. The angelologists persist, however, in speculating and forming tentative opinions about the angels on the basis of tradition, their intrinsic sense of angelological plausibility, and so on. All agree, for example, that the number of angels is not 17, but rather some more sensible number; that angels are intelligent, immaterial, etc. The facts about angels so conceived do not by hypothesis figure in the causal explanation of angelological opinion. Indeed they don't figure in the causal explanation of much of anything. This recognition may tend to cast doubt on angelological doctrine. Maybe it is ultimately unreasonable hold detailed opinions about this sort of subject matter. But even so, this has no tendency to undermine our sense that the *facts* about angels, if they exist, are thoroughly objective facts that are in no way constructed or projected by us. To take some more familiar examples, consider the moral or mathematical facts as conceived by a character we may call the moderate platonist. For the moderate platonist the facts of morality and mathematics are irreducible facts about radically transcendent entities — the good, the numbers — as they were for Plato himself. His moderation consists in rejecting the idea of a special intuitive faculty by which these objects can influence us. His view is rather that insofar as we have knowledge of them it is speculative. Our opinions are constrained by considerations of plausibility, coherence with firmly held prior doctrine, elegance, systematicity, and the like; but not by anything like an encounter with the objects or their visible traces.[34] A more peculiarly philosophical example is discourse about possible worlds as Lewis conceives them (Lewis 1986). Again, the objects here are causally isolated from us and our representations. But if they exist the facts concerning them are 'intuitively' as objective as anything can be.[35]

The conclusion towards which these examples point is that however interesting the contrast between the causally or explanatorily potent features of the world and the rest may be for epistemology, it seems to have no intrinsic connection with the metaphysical contrast between that which exists independently of us and that which does not. I concede, however, that this discussion is far too compressed to make this conclusion anything more than plausible.

### *VIII. Conclusion*

The problem was to draw a notional line between objective features of the world and those which the mind somehow constructs. This contrast is introduced to us by a series of images and metaphors which are most at home in a Kantian framework where the Mind's relation to the world of experience is problematic. The question was, what are we to make of the metaphors once this metaphysics gives way to a naturalism according to which the only minds there are are *parts* of the natural world itself? We have surveyed a number of influential answers, all of which take as their starting point some version of the contrast between primary and secondary qualities. This is promising, at least in part because this contrast pre-dates the idea of a transcendental psychology, and so promises to be compatible with the naturalism we now take for granted. In each case I suggested that the initial promise depends on a subtle conflation: a contrast drawn at the level of concepts or representations of the world is transposed into a contrast drawn at the level of properties or facts. But this move is always illegitimate without further argument, and we saw some reason to think that in these cases the argument cannot be available. As I said at the outset, this is not supposed to show all by itself that there is no genuine issue about objectivity compatible with our moderate naturalism. In the nature of the case, it is very hard to show that there is no way to sharpen a putative issue. These remarks are only a challenge, then, to say what exactly the issue is.

*Princeton University*

## NOTES

[1] I am grateful to Sally Haslanger, Richard Holton, Allan Gibbard, Mark Johnston, Jim Joyce, David Lewis, James Woodbridge, Crispin Wright, Steve Yablo and José Zalabardo for their very helpful comments on this material.

[2] For a slightly more extensive presentation of the material in this section, see Rosen and Railton (1994).

[3] For the metaphysically unambitious talk of facts, properties, correspondence, etc., see Wright (1992).

[4] Which line may of course place all of the facts on one side or other.

[5] This formulation derives from Peirce (1878), of course. For more modern versions see Wiggins (1976); Nagel (1986), ch. II; Williams (1978, ch 8).

[6] On the impossibility of presuppositionless understanding see Gadamer (1975).

[7] See Geuss (1981) for discussion.

[8] Feldman (1994) is a useful discussion of the competing conceptions of methodological objectivity in recent philosophy.

[9] What about the converse? Suppose we could convince ourselves that a judgment had resulted from a methodologically objective inquiry. Would this be sufficient to convince us that the fact it describes was really 'out there', independent of us, etc.? Not obviously. Note that the idea of methodological objectivity presupposes the idea of a purely rational inquirer. Of course it is far from clear how this should be understood. But suppose there are some opinions the failure to hold which counts as a failure of rationality in the relevant sense — elementary principles of logic are plausible candidates, along with the cogito and principles like 'Rational inquiry is

possible'. Then it might be that any methodologically objective inquiry would affirm these principles, simply because failure to affirm them would be a sign of irrationality. (See Haslanger 1992.) I take it that this sort of view about logic or the possibility of rationality does not by itself force the realist's rhetoric of metaphysical objectivity upon us. We can say that any rational creature must take a certain view without thinking of the corresponding fact as something 'out there' which rational inquiry is somehow compelled to 'discover'. On this point it is useful to note that Peirce himself is widely regarded as something less than a full blooded realist about the natural world, even though he held that scientific knowledge of that world was ideally the result of a methodologically objective inquiry.

[10] In recent writing the issue seems to have less to do with the existence of private inner *objects* — ideas, sensa — than with the existence of subjective *states* or *qualities*, known to each of us in his own case by a sort of Russellian acquaintance but somehow 'unavailable' to the objectifying third-personal gaze of natural science (Jackson 1982; Nagel 1974).

[11] The minimalist *may* of course go on to affirm the counterfactual dependence of worldly items on the existence of thought or language, and so to defend a species of empirical idealism. The point is only that minimalism by itself is neutral on the point. Thanks to Eric Bourneuf on this point.

[12] For some discussion of minimalism in this sense, see Johnston (1988), Wright (1992).

[13] Sometimes the claim that range of facts is objective is meant to contrast with claim that it is 'socially constructed', where nothing more is meant by this than a sort of causal dependence on social practice. Claims like this can be very important. But they do not imply the sort of deep metaphysical distinction with which we are currently concerned. On the senses of social construction, see Haslanger (1993).

[14] For some very suggestive but ultimately inconclusive discussion of the problem in the context of Wittgenstein's philosophy see Williams (1974) and Lear (1986).

[15] *Pace* Mackie (1976, ch. 1).

[16] The formulation in Johnston (1989) is rather different. The response-dependent concepts there are given as those for which a biconditional of the form

x is F iff in conditions C, subjects S are disposed to produce x-directed response R

may be seen to hold a priori given substantive specifications of S, R, and C. The distinction between the new formulation in terms of concept identities and the old one in terms of a priori biconditionals is subtle but important. But it does not directly affect the issues before us. For some discussion see §V below.

[17] See Curley (1972).

[18] The point about the scholastics is enough to show that it isn't. But we might have used quasi-color concepts that were explicitly response-dependent in Johnston's sense, i.e., concepts for which equations like (4) or (5) held true.

[19] It should be stressed that Johnston himself does not endorse this account of the significance of widespread response-dependence. He does say at one point that a response-dependent realism is 'if you like, a conceptual or transcendental idealism' (Johnston 1989, p. 148), and more recently:

> Descriptive Protagoreanism bids fair to be the best candidate for an appropriately qualified realism, the qualification being precisely the denial that the concepts in question are independent of subjects' responses... This independence of subjectivity is the hallmark of complete objectivity in one of its obvious senses. (Johnston 1993, p. 106)

The uses to which Johnston puts the notion are nonetheless rather different from the uses to which we imagine its being put in (8). Johnston never suggests that the facts described by means of response-dependent concepts are significantly mind-dependent in a sense that would license regarding them as constituted or constructed by our thinking,

[20] Locus classicus: Goodman (1978).

[21] There is, of course, a sense in which surface-contradictory value judgments 'disagree' on such

an analysis that does not apply in the case of 'my foot'. If you value x and I disvalue it, then our *values* conflict in the sense that they can't both be fully realized in a single world. The Lewis-style value judgment will typically *express* a state of valuing: two perfectly true such judgments may therefore disagree in the sense that the patterns of valuation they express cannot both be satisfied. The important point is that this sort of disagreement is merely disagreement in 'attitude' or desire. It is not a disagreement about what the world is like, and so cuts no metaphysical ice.

[22] Thus suppose the term 'water' had been introduced by the stipulation: Call something 'water' iff it is consubstantial with the stuff that fills up the lakes and rivers around here. It might then have been a priori (at least for those party to the stipulation) that something is water if and only if it is consubstantial with the contents of local rivers and lakes. And yet the concept of being water — the sense of the term 'water'; that which any competent speaker of the language 'grasps' — might be very different. This is suggested by the fact that later on, someone might come to understand the word 'water' perfectly well, and so to 'grasp' the associated concept, without ever having heard of rivers and lakes.

[23] For reasons not immediately relevant to the present discussion, Wright is ultimately concerned not with a priori biconditionals like (10) involving counterfactuals about the implementation of ideal conditions (so-called *basic equations*), but rather with what he calls 'provisoed equations' of the form

It is a priori that if C then (x is F iff S judges that x is F).

Here the connectives are all (presumably) material connectives. Certain difficulties to do with the conditional fallacy are thereby avoided. I have not pursued the interesting question of whether the metaphysical issues I discuss in the text are somehow affected by this shift to provisoed equations.

[24] The slogan is Putnam's. See, for example, Putnam (1981).

[25] This is not quite right, of course. It be more accurate to say that a law is constitutional if the Court would not judge it unconstitutional, since the Court routinely declines to rule.

[26] The characterization of these facts as facts about the mere responses of animals to stimuli may seem to neglect the crucial feature of definitions like (12), viz., that normative concepts like 'rational' and 'unbiased' may appear in the specification of the appropriate subjects or the appropriate conditions. Most of the interesting judgment dependent concepts will turn out to be normative in this sense: and if you think that this somehow counts against their objectivity, you may want to resist my assimilation of these cases to mere animal responses. Fair enough. My point is only that if this is right then it is the normativity of these interesting concepts and not simply their judgment dependence that is doing the work. What one wants in that case is an *argument* for the subjectivity of the normative, and in the present context it will not do simply to show, as may be the case, that the normative is in general judgment-dependent, since the subjectivity of the judgment-dependent is just what's at issue. In the absence of such an argument there is no metaphysically relevant distinction between concepts like constitutionality as defined in (12) and the concepts of dispositions to judgmental response in conditions specified in non-normative terms.

[27] Thanks to Allan Gibbard on this point.

[28] This is not necessarily a decisive objection. It could be that a successful attempt to explain the contrast between the objective facts and the rest would have the implication that every fact falls on the objective side. But this is clearly not Williams' view.

[29] I can imagine a Fregean response. The word 'F' may express one concept when employed in oratio directa, another in oratio obliqua, and it's conceivable that an anthropologist may grasp the second without being able to grasp the first.

[30] To be fair, I should say that Williams sometimes uses the phrase 'absolute conception' to refer only to the account of what the world is like that figures in the total anthropology, leaving out the story about how a world like that gives rise to the various representations. The absolute conception so conceived need not employ the array of parochial concepts, since it need not include in itself a specification of the representations whose existence is to be explained on its basis. This

brings the two accounts of the absolute conception closer together, though it by no means guaranteed their coincidence. It seems to me that we have no guarantee a priori that the features of the natural world we need to invoke in an account of the plurality of representations are available to creatures regardless of their peculiar circumstances. Suppose the scholastic realism about colors had been correct, so that the property of redness needed to be invoked in order to explain ordinary color perception; and suppose also (with Williams, I suppose) that the concept of redness would still in these circumstances have been unavailable to the color blind. Then there would have been a feature of the world that needed to be invoked in order to explain why things seem to us as they do which would not have been representable in the idiom of non-parochial concepts. We may hope that this is not the case; or we may take ourselves to have empirical evidence that it isn't. But from a philosophical point of view, we can have no guarantee.

[31] A similar application of the criterion is implicit in Harman (1977), though Harman does not employ the Kantian imagery.

[32] In fact this is only barely conceivable. It's one thing to suggest that in some level most of the causes of our opinions can be described in this austere idiom; quite another to suggest that an adequate *explanation* can be given at this level.

[33] Might one respond by pointing out that Q's lack of wide cosmological role is merely accidental? Even if as a matter of nomological fact Q's effects are all mediated by beliefs about its instantiation, we can imagine a world where the laws are slightly different in which Q's nomic connections are quite widespread. Contrast the funny. It seems odd to suppose that its merely an accident of nature that the facts about its distribution interact in the first instance only with thinking things. (Of course if one thinks of the funny as an objective dispositional property, this is not odd at all; but we are imagining that it is conceived in some other way.) This is suggestive. The response presupposes a controversial view about property individuation — if properties have their nomic roles essentially, as Shoemaker has suggested, then it is a non-starter. Nonetheless it is clearly worth exploring further.

[34] I take Quine to be a moderate platonist of this sort. See for example Quine (1961).

[35] There is a natural retrenchment in light of these arguments. Some properties have wide cosmological role, and some have none at all. The examples suggest that in both these cases we have no grounds for withholding the realist's imagery of objectivity. By contrast, properties like the funny and the interesting are *intermediate* cases: they are causally relevant, but their effects are always mediated by our representations. So consider the proposal that a discourse whose central concepts are have this intermediate status concerns a domain that is less than fully objective. The examples of angelology and moderate platonism do not tell against this proposal. What does?

The right response should begin like this. Suppose for a moment that the funny is that which is disposed to elicit in us the judgment that it is funny. The disposition to elicit such judgments is a causally relevant feature of objects. And on the face of it there is no special reason to regard the distribution of this property as a matter of less than fully objective fact. But here is a natural thing to say about dispositional properties: they are tailored to their manifestations, in the sense that any effects they have they have by first bringing about their characteristic manifestations. The fragility of the glass is not causally inert; but if it is part of the explanation for any natural event, its efficacy is mediated by its first explaining the glass's breaking. Even if the glass's fragility is constituted by the same underlying microphysical property as (say) its transparency, there is something odd about saying that Jones can see through the glass because it is fragile. Dispositions in this sense are more fine grained than their categorial bases. Suppose this is the right way to think about dispositions. Then the fact that judgment dispositional properties are effective only via judgments involving concepts which represent them is just an instance of the general feature of dispositions that they are effective only via their manifestations. In other words, the fact that judgment-dependent concepts possess intermediate cosmological role is a straightforward consequence of their being dispositional concepts. The effects of flammability are always mediated by burning; likewise, the effects of funniness are always mediated by judgments of humor. This conception of dispositional properties thus explains why the judgment dependent features of things should possess intermediate cosmological role without suggestin , that they are in any metaphysical sense peculiar.

## REFERENCES


Armstrong, D. (1989), *A Combinatorial Theory of Possibility* (Cambridge: Cambridge University Press).

Blackburn, S. (1984), *Spreading the Word* (Oxford: Oxford University Press).

Boyd, R. (1984), 'The Current Status of Scientific Realism' in J. Leplin, ed., *Scientific Realism* (California: The University of California Press).

Churchland, P. (1984), *Matter and Consciousness* (Cambridge, MA, The MIT Press).

Curley, E. (1972), 'Locke and Boyle on the Distinction between Primary and Secondary Qualities' , *The Philosophical Review.*

Feldman, H.L. (1994), 'Objectivity and Legal Judgment', *Michigan Law Review* 92 (March).

Gadamer, H-G. (1975), *Warheit und Methods*, 3rd ed. (Tübingen: J. C. B. Mohr).

Geuss, R.(1981), *The Idea of a Critical Theory* (Cambridge: Cambridge University Press).

Gibbard, A. (1990), *Wise Choices, Apt Feelings* (Cambridge: MA, Harvard University Press).

Goodman, N. (1978), *Ways of Worldmaking* (Indianapolis: Hackett).

Harman, G. (1977), *The Nature of Morality* (Oxford: Oxford University Press).

Haslanger, S. (1992), 'Ontology and Pragmatic Paradox', *Proceedings of the Aristotelian Society.*

Haslanger, S., 'On the Social Construction of Reality: Gender, Knowledge and Social Kinds', University of Michigan ms.

Jackson, F. (1982), 'Epiphenomenal Qualia', *Philosophical Quarterly* 32, pp. 127-36.

Johnston, M. (1988), 'The End of the Theory of Meaning', *Mind and Language* 3, pp. 28-42.

Johnston, M. (1989), 'Dispositional Theories of Value', *Proceedings of the Aristotelian Society.*

Johnston, M. 'Objectivity Refigured', Princeton University, ms.

Kaplan, D. (1989), 'Demonstratives', in Almog, Perry and Wettstein, eds, *Themes from Kaplan* (Oxford: Oxford University Press).

Lear, J. (1986) 'Transcendental Anthropology,' in J. McDowell and P. Petit, eds., *Subject, Thought and Context* (Oxford: Clarendon).

Lewis, D. (1986), *On the Plurality of Worlds* (Oxford: B. Blackwell).

Lewis, D. (1989), 'Dispositional Theories of Value,' *Proceedings of the Aristotelian Society,* Supplementary Volume 62.

Mackie, J. (1976), *Problems from Locke* (Oxford: Clarendon).

Mackie, J. (1977), *Ethics: Inventing Right and Wrong* (Harmondsworth: Penguin).

Nagel, T. (1974), 'What is it Like to be a Bat,' *The Philosophical Review* 83, pp. 435-450.

Nagel, T. (1986), *The View from Nowhere* (New York: Oxford University Press)..

Pettit, P. (1992), 'Realism and Response Dependence' , *Mind.*

Peirce, C. S. (1878), 'How to Make our Ideas Clear', in *The Collected Papers of Charles Sanders Peirce* (Cambridge, MA: Harvard University Press) 1931-62, v. 5.

Putnam, H. (1981), *Reason, Truth and History*, (Cambridge: Cambridge University Press).

Quine, W. V. O.(1961), 'On What There Is,' in his *From a Logical Point of View*, 2nd ed., (Cambridge, MA: Harvard University Press).

Rorty, R.(1989), *Contingency, Irony and Solidarity* (Cambridge: Cambridge University Press).

Stove, D. (1991), *The Plato Cult and other philosophical follies* (Oxford: B. Blackwell).

van Fraassen, B. (1980), *The Scientific Image* (Oxford: Clarendon).

Wiggins, D. (1976),, 'Truth, Invention and the Meaning of Life,' *Proceedings of the British Academy* LXII.

Williams, B. (1974) 'Wittgenstein and Idealism' in G. Vesey, ed., *Understanding Wittgenstein* (London: MacMillan).
Williams, B. (1978), *Descartes: The Project of Pure Inquiry* (New Jersey: Humanities Press).
Williams, B. (1986), *Ethics and the Limits of Philosophy* (Cambridge: MA, Harvard University Press).
Wright, C. (1992), *Truth and Objectivity* (Cambridge, MA: Harvard University Press).

# PHILOSOPHICAL STUDIES SERIES

*Founded by Wilfrid S. Sellars and Keith Lehrer*



1. JAY F. ROSENBERG, *Linguistic Representation,* 1974.
2. WILFRID SELLARS, *Essays in Philosophy and Its History,* 1974.
3. DICKINSON S. MILLER, *Philosophical Analysis and Human Welfare.* Selected Essays and Chapters from Six Decades. Edited with an Introduction by Lloyd D. Easton, 1975.
4. KEITH LEHRER (ed.), *Analysis and Metaphysics.* Essays in Honor of R. M. Chisholm. 1975.
5. CARL GINET, *Knowledge, Perception, and Memory,* 1975.
6. PETER H. HARE and EDWARD H. MADDEN, *Causing, Perceiving and Believing.* An Examination of the Philosophy of C. J. Ducasse, 1975.
7. HECTOR-NERI CASTAÑEDA, *Thinking and Doing.* The Philosophical Foundations of Institutions, 1975.
8. JOHN L. POLLOCK, *Subjunctive Reasoning,* 1976.
9. BRUCE AUNE, *Reason and Action,* 1977.
10. GEORGE SCHLESINGER, *Religion and Scientific Method,* 1977.
11. YIRMIAHU YOVEL (ed.), *Philosophy of History and Action.* Papers presented at the first Jerusalem Philosophical Encounter, December 1974, 1978.
12. JOSEPH C. PITT, *The Philosophy of Wilfrid Sellars: Queries and Extensions,* 1978.
13. ALVIN I. GOLDMAN and JAEGWON KIM, *Values and Morals.* Essays in Honor of William Frankena, Charles Stevenson, and Richard Brandt, 1978.
14. MICHAEL J. LOUX, *Substance and Attribute.* A Study in Ontology, 1978.
15. ERNEST SOSA (ed.), *The Philosophy of Nicholas Rescher: Discussion and Replies,* 1979.
16. JEFFRIE G. MURPHY, *Retribution, Justice, and Therapy.* Essays in the Philosophy of Law, 1979.
17. GEORGE S. PAPPAS, *Justification and Knowledge: New Studies in Epistemology,* 1979.
18. JAMES W. CORNMAN, *Skepticism, Justification, and Explanation,* 1980.
19. PETER VAN INWAGEN, *Time and Cause.* Essays presented to Richard Taylor, 1980.
20. DONALD NUTE, *Topics in Conditional Logic,* 1980.
21. RISTO HILPINEN (ed.), *Rationality in Science,* 1980.
22. GEORGES DICKER, *Perceptual Knowledge,* 1980.

23. JAY F. ROSENBERG, *One World and Our Knowledge of It,* 1980.
24. KEITH LEHRER and CARL WAGNER, *Rational Consensus in Science and Society,* 1981.
25. DAVID O'CONNOR, *The Metaphysics of G. E. Moore,* 1982.
26. JOHN D. HODSON, *The Ethics of Legal Coercion,* 1983.
27. ROBERT J. RICHMAN, *God, Free Will, and Morality,* 1983.
28. TERENCE PENELHUM, *God and Skepticism,* 1983.
29. JAMES BOGEN and JAMES E. McGUIRE (eds.), *How Things Are, Studies in Predication and the History of Philosophy of Science,* 1985.
30. CLEMENT DORE, *Theism,* 1984.
31. THOMAS L. CARSON, *The Status of Morality,* 1984.
32. MICHAEL J. WHITE, *Agency and Integrality,* 1985.
33. DONALD F. GUSTAFSON, *Intention and Agency,* 1986.
34. PAUL K. MOSER, *Empirical Justification,* 1985.
35. FRED FELDMAN, *Doing the Best We Can*, 1986.
36. G. W. FITCH, *Naming and Believing,* 1987.
37. TERRY PENNER, *The Ascent from Nominalism.* Some Existence Arguments in Plato's Middle Dialogues, 1987.
38. ROBERT G. MEYERS, *The Likelihood of Knowledge,* 1988.
39. DAVID F. AUSTIN, *Philosophical Analysis.* A Defense by Example, 1988.
40. STUART SILVERS, *Rerepresentation.* Essays in the Philosophy of Mental Rerepresentation, 1988.
41. MICHAEL P. LEVINE, *Hume and the Problem of Miracles. A Solution*, 1979.
42. MELVIN DALGARNO and ERIC MATTHEWS, *The Philosophy of Thomas Reid*, 1989.
43. KENNETH R. WESTPHAL, *Hegel's Epistemological Realism.* A Study of the Aim and Method of Hegel's *Phenomenology of Spirit*, 1989.
44. JOHN W. BENDER, *The Current State of the Coherence Theory.* Critical Essays on the Epistemic Theories of Keith Lehrer and Laurence Bonjour, with Replies, 1989.
45. ROGER D. GALLIE, *Thomas Reid and 'The Way of Ideas'*, 1989.
46. J-C. SMITH (ed.), *Historical Foundations of Cognitive Science*, 1990.
47. JOHN HEIL (ed.), *Cause, Mind, and Reality.* Essays Honoring C. B. Martin, 1990.
48. MICHAEL D. ROTH and GLENN ROSS (eds.), *Doubting.* Contemporary Perspectives on Skepticism, 1990.
49. ROD BERTOLET, *What is Said.* A Theory of Indirect Speech Reports, 1990
50. BRUCE RUSSELL (ed.), *Freedom, Rights and Pornography.* A Collection of Papers by Fred R. Berger, 1990
51. KEVIN MULLIGAN (ed.), *Language, Truth and Ontology,* 1992
52. JESÚS EZQUERRO and JESÚS M. LARRAZABAL (eds.), *Cognition, Semantics and Philosophy.* Proceedings of the First International Colloquium on Cognitive Science, 1992

53. O.H. GREEN, *The Emotions*. A Philosophical Theory, 1992
54. JEFFRIE G. MURPHY, *Retribution Reconsidered.* More Essays in the Philosophy of Law, 1992
55. PHILLIP MONTAGUE, *In the Interests of Others.* An Essay in Moral Philosophy, 1992
56. J.- P. DUBUCS (ed.), *Philosophy of Probability.* 1993
57. G.S. ROSENKRANTZ, *Haecceity:* An Ontological Essay. 1993
58. C. LANDESMAN, *The Eye and the Mind.* Reflections on Perception and the Problem of Knowledge. 1993
59. P. WEINGARTNER (ed.), *Scientific and Religious Belief.* 1994
60. MICHAELIS MICHAEL and JOHN O'LEARY-HAWTHORNE (eds.), *Philosophy in Mind.* The Place of Philosophy in the Study of Mind. 1994

9 780792 331438